Michio Masujima

Applied Mathematical Methods in Theoretical Physics

Applied Mathematical Methods in Theoretical Physics
Michio Masujima

ISBN: 978-3-527-40534-3

Michio Masujima

Applied Mathematical Methods in Theoretical Physics

WILEY-VCH Verlag GmbH & Co. KGaA

Cover Picture
K. Schmidt

1st Edition 2005
1st Reprint 2007

Library of Congress Card No.: applied for
British Library Cataloging-in-Publication Data:
A catalogue record for this book is available from the British Library

Bibliographic information published by the Deutsche Nationalbibliothek
The Deutsche Nationalbibliothek lists this publication in the Deutsche Nationalbibliografie; detailed bibliographic data are available in the Internet at http://dnb.d-nb.de.

Printed in the Federal Republic of Germany
Printed on acid-free paper

Printing Strauss GmbH, Mörlenbach
Bookbinding Litges & Dopf Buchbinderei GmbH, Heppenheim

ISBN-13: 978- 3-527-40534-3
ISBN-10: 3-527-40534-8

Contents

Preface

This book on integral equations and the calculus of variations is intended for use by senior undergraduate students and first-year graduate students in science and engineering. Basic familiarity with theories of linear algebra, calculus, differential equations, and complex analysis on the mathematics side, and classical mechanics, classical electrodynamics, quantum mechanics including the second quantization, and quantum statistical mechanics on the physics side, is assumed. Another prerequisite for this book on the mathematics side is a sound understanding of local and global analysis.

This book grew out of the course notes for the last of the three-semester sequence of *Methods of Applied Mathematics I (Local Analysis), II (Global Analysis)* and *III (Integral Equations and Calculus of Variations)* taught in the Department of Mathematics at MIT. About two-thirds of the course is devoted to integral equations and the remaining one-third to the calculus of variations. Professor Hung Cheng taught the course on integral equations and the calculus of variations every other year from the mid 1960s through the mid 1980s at MIT. Since then, younger faculty have been teaching the course in turn. The course notes evolved in the intervening years. This book is the culmination of these joint efforts.

There will be the obvious question: Why yet another book on integral equations and the calculus of variations? There are already many excellent books on the theory of integral equations. No existing book, however, discusses the singular integral equations in detail; in particular, Wiener–Hopf integral equations and Wiener–Hopf sum equations with the notion of the Wiener–Hopf index. In this book, the notion of the Wiener–Hopf index is discussed in detail.

This book is organized as follows. In Chapter 1 we discuss the notion of function space, the linear operator, the Fredholm alternative and Green's functions, to prepare the reader for the further development of the material. In Chapter 2 we discuss a few examples of integral equations and Green's functions. In Chapter 3 we discuss integral equations of the Volterra type. In Chapter 4 we discuss integral equations of the Fredholm type. In Chapter 5 we discuss the Hilbert–Schmidt theories of the symmetric kernel. In Chapter 6 we discuss singular integral equations of the Cauchy type. In Chapter 7, we discuss the Wiener–Hopf method for the mixed boundary-value problem in classical electrodynamics, Wiener–Hopf integral equations, and Wiener–Hopf sum equations; the latter two topics being discussed in terms of the notion of the index. In Chapter 8 we discuss nonlinear integral equations of the Volterra, Fredholm and Hammerstein type. In Chapter 9 we discuss the calculus of variations, in particular, the second variations, the Legendre test and the Jacobi test, and the relationship between integral equations and applications of the calculus of variations. In Chapter 10 we discuss Feynman's action principle in quantum mechanics and Feynman's variational principle, a system

Applied Mathematics in Theoretical Physics. Michio Masujima

ISBN: 3-527-40534-8

of the Schwinger–Dyson equations in quantum field theory and quantum statistical mechanics, Weyl's gauge principle and Kibble's gauge principle.

A substantial portion of Chapter 10 is taken from my monograph, "*Path Integral Quantization and Stochastic Quantization*", Vol. 165, Springer Tracts in Modern Physics, Springer, Heidelberg, published in the year 2000.

A reasonable understanding of Chapter 10 requires the reader to have a basic understanding of classical mechanics, classical field theory, classical electrodynamics, quantum mechanics including the second quantization, and quantum statistical mechanics. For this reason, Chapter 10 can be read as a side reference on theoretical physics, independently of Chapters 1 through 9.

The examples are mostly taken from classical mechanics, classical field theory, classical electrodynamics, quantum mechanics, quantum statistical mechanics and quantum field theory. Most of them are worked out in detail to illustrate the methods of the solutions. Those examples which are not worked out in detail are either intended to illustrate the general methods of the solutions or it is left to the reader to complete the solutions.

At the end of each chapter, with the exception of Chapter 1, problem sets are given for sound understanding of the content of the main text. The reader is recommended to solve all the problems at the end of each chapter. Many of the problems were created by Professor Hung Cheng over the past three decades. The problems due to him are designated by the note '(Due to H. C.)'. Some of the problems are those encountered by Professor Hung Cheng in the course of his own research activities.

Most of the problems can be solved by the direct application of the method illustrated in the main text. Difficult problems are accompanied by the citation of the original references. The problems for Chapter 10 are mostly taken from classical mechanics, classical electrodynamics, quantum mechanics, quantum statistical mechanics and quantum field theory.

A bibliography is provided at the end of the book for an in-depth study of the background materials in physics, beside the standard references on the theory of integral equations and the calculus of variations.

The instructor can cover Chapters 1 through 9 in one semester or two quarters with a choice of the topic of his or her own taste from Chapter 10.

I would like to express many heart-felt thanks to Professor Hung Cheng at MIT, who appointed me as his teaching assistant for the course when I was a graduate student in the Department of Mathematics at MIT, for his permission to publish this book under my single authorship and also for his criticism and constant encouragement without which this book would not have materialized.

I would like to thank Professor Francis E. Low and Professor Kerson Huang at MIT, who taught me many of the topics within theoretical physics. I would like to thank Professor Roberto D. Peccei at Stanford University, now at UCLA, who taught me quantum field theory and dispersion theory.

I would like to thank Professor Richard M. Dudley at MIT, who taught me real analysis and theories of probability and stochastic processes. I would like to thank Professor Herman Chernoff, then at MIT, now at Harvard University, who taught me many topics in mathematical statistics starting from multivariate normal analysis, for his supervision of my Ph. D. thesis at MIT.

I would like to thank Dr. Ali Nadim for supplying his version of the course notes and Dr. Dionisios Margetis at MIT for supplying examples and problems of integral equations from his courses at Harvard University and MIT. The problems due to him are designated by the note '(Due to D. M.)'. I would like to thank Dr. George Fikioris at the National Technical University of Athens for supplying the references on the Yagi–Uda semi-infinite arrays.

I would like to thank my parents, Mikio and Hanako Masujima, who made my undergraduate study at MIT possible by their financial support. I also very much appreciate their moral support during my graduate student days at MIT. I would like to thank my wife, Mari, and my son, Masachika, for their strong moral support, patience and encouragement during the period of the writing of this book, when the 'going got tough'.

Lastly, I would like to thank Dr. Alexander Grossmann and Dr. Ron Schulz of Wiley-VCH GmbH & Co. KGaA for their administrative and legal assistance in resolving the copyright problem with Springer.

Michio Masujima

Tokyo, Japan,
June, 2004

Introduction

Many problems within theoretical physics are frequently formulated in terms of ordinary differential equations or partial differential equations. We can often convert them into integral equations with boundary conditions or with initial conditions built in. We can formally develop the perturbation series by iterations. A good example is the Born series for the potential scattering problem in quantum mechanics. In some cases, the resulting equations are nonlinear integro-differential equations. A good example is the Schwinger–Dyson equation in quantum field theory and quantum statistical mechanics. It is the nonlinear integro-differential equation, and is exact and closed. It provides the starting point of Feynman–Dyson type perturbation theory in configuration space and in momentum space. In some singular cases, the resulting equations are Wiener–Hopf integral equations. These originate from research on the radiative equilibrium on the surface of a star. In the two-dimensional Ising model and the analysis of the Yagi–Uda semi-infinite arrays of antennas, among others, we have the Wiener–Hopf sum equation.

The theory of integral equations is best illustrated by the notion of functionals defined on some function space. If the functionals involved are quadratic in the function, the integral equations are said to be linear integral equations, and if they are higher than quadratic in the function, the integral equations are said to be nonlinear integral equations. Depending on the form of the functionals, the resulting integral equations are said to be of the first kind, of the second kind, or of the third kind. If the kernels of the integral equations are square-integrable, the integral equations are said to be nonsingular, and if the kernels of the integral equations are not square-integrable, the integral equations are then said to be singular. Furthermore, depending on whether or not the endpoints of the kernel are fixed constants, the integral equations are said to be of the Fredholm type, Volterra type, Cauchy type, or Wiener–Hopf types, etc. By the discussion of the variational derivative of the quadratic functional, we can also establish the relationship between the theory of integral equations and the calculus of variations. The integro-differential equations can best be formulated in this manner. Analogies of the theory of integral equations with the system of linear algebraic equations are also useful.

The integral equation of Cauchy type has an interesting application to classical electrodynamics, namely, dispersion relations. Dispersion relations were derived by Kramers in 1927 and Kronig in 1926, for X-ray dispersion and optical dispersion, respectively. Kramers-Kronig dispersion relations are of very general validity which only depends on the assumption of the causality. The requirement of the causality alone determines the region of analyticity of dielectric constants. In the mid 1950s, these dispersion relations were also derived from quantum field theory and applied to strong interaction physics. The application of the covariant perturbation theory to strong interaction physics was impossible due to the large coupling

Applied Mathematics in Theoretical Physics. Michio Masujima

ISBN: 3-527-40534-8

constant. From the mid 1950s to the 1960s, the dispersion-theoretic approach to strong interaction physics was the only realistic approach that provided many sum rules. To cite a few, we have the Goldberger–Treiman relation, the Goldberger–Miyazawa–Oehme formula and the Adler–Weisberger sum rule. In the dispersion-theoretic approach to strong interaction physics, experimentally observed data were directly used in the sum rules. The situation changed dramatically in the early 1970s when quantum chromodynamics, the relativistic quantum field theory of strong interaction physics, was invented by the use of asymptotically-free non-Abelian gauge field theory.

The region of analyticity of the scattering amplitude in the upper-half k-plane in quantum field theory, when expressed in terms of the Fourier transform, is immediate since quantum field theory has microscopic causality. But, the region of analyticity of the scattering amplitude in the upper-half k-plane in quantum mechanics, when expressed in terms of the Fourier transform, is not immediate since quantum mechanics does not have microscopic causality. We shall invoke the generalized triangular inequality to derive the region of analyticity of the scattering amplitude in the upper-half k-plane in quantum mechanics. This region of analyticity of the scattering amplitudes in the upper-half k-plane in quantum mechanics and quantum field theory strongly depends on the fact that the scattering amplitudes are expressed in terms of the Fourier transform. When another expansion basis is chosen, such as the Fourier–Bessel series, the region of analyticity drastically changes its domain.

In the standard application of the calculus of variations to the variety of problems in theoretical physics, we simply write the Euler equation and are rarely concerned with the second variations; the Legendre test and the Jacobi test. Examination of the second variations and the application of the Legendre test and the Jacobi test becomes necessary in some cases of the application of the calculus of variations theoretical physics problems. In order to bring the development of theoretical physics and the calculus of variations much closer, some historical comments are in order here.

Euler formulated Newtonian mechanics by the variational principle; the Euler equation. Lagrange began the whole field of the calculus of variations. He also introduced the notion of generalized coordinates into classical mechanics and completely reduced the problem to that of differential equations, which are presently known as Lagrange equations of motion, with the Lagrangian appropriately written in terms of kinetic energy and potential energy. He successfully converted classical mechanics into analytical mechanics using the variational principle. Legendre constructed the transformation methods for thermodynamics which are presently known as the Legendre transformations. Hamilton succeeded in transforming the Lagrange equations of motion, which are of the second order, into a set of first-order differential equations with twice as many variables. He did this by introducing the canonical momenta which are conjugate to the generalized coordinates. His equations are known as Hamilton's canonical equations of motion. He successfully formulated classical mechanics in terms of the principle of least action. The variational principles formulated by Euler and Lagrange apply only to the conservative system. Hamilton recognized that the principle of least action in classical mechanics and Fermat's principle of shortest time in geometrical optics are strikingly analogous, permitting the interpretation of optical phenomena in mechanical terms and vice versa. Jacobi quickly realized the importance of the work of Hamilton. He noted that Hamilton was using just one particular set of the variables to describe the mechanical system and formulated the canonical transformation theory using the Legendre transformation. He duly arrived at what is presently known as the Hamilton–Jacobi equation. He formulated his version of the principle of least action for the time-independent case.

Path integral quantization procedure, invented by Feynman in 1942 in the Lagrangian formalism, is usually justified by the Hamiltonian formalism. We deduce the canonical formalism of quantum mechanics from the path integral formalism. As a byproduct of the discussion of the Schwinger–Dyson equation, we deduce the path integral formalism of quantum field theory from the canonical formalism of quantum field theory.

Weyl's gauge principle also attracts considerable attention due to the fact that all forces in nature; the electromagnetic force, the weak force and the strong force, can be unified with Weyl's gauge principle by the appropriate choice of the grand unifying Lie groups as the gauge group. Inclusion of the gravitational force requires the use of superstring theory.

Basic to these are the integral equations and the calculus of variations.

1 Function Spaces, Linear Operators and Green's Functions

1.1 Function Spaces

Consider the set of all complex valued functions of the real variable x, denoted by $f(x), g(x), \ldots$ and defined on the interval (a, b). We shall restrict ourselves to those functions which are *square-integrable*. Define the *inner product* of any two of the latter functions by

$$(f, g) \equiv \int_a^b f^*(x) g(x)\, dx, \tag{1.1.1}$$

in which $f^*(x)$ is the complex conjugate of $f(x)$. The following properties of the inner product follow from the definition (1.1.1).

$$\begin{aligned}
(f, g)^* &= (g, f), \\
(f, g+h) &= (f, g) + (f, h), \\
(f, \alpha g) &= \alpha(f, g), \\
(\alpha f, g) &= \alpha^*(f, g),
\end{aligned} \tag{1.1.2}$$

with α a complex scalar.

While the inner product of any two functions is in general a complex number, the inner product of a function with itself is a real number and is non-negative. This prompts us to define the *norm of a function* by

$$\|f\| \equiv \sqrt{(f, f)} = \left[\int_a^b f^*(x) f(x)\, dx\right]^{\frac{1}{2}}, \tag{1.1.3}$$

provided that f is *square-integrable*, i.e., $\|f\| < \infty$. Equation (1.1.3) constitutes a proper definition for a norm since it satisfies the following conditions,

$$\begin{array}{llll}
\text{(i)} & \textit{scalar multiplication} & \|\alpha f\| = |\alpha| \cdot \|f\|, & \text{for all complex } \alpha, \\
\text{(ii)} & \textit{positivity} & \|f\| > 0, & \text{for all } f \neq 0, \\
 & & \|f\| = 0, & \text{if and only if } f = 0, \\
\text{(iii)} & \textit{triangular inequality} & \|f + g\| \leq \|f\| + \|g\|. &
\end{array} \tag{1.1.4}$$

Applied Mathematics in Theoretical Physics. Michio Masujima

ISBN: 3-527-40534-8

A very important inequality satisfied by the *inner product* (1.1.1) is the so-called *Schwarz inequality* which says

$$|(f,g)| \leq \|f\| \cdot \|g\| . \tag{1.1.5}$$

To prove the latter, start with the trivial inequality $\|(f+\alpha g)\|^2 \geq 0$, which holds for any $f(x)$ and $g(x)$ and for any complex number α. With a little algebra, the left-hand side of this inequality may be expanded to yield

$$(f,f) + \alpha^*(g,f) + \alpha(f,g) + \alpha\alpha^*(g,g) \geq 0. \tag{1.1.6}$$

The latter inequality is true for any α, and is thus true for the value of α which minimizes the left-hand side. This value can be found by writing α as $a + ib$ and minimizing the left-hand side of Eq. (1.1.6) with respect to the real variables a and b. A quicker way would be to treat α and α^* as independent variables and requiring $\partial/\partial\alpha$ and $\partial/\partial\alpha^*$ of the left hand side of Eq. (1.1.6) to vanish. This immediately yields $\alpha = -(g,f)/(g,g)$ as the value of α at which the minimum occurs. Evaluating the left-hand side of Eq. (1.1.6) at this minimum then yields

$$\|f\|^2 \geq \frac{|(f,g)|^2}{\|g\|^2}, \tag{1.1.7}$$

which proves the Schwarz inequality (1.1.5).

Once the Schwarz inequality has been established, it is relatively easy to prove the *triangular inequality* (1.1.4)(iii). To do this, we simply begin from the definition

$$\|f+g\|^2 = (f+g, f+g) = (f,f) + (f,g) + (g,f) + (g,g). \tag{1.1.8}$$

Now the right-hand side of Eq. (1.1.8) is a sum of complex numbers. Applying the usual triangular inequality for complex numbers to the right-hand side of Eq. (1.1.8) yields

$$\begin{aligned} |\text{Right-hand side of Eq. (1.1.8)}| &\leq \|f\|^2 + |(f,g)| + |(g,f)| + \|g\|^2 \\ &= (\|f\| + \|g\|)^2. \end{aligned} \tag{1.1.9}$$

Combining Eqs. (1.1.8) and (1.1.9) finally proves the triangular inequality (1.1.4)(iii).

We remark finally that the set of functions $f(x)$, $g(x)$, ... is an example of a *linear vector space*, equipped with an inner product and a norm based on that inner product. A similar set of properties, including the Schwarz and triangular inequalities, can be established for other linear vector spaces. For instance, consider the set of all complex column vectors $\vec{u}$, $\vec{v}$, $\vec{w}$, ... of finite dimension n. If we define the inner product

$$(\vec{u}, \vec{v}) \equiv (\vec{u}^*)^{\mathrm{T}}\vec{v} = \sum_{k=1}^{n} u_k^* v_k, \tag{1.1.10}$$

and the related norm

$$\|\vec{u}\| \equiv \sqrt{(\vec{u}, \vec{u})}, \tag{1.1.11}$$

then the corresponding Schwarz and triangular inequalities can be proven in an identical manner yielding

$$|(\vec{u}, \vec{v})| \leq \|\vec{u}\| \, \|\vec{v}\| , \tag{1.1.12}$$

and

$$\|\vec{u} + \vec{v}\| \leq \|\vec{u}\| + \|\vec{v}\| . \tag{1.1.13}$$

1.2 Orthonormal System of Functions

Two functions $f(x)$ and $g(x)$ are said to be *orthogonal* if their inner product vanishes, i.e.,

$$(f, g) = \int_a^b f^*(x) g(x) \, dx = 0. \tag{1.2.1}$$

A function is said to be *normalized* if its norm equals to unity, i.e.,

$$\|f\| = \sqrt{(f, f)} = 1. \tag{1.2.2}$$

Consider now a set of normalized functions $\{\phi_1(x), \phi_2(x), \phi_3(x), \ldots\}$ which are mutually orthogonal. Such a set is called an *orthonormal set of functions*, satisfying the orthonormality condition

$$(\phi_i, \phi_j) = \delta_{ij} = \begin{cases} 1, & \text{if } i = j, \\ 0, & \text{otherwise,} \end{cases} \tag{1.2.3}$$

where δ_{ij} is the *Kronecker delta symbol* itself defined by Eq. (1.2.3).

An orthonormal set of functions $\{\phi_n(x)\}$is said to form a *basis for a function space*, or to be *complete*, if any function $f(x)$ in that space can be expanded in a series of the form

$$f(x) = \sum_{n=1}^{\infty} a_n \phi_n(x). \tag{1.2.4}$$

(This is not the exact definition of a complete set but it will do for our purposes.) To find the coefficients of the expansion in Eq. (1.2.4), we take the inner product of both sides with $\phi_m(x)$ from the left to obtain

$$\begin{aligned} (\phi_m, f) &= \sum_{n=1}^{\infty} (\phi_m, a_n \phi_n) \\ &= \sum_{n=1}^{\infty} a_n \, (\phi_m, \phi_n) \\ &= \sum_{n=1}^{\infty} a_n \delta_{mn} = a_m. \end{aligned} \tag{1.2.5}$$

In other words, for any n,

$$a_n = (\phi_n, f) = \int_a^b \phi_n^* (x) f (x) \ dx. \tag{1.2.6}$$

An example of an orthonormal system of functions on the interval $(-l, l)$ is the infinite set

$$\phi_n (x) = \frac{1}{\sqrt{2l}} \exp\left[\frac{in\pi x}{l}\right], \quad n = 0, \pm 1, \pm 2, \ldots \tag{1.2.7}$$

with which the expansion of a *square-integrable* function $f(x)$ on $(-l, l)$ takes the form

$$f(x) = \sum_{n=-\infty}^{\infty} c_n \exp\left[\frac{in\pi x}{l}\right], \tag{1.2.8a}$$

with

$$c_n = \frac{1}{2l} \int_{-l}^{+l} f(x) \exp\left[-\frac{in\pi x}{l}\right], \tag{1.2.8b}$$

which is the familiar complex form of the *Fourier series* of $f(x)$.

Finally the *Dirac delta function* $\delta (x - x')$, defined with x and x' in (a, b), can be expanded in terms of a complete set of orthonormal functions $\phi_n(x)$ in the form

$$\delta (x - x') = \sum_n a_n \phi_n(x)$$

with

$$a_n = \int_a^b \phi_n^*(x) \delta(x - x') \ dx = \phi_n^* (x') .$$

That is,

$$\delta(x - x') = \sum_n \phi_n^*(x') \phi_n(x). \tag{1.2.9}$$

The expression (1.2.9) is sometimes taken as the statement which implies the *completeness of an orthonormal system of functions*.

1.3 Linear Operators

An *operator* can be thought of as a mapping or a transformation which acts on a member of the function space (i.e., a function) to produce another member of that space (i.e., another function). The operator, typically denoted by a symbol such as L, is said to be *linear* if it satisfies

$$L(\alpha f + \beta g) = \alpha L f + \beta L g, \tag{1.3.1}$$

where α and β are complex numbers, and f and g are members of that function space.

Some trivial examples of linear operators L are

(i) multiplication by a constant scalar, i.e.,

$$L\phi = a\phi,$$

(ii) taking the third derivative of a function, i.e.,

$$L\phi = \frac{d^3}{dx^3}\phi \quad \text{or} \quad L = \frac{d^3}{dx^3},$$

which is a differential operator, or,

(iii) multiplying a function by the kernel, $K(x, x')$, and integrating over (a, b) with respect to x', i.e.,

$$L\phi(x) = \int_a^b K(x, x')\phi(x')\, dx',$$

which is an integral operator.

An important concept in the theory of the linear operator is that of the *adjoint* of the operator which is defined as follows. Given the operator L, together with an inner product defined on a vector space, the adjoint L^{adj} of the operator L is that operator for which

$$(\psi, L\phi) = (L^{\text{adj}}\psi, \phi), \tag{1.3.2}$$

is an identity for any two members ϕ and ψ of the vector space. Actually, as we shall see later, in the case of the differential operators, we frequently need to worry to some extent about the boundary conditions associated with the original and the adjoint problems. Indeed, there often arise additional terms on the right-hand side of Eq. (1.3.2) which involve the boundary points, and a prudent choice of the adjoint boundary conditions will need to be made in order to avoid unnecessary difficulties. These issues will be raised in connection with Green's functions for differential equations.

As our first example of the adjoint operator, consider the liner vector space of n-dimensional complex column vectors $\vec{u}$, $\vec{v}$, ... with their associated inner product (1.1.10). In this space, $n\times n$ square matrices A, B, ... with complex entries are linear operators when multiplied by the n-dimensional complex column vectors according to the usual rules of matrix multiplication. Consider now the problem of finding the adjoint A^{adj} of the matrix A. According to the definition (1.3.2) of the adjoint operator, we search for the matrix A^{adj} satisfying

$$(\vec{u}, A\vec{v}) = (A^{\text{adj}}\vec{u}, \vec{v}). \tag{1.3.3}$$

Now, from the definition of the inner product (1.1.10), we must have

$$\vec{u}^{*\text{T}}(A^{\text{adj}})^{*\text{T}}\vec{v} = \vec{u}^{*\text{T}}A\vec{v},$$

i.e.,

$$(A^{\text{adj}})^{*\text{T}} = A \quad \text{or} \quad A^{\text{adj}} = A^{*\text{T}}. \tag{1.3.4}$$

That is, the adjoint A^{adj} of a matrix A is equal to the complex conjugate of its transpose, which is also known as its *Hermitian transpose*,

$$A^{\mathrm{adj}} = A^{*\mathrm{T}} \equiv A^{\mathrm{H}}. \tag{1.3.5}$$

As a second example, consider the problem of finding the adjoint of the linear integral operator

$$L = \int_a^b dx' K(x, x'), \tag{1.3.6}$$

on our function space. By definition, the adjoint L^{adj} of L is the operator which satisfies Eq. (1.3.2). Upon expressing the left-hand side of Eq. (1.3.2) explicitly with the operator L given by Eq. (1.3.6), we find

$$(\psi, L\phi) = \int_a^b dx\, \psi^*(x) L\phi(x) = \int_a^b dx' \left[\int_a^b dx K(x, x') \psi^*(x) \right] \phi(x'). \tag{1.3.7}$$

Requiring Eq. (1.3.7) to be equal to

$$(L^{\mathrm{adj}}\psi, \phi) = \int_a^b dx (L^{\mathrm{adj}}\psi(x))^* \phi(x)$$

necessitates defining

$$L^{\mathrm{adj}}\psi(x) = \int_a^b d\xi K^*(\xi, x)\psi(\xi).$$

Hence the adjoint of integral operator (1.3.6) is found to be

$$L^{\mathrm{adj}} = \int_a^b dx' K^*(x', x). \tag{1.3.8}$$

Note that, aside from the complex conjugation of the kernel $K(x, x')$, the integration in Eq. (1.3.6) is carried out with respect to the second argument of $K(x, x')$ while that in Eq. (1.3.8) is carried out with respect to the first argument of $K^*(x', x)$. Also, be careful to note which of the variables throughout the above is the dummy variable of integration.

Before we end this section, let us define what is meant by a *self-adjoint* operator. An operator L is said to be self-adjoint (or *Hermitian*) if it is equal to its own adjoint L^{adj}. Hermitian operators have very nice properties which will be discussed in Section 1.6. Not the least of these is that their eigenvalues are real. (Eigenvalue problems are discussed in the next section.)

Examples of self-adjoint operators are Hermitian matrices, i.e., matrices which satisfies

$$A = A^{\mathrm{H}},$$

and linear integral operators of the type (1.3.6) whose kernel satisfy

$$K(x, x') = K^*(x', x),$$

each on their respective linear spaces and with their respective inner products.

1.4 Eigenvalues and Eigenfunctions

Given a linear operator L on a linear vector space, we can set up the following eigenvalue problem

$$L\phi_n = \lambda_n \phi_n \quad (n = 1, 2, 3, \ldots). \tag{1.4.1}$$

Obviously the trivial solution $\phi(x) = 0$ always satisfies this equation, but it also turns out that for some particular values of λ (called the *eigenvalues* and denoted by λ_n), nontrivial solutions to Eq. (1.4.1) also exist. Note that for the case of the differential operators on bounded domains, we must also specify an appropriate homogeneous boundary condition (such that $\phi = 0$ satisfies those boundary conditions) for the *eigenfunctions* $\phi_n(x)$. We have affixed the subscript n to the eigenvalues and eigenfunctions under the assumption that the eigenvalues are discrete and that they can be counted (i.e., with $n = 1, 2, 3, \ldots$). This is not always the case. The conditions which guarantee the existence of a discrete (and complete) set of eigenfunctions are beyond the scope of this introductory chapter and will not be discussed.

So, for the moment, let us tacitly assume that the eigenvalues λ_n of Eq. (1.4.1) are discrete and that their eigenfunctions ϕ_n form a basis (i.e., a complete set) for their space.

Similarly the adjoint L^{adj} of the operator L would posses a set of eigenvalues and eigenfunctions satisfying

$$L^{\text{adj}}\psi_m = \mu_m \psi_m \quad (m = 1, 2, 3, \ldots). \tag{1.4.2}$$

It can be shown that the eigenvalues μ_m of the adjoint problem are equal to complex conjugates of the eigenvalues λ_n of the original problem. (We will prove this only for matrices but it remains true for general operators.) That is, if λ_n is an eigenvalue of L, λ_n^* is an eigenvalue of L^{adj}. This prompts us to rewrite Eq. (1.4.2) as

$$L^{\text{adj}}\psi_m = \lambda_m^* \psi_m, \quad (m = 1, 2, 3, \ldots). \tag{1.4.3}$$

It is then a trivial matter to show that the eigenfunctions of the adjoint and original operators are all orthogonal, except those corresponding to the same index ($n = m$). To do this, take the inner product of Eq. (1.4.1) with ψ_m from the left, and the inner product of Eq. (1.4.3) with ϕ_n from the right, to find

$$(\psi_m, L\phi_n) = (\psi_m, \lambda_n \phi_n) = \lambda_n (\psi_m, \phi_n) \tag{1.4.4}$$

and

$$(L^{\text{adj}}\psi_m, \phi_n) = (\lambda_m^* \psi_m, \phi_n) = \lambda_m (\psi_m, \phi_n). \tag{1.4.5}$$

Subtract the latter two equations and note that their left-hand sides are equal because of the definition of the adjoint, to get

$$0 = (\lambda_n - \lambda_m)(\psi_m, \phi_n). \tag{1.4.6}$$

This implies

$$(\psi_m, \phi_n) = 0 \quad \text{if} \quad \lambda_n \neq \lambda_m, \tag{1.4.7}$$

which proves the desired result. Also, since each ϕ_n and ψ_m is determined to within a multiplicative constant (e.g., if ϕ_n satisfies Eq. (1.4.1) so does $\alpha\phi_n$), the normalization for the latter can be chosen such that

$$(\psi_m, \phi_n) = \delta_{mn} = \begin{cases} 1, & \text{for } n = m, \\ 0, & \text{otherwise.} \end{cases} \tag{1.4.8}$$

Now, if the set of eigenfunctions ϕ_n $(n = 1, 2, \ldots)$ forms a complete set, any arbitrary function $f(x)$ in the space may be expanded as

$$f(x) = \sum_n a_n \phi_n(x), \tag{1.4.9}$$

and to find the coefficients a_n, we simply take the inner product of both sides with ψ_k to get

$$\begin{aligned}(\psi_k, f) = \sum_n (\psi_k, a_n \phi_n) &= \sum_n a_n (\psi_k, \phi_n) \\ &= \sum_n a_n \delta_{kn} = a_k,\end{aligned}$$

i.e.,

$$a_n = (\psi_n, f), \quad (n = 1, 2, 3, \ldots). \tag{1.4.10}$$

Note the difference between Eqs. (1.4.9) and (1.4.10) and the corresponding formulas (1.2.4) and (1.2.6) for an orthonormal system of functions. In the present case, neither $\{\phi_n\}$ nor $\{\psi_n\}$ form an orthonormal system, but they are orthogonal to one another.

Proof that the eigenvalues of the adjoint matrix are complex conjugates of the eigenvalues of the original matrix.

Above, we claimed without justification that the eigenvalues of the adjoint of an operator are complex conjugates of those of the original operator. Here we show this for the matrix case. The eigenvalues of a matrix A are given by

$$\det(A - \lambda I) = 0. \tag{1.4.11}$$

The latter is the characteristic equation whose n solutions for λ are the desired eigenvalues. On the other hand, the eigenvalues of A^{adj} are determined by setting

$$\det(A^{\text{adj}} - \mu I) = 0. \tag{1.4.12}$$

Since the determinant of a matrix is equal to that of its transpose, we easily conclude that the eigenvalues of A^{adj} are the complex conjugates of λ_n. □

1.5 The Fredholm Alternative

The *Fredholm Alternative*, which may be also called the *Fredholm solvability condition*, is concerned with the existence of the solution $y(x)$ of the inhomogeneous problem

$$Ly(x) = f(x), \tag{1.5.1}$$

where L is a given linear operator and $f(x)$ a known forcing term. As usual, if L is a differential operator, additional boundary or initial conditions must also be specified.

The Fredholm Alternative *states that the unknown function $y(x)$ can be determined uniquely if the corresponding homogeneous problem*

$$L\phi_H(x) = 0 \tag{1.5.2}$$

with homogeneous boundary conditions, has no nontrivial solutions. On the other hand, if the homogeneous problem (1.5.2) *does possess a nontrivial solution, then the inhomogeneous problem* (1.5.1) *has either no solution or infinitely many solutions.*

What determines the latter is the homogeneous solution ψ_H to the adjoint problem

$$L^{\text{adj}}\psi_H = 0. \tag{1.5.3}$$

Taking the inner product of Eq. (1.5.1) with ψ_H from the left,

$$(\psi_H, Ly) = (\psi_H, f).$$

Then, by the definition of the adjoint operator (excluding the case wherein L is a differential operator, to be discussed in Section 1.7.), we have

$$(L^{\text{adj}}\psi_H, y) = (\psi_H, f).$$

The left-hand side of the equation above is zero by the definition of ψ_H, Eq. (1.5.3).

Thus the criteria for the solvability of the inhomogeneous problem Eq. (1.5.1) *is given by*

$$(\psi_H, f) = 0.$$

If these criteria are satisfied, there will be an infinity of solutions to Eq. (1.5.1), *otherwise Eq.* (1.5.1) *will have no solution.*

To understand the above claims, let us suppose that L and L^{adj} possess complete sets of eigenfunctions satisfying

$$L\phi_n = \lambda_n\phi_n \quad (n = 0, 1, 2, \ldots), \tag{1.5.4a}$$

$$L^{\text{adj}}\psi_n = \lambda_n^*\psi_n \quad (n = 0, 1, 2, \ldots), \tag{1.5.4b}$$

with

$$(\psi_m, \phi_n) = \delta_{mn}. \tag{1.5.5}$$

The existence of a nontrivial homogeneous solution $\phi_H(x)$ to Eq. (1.5.2), as well as $\psi_H(x)$ to Eq. (1.5.3), is the same as having one of the eigenvalues λ_n in Eqs. (1.5.4a), (1.5.4b) to be zero. If this is the case, i.e., if zero is an eigenvalue of Eq. (1.5.4a) and hence Eq. (1.5.4b), we shall choose the subscript $n = 0$ to signify that eigenvalue ($\lambda_0 = 0$), and in that case

ϕ_0 and ψ_0 are the same as ϕ_H and ψ_H. The two circumstances in the Fredholm Alternative correspond to cases where zero is an eigenvalue of Eqs. (1.5.4a), (1.5.4b) and where it is not.

Let us proceed formally with the problem of solving the inhomogeneous problem Eq. (1.5.1). Since the set of eigenfunctions ϕ_n of Eq. (1.5.4a) is assumed to be complete, both the known function $f(x)$ and the unknown function $y(x)$ in Eq. (1.5.1) can presumably be expanded in terms of $\phi_n(x)$:

$$f(x) = \sum_{n=0}^{\infty} \alpha_n \phi_n(x), \tag{1.5.6}$$

$$y(x) = \sum_{n=0}^{\infty} \beta_n \phi_n(x), \tag{1.5.7}$$

where the α_n are known (since $f(x)$ is known), i.e., according to Eq. (1.4.10)

$$\alpha_n = (\psi_n, f), \tag{1.5.8}$$

while the β_n are unknown. Thus, if all the β_n can be determined, then the solution $y(x)$ to Eq. (1.5.1) is regarded as having been found.

To try to determine the β_n, substitute both Eqs. (1.5.6) and (1.5.7) into Eq. (1.5.1) to find

$$\sum_{n=0}^{\infty} \lambda_n \beta_n \phi_n = \sum_{k=0}^{\infty} \alpha_k \phi_k, \tag{1.5.9}$$

where different summation indices have been used on the two sides to remind the reader that the latter are dummy indices of summation. Next, take the inner product of both sides with ψ_m (with an index which must be different from the two above) to get

$$\sum_{n=0}^{\infty} \lambda_n \beta_n (\psi_m, \phi_n) = \sum_{k=0}^{\infty} \alpha_k (\psi_m, \phi_k),$$

or

$$\sum_{n=0}^{\infty} \lambda_n \beta_n \delta_{mn} = \sum_{k=0}^{\infty} \alpha_k \delta_{mk},$$

i.e.,

$$\lambda_m \beta_m = \alpha_m. \tag{1.5.10}$$

Thus, for any $m = 0, 1, 2, \ldots$, we can solve Eq. (1.5.10) for the unknowns β_m to get

$$\beta_n = \alpha_n / \lambda_n \quad (n = 0, 1, 2, \ldots), \tag{1.5.11}$$

provided that λ_n is not equal to zero. Obviously the only possible difficulty occurs if one of the eigenvalues (which we take to be λ_0) is equal to zero. In that case, equation (1.5.10) with $m = 0$ reads

$$\lambda_0 \beta_0 = \alpha_0 \quad (\lambda_0 = 0). \tag{1.5.12}$$

Now if $\alpha_0 \neq 0$, then we cannot solve for β_0 and thus the problem $Ly = f$ has no solution. On the other hand if $\alpha_0 = 0$, i.e., if

$$(\psi_0, f) = (\psi_H, f) = 0, \tag{1.5.13}$$

implying that f is orthogonal to the homogeneous solution to the adjoint problem, then Eq. (1.5.12) is satisfied by any choice of β_0. All the other β_n ($n = 1, 2, \ldots$) are uniquely determined but there are infinitely many solutions $y(x)$ to Eq. (1.5.1) corresponding to the infinitely many values possible for β_0. The reader must make certain that he or she understands the equivalence of the above with the original statement of the Fredholm Alternative.

1.6 Self-adjoint Operators

Operators which are self-adjoint or Hermitian form a very useful class of operators. They possess a number of special properties, some of which are described in this section.

The first important property of self-adjoint operators is that their *eigenvalues are real.* To prove this, begin with

$$\begin{aligned} L\phi_n &= \lambda_n \phi_n, \\ L\phi_m &= \lambda_m \phi_m, \end{aligned} \tag{1.6.1}$$

and take the inner product of both sides of the former with ϕ_m from the left, and the latter with ϕ_n from the right, to obtain

$$\begin{aligned} (\phi_m, L\phi_n) &= \lambda_n (\phi_m, \phi_n), \\ (L\phi_m, \phi_n) &= \lambda_m^* (\phi_m, \phi_n). \end{aligned} \tag{1.6.2}$$

For a self-adjoint operator $L = L^{\text{adj}}$, the two left-hand sides of Eq. (1.6.2) are equal and hence, upon subtraction of the latter from the former, we find

$$0 = (\lambda_n - \lambda_m^*)(\phi_m, \phi_n). \tag{1.6.3}$$

Now, if $m = n$, the inner product $(\phi_n, \phi_n) = \|\phi_n\|^2$ is nonzero and Eq. (1.6.3) implies

$$\lambda_n = \lambda_n^*, \tag{1.6.4}$$

proving that all the eigenvalues are real. Thus Eq. (1.6.3) can be rewritten as

$$0 = (\lambda_n - \lambda_m)(\phi_m, \phi_n), \tag{1.6.5}$$

indicating that if $\lambda_n \neq \lambda_m$, then the eigenfunctions ϕ_m and ϕ_n are orthogonal. Thus, upon normalizing each ϕ_n, we verify a second important property of self-adjoint operators that (upon normalization) the *eigenfunctions of a self-adjoint operator form an orthonormal set.*

The Fredholm Alternative can also be restated for a self-adjoint operator L in the following form: The inhomogeneous problem $Ly = f$ (with L self-adjoint) is solvable for y, if f is orthogonal to all eigenfunctions ϕ_0 of L with eigenvalue zero (if indeed any exist). If zero is not an eigenvalue of L, the solution is unique. Otherwise, there is no solution if $(\phi_0, f) \neq 0$, and an infinite number of solutions if $(\phi_0, f) = 0$.

Diagonalization of Self-adjoint Operators: Any linear operator can be expanded in some sense in terms of any orthonormal basis set. To elaborate on this, suppose that the orthonormal system $\{e_i(x)\}_i$, with $(e_i, e_j) = \delta_{ij}$ forms a complete set. Any function $f(x)$ can be expanded as

$$f(x) = \sum_{j=1}^{\infty} \alpha_j e_j(x), \quad \alpha_j = (e_j, f). \tag{1.6.6}$$

Thus the function $f(x)$ can be thought of as an infinite dimensional vector with components α_j. Now consider the action of an arbitrary linear operator L on the function $f(x)$. Obviously

$$Lf(x) = \sum_{j=1}^{\infty} \alpha_j L e_j(x). \tag{1.6.7}$$

But L acting on $e_j(x)$ is itself a function of x which can be expanded in the orthonormal basis $\{e_i(x)\}_i$. Thus we write

$$Le_j(x) = \sum_{i=1}^{\infty} l_{ij} e_i(x), \tag{1.6.8}$$

wherein the coefficients l_{ij} of the expansion are found to be $l_{ij} = (e_i, Le_j)$. Substitution of Eq. (1.6.8) into Eq. (1.6.7) then shows

$$Lf(x) = \sum_{i=1}^{\infty} \left(\sum_{j=1}^{\infty} l_{ij} \alpha_j \right) e_i(x). \tag{1.6.9}$$

We discover that just as we can think of $f(x)$ as the infinite dimensional vector with components α_j, we can consider L to be equivalent to an infinite dimensional matrix with components l_{ij}, and we can regard Eq. (1.6.9) as a regular multiplication of the matrix L (components l_{ij}) with the vector f (components α_j). However, this equivalence of the operator L with the matrix whose components are l_{ij}, i.e., $L \Leftrightarrow l_{ij}$, depends on the choice of the orthonormal set.

For a self-adjoint operator $L = L^{\text{adj}}$, the most natural choice of the basis set is the set of eigenfunctions of L. Denoting these by $\{\phi_i(x)\}_i$, the components of the equivalent matrix for L take the form

$$l_{ij} = (\phi_i, L\phi_j) = (\phi_i, \lambda_j \phi_j) = \lambda_j (\phi_i, \phi_j) = \lambda_j \delta_{ij}. \tag{1.6.10}$$

1.7 Green's Functions for Differential Equations

In this section, we describe the conceptual basis of the theory of *Green's functions*. We do this by first outlining the abstract themes involved and then by presenting a simple example. More complicated examples will appear in later chapters.

Prior to discussing Green's functions, recall some of the elementary properties of the so-called Dirac delta function $\delta(x - x')$. In particular, remember that if x' is inside the domain of integration (a, b), for any well-behaved function $f(x)$, we have

$$\int_a^b \delta(x - x')f(x)\,dx = f(x'), \tag{1.7.1}$$

which can be written as

$$(\delta(x - x'), f(x)) = f(x'), \tag{1.7.2}$$

with the inner product taken with respect to x. Also remember that $\delta(x - x')$ is equal to zero for any $x \neq x'$.

Suppose now that we wish to solve a differential equation

$$Lu(x) = f(x), \tag{1.7.3}$$

on the domain $x \in (a, b)$ and subject to given boundary conditions, with L a differential operator. Consider what happens when a function $g(x, x')$ (which is as yet unknown but will end up being the Green's function) is multiplied on both sides of Eq. (1.7.3) followed by integration of both sides with respect to x from a to b. That is, consider taking the inner product of both sides of Eq. (1.7.3) with $g(x, x')$ with respect to x. (We suppose everything is real in this section so that no complex conjugation is necessary.) This yields

$$(g(x, x'), Lu(x)) = (g(x, x'), f(x)). \tag{1.7.4}$$

Now by definition of the adjoint L^{adj} of L, the left-hand side of Eq. (1.7.4) can be written as

$$(g(x, x'), Lu(x)) = (L^{\text{adj}}g(x, x'), u(x)) + \text{boundary terms}, \tag{1.7.5}$$

in which, for the first time, we explicitly recognize the terms involving the boundary points which arise when L is a differential operator. The *boundary terms* on the right-hand side of Eq. (1.7.5) emerge when we integrate by parts. It is difficult to be more specific than this when we work in the abstract, but our example should clarify what we mean shortly. If Eq. (1.7.5) is substituted back into Eq. (1.7.4), it provides

$$(L^{\text{adj}}g(x, x'), u(x)) = (g(x, x'), f(x)) + \text{boundary terms}. \tag{1.7.6}$$

So far we have not discussed what function $g(x, x')$ to choose. Suppose we choose that $g(x, x')$ which satisfies

$$L^{\text{adj}}g(x, x') = \delta(x - x'), \tag{1.7.7}$$

subject to appropriately selected boundary conditions which eliminate all the unknown terms within the boundary terms. This function $g(x, x')$ is known as Green's function. Substituting Eq. (1.7.7) into Eq. (1.7.6) and using property (1.7.2) then yields

$$u(x') = (g(x, x'), f(x)) + \text{known boundary terms}, \tag{1.7.8}$$

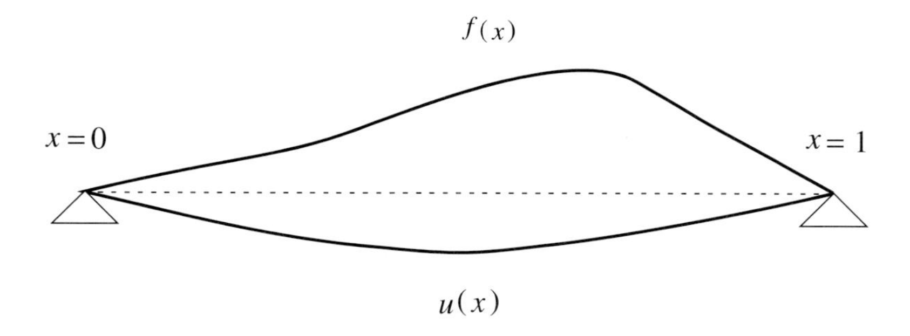

Fig. 1.1: Displacement $u(x)$ of a taut string under the distributed load $f(x)$ with $x \in (0,1)$.

which is the solution to the differential equation since everything on the right-hand side is known once $g(x,x')$ has been found. More accurately, if we change x' to x in the above and use a different dummy variable ξ of integration in the inner product, we have

$$u(x) = \int_a^b g(\xi,x)f(\xi)\,d\xi + \text{known boundary terms}. \tag{1.7.9}$$

In summary, to solve the linear inhomogeneous differential equation

$$Lu(x) = f(x)$$

using Green's function, we first solve the equation

$$L^{\text{adj}}g(x,x') = \delta(x-x')$$

for Green's function $g(x,x')$, subject to the appropriately selected boundary conditions, and immediately obtain the solution to our differential equation given by Eq. (1.7.9).

The above will we hope become more clear in the context of the following simple example.

❑ **Example 1.1.** Consider the problem of finding the displacement $u(x)$ of a taut string under the *distributed load* $f(x)$ as in Figure 1.1.

Solution. The governing ordinary differential equation for the vertical displacement $u(x)$ has the form

$$\frac{d^2u}{dx^2} = f(x) \quad \text{for} \quad x \in (0,1) \tag{1.7.10}$$

subject to boundary conditions

$$u(0) = 0 \quad \text{and} \quad u(1) = 0. \tag{1.7.11}$$

To proceed formally, multiply both sides of Eq. (1.7.10) by $g(x,x')$ and integrate from 0 to 1 with respect to x to find

$$\int_0^1 g(x,x')\frac{d^2u}{dx^2}\,dx = \int_0^1 g(x,x')f(x)\,dx.$$

Integrate the left-hand side by parts twice to obtain

$$\int_0^1 \frac{d^2}{dx^2} g(x,x')u(x)\,dx$$

$$+\left[g(1,x')\frac{du}{dx}\bigg|_{x=1} - g(0,x')\frac{du}{dx}\bigg|_{x=0} - u(1)\frac{dg(1,x')}{dx} + u(0)\frac{dg(0,x')}{dx}\right]$$

$$= \int_0^1 g(x,x')f(x)\,dx. \tag{1.7.12}$$

The terms contained within the square brackets on the left-hand side of (1.7.12) are the boundary terms. In consequence of the boundary conditions (1.7.11), the last two terms therein vanish. Hence a prudent choice of boundary conditions for $g(x,x')$ would be to set

$$g(0,x') = 0 \quad \text{and} \quad g(1,x') = 0. \tag{1.7.13}$$

With that choice, all the boundary terms vanish (this does not necessarily happen for other problems). Now suppose that $g(x,x')$ satisfies

$$\frac{d^2 g(x,x')}{dx^2} = \delta(x-x'), \tag{1.7.14}$$

subject to the boundary conditions (1.7.13). Use of Eqs. (1.7.14) and (1.7.13) in Eq. (1.7.12) yields

$$u(x') = \int_0^1 g(x,x')f(x)\,dx, \tag{1.7.15}$$

as our solution, once $g(x,x')$ has been obtained. Note that, if the original differential operator d^2/dx^2 is denoted by L, its adjoint L^{adj} is also d^2/dx^2 as found by twice integrating by parts. Hence the latter operator is indeed self-adjoint.

The last step involves the actual solution of (1.7.14) subject to (1.7.13). The variable x' plays the role of a parameter throughout. With x' somewhere between 0 and 1, Eq. (1.7.14) can actually be solved separately in each domain $0 < x < x'$ and $x' < x < 1$. For each of these, we have

$$\frac{d^2 g(x,x')}{dx^2} = 0 \quad \text{for} \quad 0 < x < x', \tag{1.7.16a}$$

$$\frac{d^2 g(x,x')}{dx^2} = 0 \quad \text{for} \quad x' < x < 1. \tag{1.7.16b}$$

The general solution in each subdomain is easily written down as

$$g(x,x') = Ax + B \quad \text{for} \quad 0 < x < x', \tag{1.7.17a}$$

$$g(x,x') = Cx + D \quad \text{for} \quad x' < x < 1, \tag{1.7.17b}$$

involving the four unknown constants A, B, C and D. Two relations for the constants are found using the two boundary conditions (1.7.13). In particular, we have

$$g(0, x') = 0 \rightarrow B = 0, \tag{1.7.18a}$$

$$g(1, x') = 0 \rightarrow C + D = 0. \tag{1.7.18b}$$

To provide two more relations which are needed to permit all four of the constants to be determined, we return to the governing equation (1.7.14). Integrate both sides of the latter with respect to x from $x' - \varepsilon$ to $x' + \varepsilon$ and take the limit as $\varepsilon \rightarrow 0$ to find

$$\lim_{\varepsilon \rightarrow 0} \int_{x'-\varepsilon}^{x'+\varepsilon} \frac{d^2 g(x, x')}{dx^2}\, dx = \lim_{\varepsilon \rightarrow 0} \int_{x'-\varepsilon}^{x'+\varepsilon} \delta(x - x')\, dx,$$

from which, we obtain

$$\left.\frac{dg(x, x')}{dx}\right|_{x=x'^+} - \left.\frac{dg(x, x')}{dx}\right|_{x=x'^-} = 1. \tag{1.7.19}$$

Thus the first derivative of $g(x, x')$ undergoes a jump discontinuity as x passes through x'. But we can expect $g(x, x')$ itself to be continuous across x', i.e.,

$$\left. g(x, x') \right|_{x=x'^+} = \left. g(x, x') \right|_{x=x'^-}. \tag{1.7.20}$$

In the above, x'^+ and x'^- denote points infinitesimally to the right and the left of x', respectively. Using solutions (1.7.17a) and (1.7.17b) for $g(x, x')$ in each subdomain, we find that Eqs. (1.7.19) and (1.7.20), respectively, imply

$$C - A = 1, \tag{1.7.21a}$$

$$Cx' + D = Ax' + B. \tag{1.7.21b}$$

Equations (1.7.18a), (1.7.18b) and (1.7.21a), (1.7.21b) can be used to solve for the four constants A, B, C and D to yield

$$A = x' - 1, \quad B = 0, \quad C = x', \quad D = -x',$$

from whence our solution (1.7.17) takes the form

$$g(x, x') = \begin{cases} (x' - 1)x & \text{for} \quad x < x', \\ x'(x - 1) & \text{for} \quad x > x', \end{cases} \tag{1.7.22a}$$

$$= x_<(x_> - 1) \quad \text{for} \quad \begin{cases} x_< = \frac{(x+x')}{2} - \frac{|x-x'|}{2}, \\ x_> = \frac{(x+x')}{2} + \frac{|x-x'|}{2}. \end{cases} \tag{1.7.22b}$$

Physically the Green's function (1.7.22) represents the displacement of the string subject to a *concentrated load* $\delta(x - x')$ at $x = x'$ as in Figure 1.2. For this reason, it is also called the *influence function*.

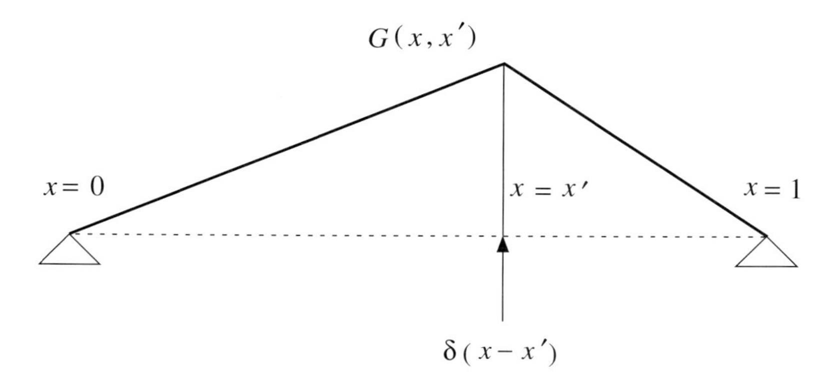

Fig. 1.2: Displacement $u(x)$ of a taut string under the concentrated load $\delta(x-x')$ at $x=x'$.

Having found the influence function above for a concentrated load, the solution with any given distributed load $f(x)$ is given by Eq. (1.7.15) as

$$\begin{aligned} u(x) &= \int_0^1 g(\xi, x) f(\xi)\, d\xi \\ &= \int_0^x (x-1)\xi f(\xi)\, d\xi + \int_x^1 x(\xi-1) f(\xi)\, d\xi \\ &= (x-1)\int_0^x \xi f(\xi)\, d\xi + x\int_x^1 (\xi-1) f(\xi)\, d\xi. \end{aligned} \tag{1.7.23}$$

Although this example has been rather elementary, we hope it has provided the reader with a basic understanding of what the Green's function is. More complex, and hence more interesting examples, are encountered in later chapters.

1.8 Review of Complex Analysis

Let us review some important results from complex analysis.

Cauchy Integral Formula. *Let $f(z)$ be analytic on and inside the closed, positively oriented contour C. Then we have*

$$f(z) = \frac{1}{2\pi i}\oint_C \frac{f(\zeta)}{\zeta - z} d\zeta. \tag{1.8.1}$$

Differentiate this formula with respect to z to obtain

$$\frac{d}{dz} f(z) = \frac{1}{2\pi i}\oint_C \frac{f(\zeta)}{(\zeta - z)^2}\, d\zeta, \text{ and } \left(\frac{d}{dz}\right)^n f(z) = \frac{n!}{2\pi i}\oint_C \frac{f(\zeta)}{(\zeta - z)^{n+1}}\, d\zeta. \tag{1.8.2}$$

Liouville's theorem. *The only entire functions which are bounded (at infinity) are constants.*

Proof. Suppose that $f(z)$ is entire. Then it can be represented by the Taylor series,

$$f(z) = f(0) + f^{(1)}(0)z + \frac{1}{2!}f^{(2)}(0)z^2 + \cdots .$$

Now consider $f^{(n)}(0)$. By the Cauchy Integral Formula, we have

$$f^{(n)}(0) = \frac{n!}{2\pi i}\oint_C \frac{f(\zeta)}{\zeta^{n+1}}d\zeta.$$

Since $f(\zeta)$ is bounded, we have

$$|f(\zeta)| \leq M.$$

Consider C to be a circle of radius R, centered at the origin. Then we have

$$\left|f^{(n)}(0)\right| \leq \frac{n!}{2\pi}\cdot\frac{2\pi RM}{R^{n+1}} = n!\cdot\frac{M}{R^n} \to 0 \quad \text{as} \quad R\to\infty.$$

Thus

$$f^{(n)}(0) = 0 \quad \text{for} \quad n = 1, 2, 3, \ldots .$$

Hence

$$f(z) = \text{constant},$$

completing the proof. □

More generally,

(i) Suppose that $f(z)$ is entire and we know $|f(z)| \leq |z|^a$ as $R\to\infty$, with $0 < a < 1$. We still find $f(z) = $ constant.

(ii) Suppose that $f(z)$ is entire and we know $|f(z)| \leq |z|^a$ as $R\to\infty$, with $n-1 \leq a < n$. Then $f(z)$ is at most a polynomial of degree $n-1$.

Discontinuity theorem. *Suppose that $f(z)$ has a branch cut on the real axis from a to b. It has no other singularities and it vanishes at infinity. If we know the difference between the value of $f(z)$ above and below the cut,*

$$D(x) \equiv f(x+i\varepsilon) - f(x-i\varepsilon), \quad (a \leq x \leq b), \tag{1.8.3}$$

with ε positive infinitesimal, then

$$f(z) = \frac{1}{2\pi i}\int_a^b \frac{D(x)}{(x-z)}\,dx. \tag{1.8.4}$$

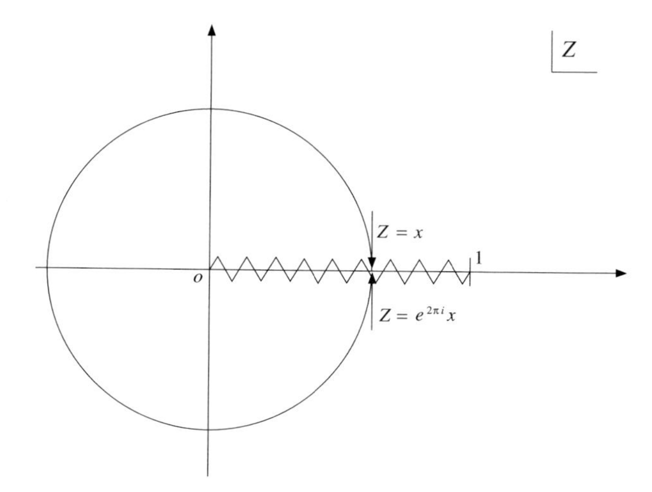

Fig. 1.3: The contours of the integration for $f(z)$. C_R is the circle of radius R centered at the origin.

Proof. By the Cauchy Integral Formula, we know

$$f(z) = \frac{1}{2\pi i}\oint_\Gamma \frac{f(\zeta)}{\zeta - z} d\zeta,$$

where Γ consists of the following pieces (see Figure 1.3),

$$\Gamma = \Gamma_1 + \Gamma_2 + \Gamma_3 + \Gamma_4 + C_R.$$

The contribution from C_R vanishes since $|f(z)| \to 0$ as $R \to \infty$. Contributions from Γ_3 and Γ_4 cancel each other. Hence we have

$$f(z) = \frac{1}{2\pi i}\left(\int_{\Gamma_1} + \int_{\Gamma_2}\right)\frac{f(\zeta)}{\zeta - z} d\zeta.$$

On Γ_1, we have

$$\zeta = x + i\varepsilon \quad \text{with} \quad x : a \to b, \quad f(\zeta) = f(x + i\varepsilon),$$

$$\int_{\Gamma_1} \frac{f(\zeta)}{\zeta - z} d\zeta = \int_a^b \frac{f(x + i\varepsilon)}{x - z + i\varepsilon}\, dx \to \int_a^b \frac{f(x + i\varepsilon)}{x - z}\, dx \quad \text{as} \quad \varepsilon \to 0^+.$$

On Γ_2, we have

$$\zeta = x - i\varepsilon \quad \text{with} \quad x : b \to a, \quad f(\zeta) = f(x - i\varepsilon),$$

$$\int_{\Gamma_2} \frac{f(\zeta)}{\zeta - z} d\zeta = \int_b^a \frac{f(x - i\varepsilon)}{x - z - i\varepsilon}\, dx \to -\int_a^b \frac{f(x - i\varepsilon)}{x - z}\, dx \quad \text{as} \quad \varepsilon \to 0^+.$$

Thus we obtain

$$f(z) = \frac{1}{2\pi i}\int_a^b \frac{f(x + i\varepsilon) - f(x - i\varepsilon)}{x - z}\, dx = \frac{1}{2\pi i}\int_a^b \frac{D(x)}{x - z}\, dx,$$

completing the proof. □

If, in addition, $f(z)$ is known to have other singularities elsewhere, or may possibly be nonzero as $|z| \to \infty$, then it is of the form

$$f(z) = \frac{1}{2\pi i} \int_a^b \frac{D(x)}{x-z}\,dx + g(z), \tag{1.8.5}$$

with $g(z)$ free of cut on $[a, b]$. This is a very important result. Memorizing it will give a better understanding of the subsequent sections.

Behavior near the endpoints. *Consider the case when z is in the vicinity of the endpoint a. The behavior of $f(z)$ as $z \to a$ is related to the form of $D(x)$ as $x \to a$. Suppose that $D(x)$ is finite at $x = a$, say $D(a)$. Then we have*

$$\begin{aligned} f(z) &= \frac{1}{2\pi i} \int_a^b \frac{D(a) + D(x) - D(a)}{x-z}\,dx \\ &= \frac{D(a)}{2\pi i} \ln\left(\frac{b-z}{a-z}\right) + \frac{1}{2\pi i} \int_a^b \frac{D(x) - D(a)}{x-z}\,dx. \end{aligned} \tag{1.8.6}$$

The second integral above converges as $z \to a$ as long as $D(x)$ satisfies a *Hölder condition* (which is implicitly assumed) requiring

$$|D(x) - D(a)| < A\,|x-a|^{\mu}, \quad A, \mu > 0. \tag{1.8.7}$$

Thus the endpoint behavior of $f(z)$ as $z \to a$ is of the form

$$f(z) = O\left(\ln(a-z)\right) \quad \text{as} \quad z \to a, \tag{1.8.8}$$

if

$$D(x) \quad \text{finite} \quad \text{as} \quad x \to a. \tag{1.8.9}$$

Another possibility is for $D(x)$ to be of the form

$$D(x) \to 1/(x-a)^{\alpha} \quad \text{with} \quad \alpha < 1 \quad \text{as} \quad x \to a, \tag{1.8.10}$$

since, even with such a singularity in $D(x)$, the integral defining $f(z)$ is well-defined. We claim that in that case, $f(z)$ also behaves as

$$f(z) = O\left(1/(z-a)^{\alpha}\right) \quad \text{as} \quad z \to a, \quad \text{with} \quad \alpha < 1, \tag{1.8.11}$$

that is, $f(z)$ is less singular than a simple pole.

Proof of the claim. Using the Cauchy Integral Formula, we have

$$1/(z-a)^{\alpha} = \frac{1}{2\pi i} \int_{\Gamma} \frac{d\zeta}{(\zeta - a)^{\alpha}(\zeta - z)},$$

where Γ consists of the following paths (see Figure 1.4)

$$\Gamma = \Gamma_1 + \Gamma_2 + C_R.$$

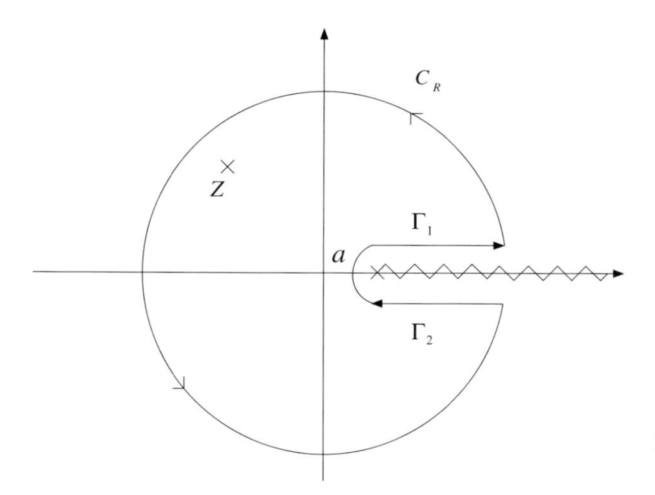

Fig. 1.4: The contour Γ of the integration for $1/(z-a)^\alpha$.

The contribution from C_R vanishes as $R \to \infty$.
On Γ_1, we set

$$\zeta - a = r, \quad \text{and} \quad (\zeta - a)^\alpha = r^\alpha,$$

$$\frac{1}{2\pi i}\int_{\Gamma_1} \frac{d\zeta}{(\zeta-a)^\alpha(\zeta-z)} = \frac{1}{2\pi i}\int_0^{+\infty} \frac{dr}{r^\alpha(r+a-z)}.$$

On Γ_2, we set

$$\zeta - a = re^{2\pi i}, \quad \text{and} \quad (\zeta - a)^\alpha = r^\alpha e^{2\pi i\alpha},$$

$$\frac{1}{2\pi i}\int_{\Gamma_2} \frac{d\zeta}{(\zeta-a)^\alpha(\zeta-z)} = \frac{e^{-2\pi i\alpha}}{2\pi i}\int_{+\infty}^{0} \frac{dr}{r^\alpha(r+a-z)}.$$

Thus we obtain

$$1/(z-a)^\alpha = \frac{1-e^{-2\pi i\alpha}}{2\pi i}\int_a^{+\infty} \frac{dx}{(x-a)^\alpha(x-z)},$$

which may be written as

$$1/(z-a)^\alpha = \frac{1-e^{-2\pi i\alpha}}{2\pi i}\left[\int_a^b \frac{dx}{(x-a)^\alpha(x-z)} + \int_b^{+\infty} \frac{dx}{(x-a)^\alpha(x-z)}\right].$$

The second integral above is convergent for z close to a. Obviously then, we have

$$\frac{1}{2\pi i}\int_a^b \frac{dx}{(x-a)^\alpha(x-z)} = O\left(\frac{1}{(z-a)^\alpha}\right) \quad \text{as} \quad z \to a.$$

A similar analysis can be done as $z \to b$, completing the proof. □

Summary of behavior near the endpoints

$$f(z) = \frac{1}{2\pi i} \int_a^b \frac{D(x)dx}{x - z},$$

$$\begin{cases} \text{if } D(x \to a) = D(a), & & \text{then} \quad f(z) = O(\ln(a - z)), \\ \text{if } D(x \to a) = \frac{1}{(x-a)^\alpha}, & (0 < \alpha < 1), & \text{then} \quad f(z) = O\left(\frac{1}{(z-a)^\alpha}\right), \end{cases} \tag{1.8.12a}$$

$$\begin{cases} \text{if } D(x \to b) = D(b), & & \text{then} \quad f(z) = O(\ln(b - z)), \\ \text{if } D(x \to b) = \frac{1}{(x-b)^\beta}, & (0 < \beta < 1), & \text{then} \quad f(z) = O\left(\frac{1}{(z-b)^\beta}\right). \end{cases} \tag{1.8.12b}$$

Principal Value Integrals. We define the principal value integral by

$$\mathrm{P}\int_a^b \frac{f(x)}{x - y}\, dx \equiv \lim_{\varepsilon \to 0+} \left[\int_a^{y-\varepsilon} \frac{f(x)}{x - y}\, dx + \int_{y+\varepsilon}^b \frac{f(x)}{x - y}\, dx \right]. \tag{1.8.13}$$

Graphically expressed, the principal value integral contour is as in Figure 1.5. As such, to evaluate a principal value integral by doing complex integration, we usually make use of either of the two contours as in Figure 1.6.

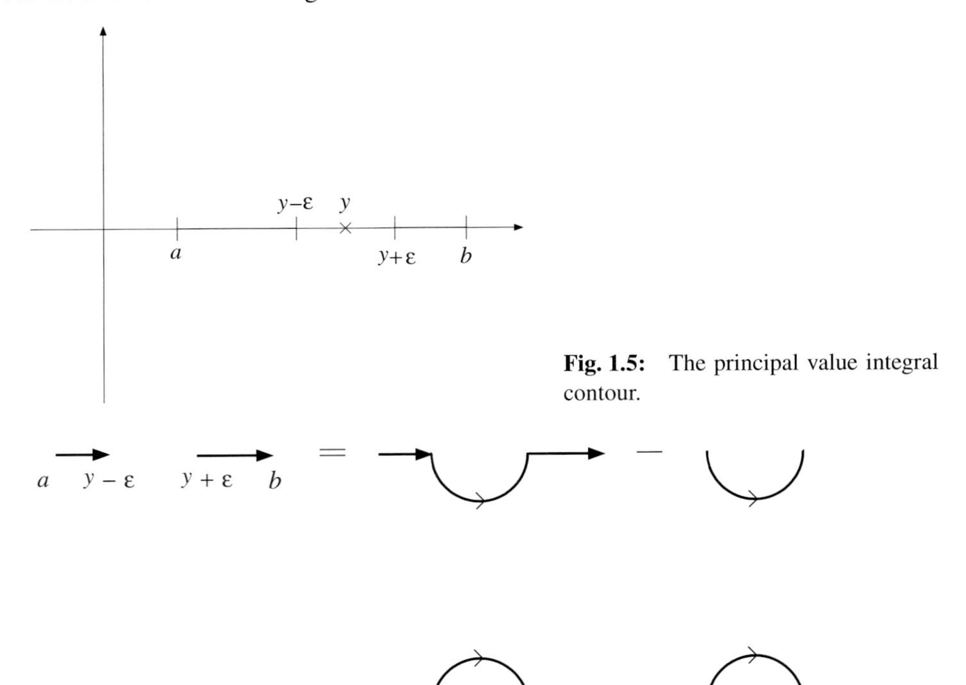

Fig. 1.5: The principal value integral contour.

Fig. 1.6: Two contours for the principal value integral (1.8.13).

Now, the contour integrals on the right of Figure 1.6 are usually possible and hence the principal value integral can be evaluated. Also, the contributions from the lower semicircle C_- and the upper semicircle C_+ take the forms,

$$\int_{C_-} \frac{f(z)}{z-y}\,dz = i\pi f(y), \qquad \int_{C_+} \frac{f(z)}{z-y}\,dz = -i\pi f(y),$$

as $\varepsilon \to 0^+$, as long as $f(z)$ is not singular at y. Mathematically expressed, the principal value integral is given by either of the following two formulae, known as the *Plemelj formula*,

$$\frac{1}{2\pi i}\mathrm{P}\int_a^b \frac{f(x)}{x-y}\,dx = \lim_{\varepsilon\to 0+}\frac{1}{2\pi i}\int_a^b \frac{f(x)}{x-y-i\varepsilon}\,dx - \frac{1}{2}f(y), \tag{1.8.14a}$$

$$\frac{1}{2\pi i}\mathrm{P}\int_a^b \frac{f(x)}{x-y}\,dx = \lim_{\varepsilon\to 0+}\frac{1}{2\pi i}\int_a^b \frac{f(x)}{x-y+i\varepsilon}\,dx + \frac{1}{2}f(y). \tag{1.8.14b}$$

These are customarily written as

$$\lim_{\varepsilon\to 0+}\frac{1}{x-y\mp i\varepsilon} = \mathrm{P}\left(\frac{1}{x-y}\right) \pm i\pi\delta(x-y), \tag{1.8.15a}$$

or may equivalently be written as

$$\mathrm{P}\left(\frac{1}{x-y}\right) = \lim_{\varepsilon\to 0+}\frac{1}{x-y\mp i\varepsilon} \mp i\pi\delta(x-y). \tag{1.8.15b}$$

Then we interchange the order of the limit $\varepsilon \to 0^+$ and the integration over x. The principal value integrand seems to diverge at $x = y$, but it is actually finite at $x = y$ as long as $f(x)$ is not singular at $x = y$. This comes about as follows;

$$\begin{aligned}\frac{1}{x-y\mp i\varepsilon} &= \frac{(x-y)\pm i\varepsilon}{(x-y)^2+\varepsilon^2} = \frac{(x-y)}{(x-y)^2+\varepsilon^2} \pm i\pi\cdot\frac{1}{\pi}\frac{\varepsilon}{(x-y)^2+\varepsilon^2} \\ &= \frac{(x-y)}{(x-y)^2+\varepsilon^2} \pm i\pi\delta_\varepsilon(x-y),\end{aligned} \tag{1.8.16}$$

where $\delta_\varepsilon(x-y)$ is defined by

$$\delta_\varepsilon(x-y) \equiv \frac{1}{\pi}\frac{\varepsilon}{(x-y)^2+\varepsilon^2}, \tag{1.8.17}$$

with the following properties,

$$\delta_\varepsilon(x \neq y) \to 0^+ \quad \text{as} \quad \varepsilon \to 0^+,$$

$$\delta_\varepsilon(x = y) = \frac{1}{\pi}\frac{1}{\varepsilon} \to +\infty \quad \text{as} \quad \varepsilon \to 0^+,$$

and

$$\int_{-\infty}^{+\infty} \delta_\varepsilon(x-y)\,dx = 1.$$

The first term on the right-hand side of Eq. (1.8.16) vanishes at $x = y$ before we take the limit $\varepsilon \to 0^+$, while the second term $\delta_\varepsilon(x-y)$ approaches the Dirac delta function, $\delta(x-y)$, as $\varepsilon \to 0^+$. This is the content of Eq. (1.8.15a).

1.9 Review of Fourier Transform

The Fourier transform of a function $f(x)$, where $-\infty < x < \infty$, is defined as

$$\tilde{f}(k) = \int_{-\infty}^{\infty} dx \exp[-ikx] f(x). \tag{1.9.1}$$

There are two distinct theories of the Fourier transforms.

(I) Fourier transform of square-integrable functions

It is assumed that

$$\int_{-\infty}^{\infty} dx \left| f(x) \right|^2 < \infty. \tag{1.9.2}$$

The inverse Fourier transform is given by

$$f(x) = \int_{-\infty}^{\infty} \frac{dk}{2\pi} \exp[ikx] \tilde{f}(k). \tag{1.9.3}$$

We note that, in this case, $\tilde{f}(k)$ is defined for real k. Accordingly, the inversion path in Eq. (1.9.3) coincides with the entire real axis. It should be borne in mind that Eq. (1.9.1) is meaningful in the sense of the convergence in the mean, namely, Eq. (1.9.1) means that there exists $\tilde{f}(k)$ for all real k such that

$$\lim_{R\to\infty} \int_{-\infty}^{\infty} dk \left| \tilde{f}(k) - \int_{-R}^{R} dx \exp[-ikx] f(x) \right|^2 = 0. \tag{1.9.4}$$

Symbolically we write

$$\tilde{f}(k) = \text{l.i.m.}_{R\to\infty} \int_{-R}^{R} dx \exp[-ikx] f(x). \tag{1.9.5}$$

Similarly in Eq. (1.9.3), we mean that, given $\tilde{f}(k)$, there exists an $f(x)$ such that

$$\lim_{R\to\infty} \int_{-\infty}^{\infty} dx \left| f(f) - \int_{-R}^{R} \frac{dk}{2\pi} \exp[ikx] \tilde{f}(k) \right|^2 = 0. \tag{1.9.6}$$

We can then prove that

$$\int_{-\infty}^{\infty} dk \left| \tilde{f}(k) \right|^2 = 2\pi \int_{-\infty}^{\infty} dx \left| f(x) \right|^2, \tag{1.9.7}$$

which is Parseval's identity for the *square-integrable* functions. We see that the pair $(f(x), \tilde{f}(k))$ defined in this way, consists of two functions with very similar properties. We shall find that this situation may change drastically if the condition (1.9.2) is relaxed.

(II) Fourier transform of integrable functions

We relax the condition on the function $f(x)$ as

$$\int_{-\infty}^{\infty} dx\,|f(x)| < \infty. \tag{1.9.8}$$

Then we can still define $\tilde{f}(k)$ for real k. Indeed, from Eq. (1.9.1), we obtain

$$\begin{aligned}\left|\tilde{f}(k\text{: real})\right| &= \left|\int_{-\infty}^{\infty} dx \exp[-ikx] f(x)\right| \\ &\leq \int_{-\infty}^{\infty} dx\,|\exp[-ikx] f(x)| = \int_{-\infty}^{\infty} dx\,|f(x)| < \infty.\end{aligned} \tag{1.9.9}$$

We can further show that the function defined by

$$\tilde{f}_{+}(k) = \int_{-\infty}^{0} dx \exp[-ikx] f(x) \tag{1.9.10}$$

is analytic in the *upper* half-plane of the complex k plane, and

$$\tilde{f}_{+}(k) \to 0 \quad \text{as} \quad |k| \to \infty \quad \text{with} \quad \operatorname{Im} k > 0. \tag{1.9.11}$$

Similarly, we can show that the function defined by

$$\tilde{f}_{-}(k) = \int_{0}^{\infty} dx \exp[-ikx] f(x) \tag{1.9.12}$$

is analytic in the *lower* half-plane of the complex k plane, and

$$\tilde{f}_{-}(k) \to 0 \quad \text{as} \quad |k| \to \infty \quad \text{with} \quad \operatorname{Im} k < 0. \tag{1.9.13}$$

Clearly we have

$$\tilde{f}(k) = \tilde{f}_{+}(k) + \tilde{f}_{-}(k), \quad k\text{: real}. \tag{1.9.14}$$

We can show that

$$\tilde{f}(k) \to 0 \quad \text{as} \quad k \to \pm\infty, \quad k\text{: real}. \tag{1.9.15}$$

This is a property in common with the Fourier transform of the *square-integrable* functions.

❑ **Example 1.2.** Find the Fourier transform of the following function,

$$f(x) = \frac{\sin(ax)}{x}, \quad a > 0, \quad -\infty < x < \infty. \tag{1.9.16}$$

Solution. The Fourier transform $\tilde{f}(k)$ is given by

$$\tilde{f}(k) = \int_{-\infty}^{\infty} dx \exp[ikx]\frac{\sin(ax)}{x} = \int_{-\infty}^{\infty} dx \exp[ikx]\frac{\exp[iax] - \exp[-iax]}{2ix}$$
$$= \int_{-\infty}^{\infty} dx \frac{\exp[i(k+a)x] - \exp[i(k-a)x]}{2ix} = I(k+a) - I(k-a),$$

where we define the integral $I(b)$ by

$$I(b) \equiv \int_{-\infty}^{\infty} dx \frac{\exp[ibx]}{2ix} = \int_{\Gamma} dx \frac{\exp[ibx]}{2ix}.$$

The contour Γ extends from $x = -\infty$ to $x = \infty$ with the infinitesimal indent below the real x-axis at the pole $x = 0$. Noting $x = \mathrm{Re}\, x + i\, \mathrm{Im}\, x$ for the complex x, we have

$$I(b) = \begin{cases} 2\pi i \cdot \mathrm{Res}\left[\frac{\exp[ibx]}{2ix}\right]_{x=0} = \pi, & b > 0, \\ 0, & b < 0. \end{cases}$$

Thus we have

$$\tilde{f}(k) = I(k+a) - I(k-a)$$
$$= \int_{-\infty}^{\infty} dx \exp[ikx]\frac{\sin(ax)}{x} = \begin{cases} \pi & \text{for} \quad |k| < a, \\ 0 & \text{for} \quad |k| > a, \end{cases} \tag{1.9.17}$$

while at $k = \pm a$, we have

$$\tilde{f}(k = \pm a) = \frac{\pi}{2},$$

which is equal to

$$\frac{1}{2}[\tilde{f}(k = \pm a^+) + \tilde{f}(k = \pm a^-)].$$

❑ **Example 1.3.** Find the Fourier transform of the following function,

$$f(x) = \frac{\sin(ax)}{x(x^2 + b^2)}, \quad a, b > 0, \quad -\infty < x < \infty. \tag{1.9.18}$$

Solution. The Fourier transform $\tilde{f}(k)$ is given by

$$\tilde{f}(k) = \int_{\Gamma} dz \frac{\exp[i(k+a)z]}{2iz(z^2 + b^2)} - \int_{\Gamma} dz \frac{\exp[i(k-a)z]}{2iz(z^2 + b^2)} = I(k+a) - I(k-a), \tag{1.9.19}$$

where we define the integral $I(c)$ by

$$I(c) \equiv \int_{-\infty}^{\infty} dz \frac{\exp[icz]}{2iz(z^2 + b^2)} = \int_{\Gamma} dz \frac{\exp[icz]}{2iz(z^2 + b^2)}, \tag{1.9.20}$$

where the contour Γ is the same as in Example 1.2. The integrand has simple poles at

$$z = 0 \quad \text{and} \quad z = \pm ib.$$

Noting $z = \operatorname{Re} z + i \operatorname{Im} z$, we have

$$I(c) = \begin{cases} 2\pi i \cdot \operatorname{Res}\left[\frac{\exp[icz]}{2iz(z^2+b^2)}\right]_{z=0} + 2\pi i \cdot \operatorname{Res}\left[\frac{\exp[icz]}{2iz(z^2+b^2)}\right]_{z=ib}, & c > 0, \\ -2\pi i \cdot \operatorname{Res}\left[\frac{\exp[icz]}{2iz(z^2+b^2)}\right]_{z=-ib}, & c < 0, \end{cases}$$

or

$$I(c) = \begin{cases} (\pi/2b^2)(2 - \exp[-bc]), & c > 0, \\ (\pi/2b^2)\exp[bc], & c < 0. \end{cases}$$

Thus we have

$$\tilde{f}(k) = I(k+a) - I(k-a)$$
$$= \begin{cases} (\pi/b^2)\sinh(ab)\exp[bk], & k < -a, \\ (\pi/b^2)\{1 - \exp[-ab]\cosh(bk)\}, & |k| < a, \\ (\pi/b^2)\sinh(ab)\exp[-bk], & k > a. \end{cases} \tag{1.9.21}$$

We note that $\tilde{f}(k)$ is *step-discontinuous* at $k = \pm a$ in Example 1.2. We also note that $\tilde{f}(k)$ and $\tilde{f}'(k)$ are *continuous* for real k, while $\tilde{f}''(k)$ is *step-discontinuous* at $k = \pm a$ in Example 1.3.

We note the rate with which

$$f(x) \to 0 \quad \text{as} \quad |x| \to +\infty$$

affects the degree of smoothness of $\tilde{f}(k)$. For the square-integrable functions, we usually have

$$f(x) = O\left(\frac{1}{x}\right) \text{ as } |x| \to +\infty \Rightarrow \tilde{f}(k) \text{ step-discontinuous},$$

$$f(x) = O\left(\frac{1}{x^2}\right) \text{ as } |x| \to +\infty \Rightarrow \begin{cases} \tilde{f}(k) \text{ continuous}, \\ \tilde{f}'(k) \text{ step-discontinuous}, \end{cases}$$

$$f(x) = O\left(\frac{1}{x^3}\right) \text{ as } |x| \to +\infty \Rightarrow \begin{cases} \tilde{f}(k), \tilde{f}'(k) \text{ continuous}, \\ \tilde{f}''(k) \text{ step-discontinuous}, \end{cases}$$

and so on.

* * *

Having learned in the above the abstract notions relating to linear space, inner product, the operator and its adjoint, eigenvalue and eigenfunction, Green's function, and having reviewed the Fourier transform and complex analysis, we are now ready to embark on our study of integral equations. We encourage the reader to make an effort to connect the concrete example that will follow with the abstract idea of linear function space and the linear operator. This will not be possible in all circumstances.

The abstract idea of function space is also useful in the discussion of the calculus of variations where a piecewise continuous, but nowhere differentiable, function and a discontinuous function appear as the solution of the problem.

We present the applications of the calculus of variations to theoretical physics specifically, classical mechanics, canonical transformation theory, the Hamilton–Jacobi equation, classical electrodynamics, quantum mechanics, quantum field theory and quantum statistical mechanics.

The mathematically oriented reader is referred to the monographs by R. Kress, and I. M. Gelfand and S. V. Fomin for details of the theories of integral equations and the calculus of variations.

2 Integral Equations and Green's Functions

2.1 Introduction to Integral Equations

An integral equation is the equation in which the function to be determined appears in an integral. There exist several types of integral equations:

Fredholm Integral Equation of the second kind:

$$\phi(x) = F(x) + \lambda \int_a^b K(x,y)\phi(y)\,dy \qquad (a \le x \le b),$$

Fredholm Integral Equation of the first kind:

$$F(x) = \int_a^b K(x,y)\phi(y)\,dy \qquad (a \le x \le b),$$

Volterra Integral Equation of the second kind:

$$\phi(x) = F(x) + \lambda \int_0^x K(x,y)\phi(y)\,dy \quad \text{with} \quad K(x,y) = 0 \quad \text{for} \quad y > x,$$

Volterra Integral Equation of the first kind:

$$F(x) = \int_0^x K(x,y)\phi(y)\,dy \quad \text{with} \quad K(x,y) = 0 \quad \text{for} \quad y > x.$$

In the above, $K(x,y)$ is the *kernel* of the integral equation and $\phi(x)$ is the unknown function. If $F(x) = 0$, the equations are said to be *homogeneous*, and if $F(x) \neq 0$, they are said to be *inhomogeneous*.

Now, begin with some simple examples of Fredholm Integral Equations.

❑ **Example 2.1.** Inhomogeneous Fredholm Integral Equation of the second kind.

$$\phi(x) = x + \lambda \int_{-1}^{1} xy\phi(y)\,dy, \qquad -1 \le x \le 1. \tag{2.1.1}$$

Applied Mathematics in Theoretical Physics. Michio Masujima

ISBN: 3-527-40534-8

Solution. Since $\int_{-1}^{1} y\phi(y)\,dy$ is some constant, define

$$A = \int_{-1}^{1} y\phi(y)\,dy. \tag{2.1.2}$$

Then Eq. (2.1.1) takes the form

$$\phi(x) = x(1 + \lambda A). \tag{2.1.3}$$

Substituting Eq. (2.1.3) into the right-hand side of Eq. (2.1.2), we obtain

$$A = \int_{-1}^{1} (1 + \lambda A)y^2\,dy = \frac{2}{3}(1 + \lambda A).$$

Solving for A, we obtain

$$\left(1 - \frac{2}{3}\lambda\right) A = \frac{2}{3}.$$

If $\lambda = \frac{3}{2}$, no such A exists. Otherwise A is uniquely determined to be

$$A = \frac{2}{3}\left(1 - \frac{2}{3}\lambda\right)^{-1}. \tag{2.1.4}$$

Thus, if $\lambda = \frac{3}{2}$, no solution exists. Otherwise, a unique solution exists and is given by

$$\phi(x) = x\left(1 - \frac{2}{3}\lambda\right)^{-1}. \tag{2.1.5}$$

We shall now consider the homogeneous counter part of the inhomogeneous Fredholm integral equation of the second kind, considered in Example 2.1.

❑ **Example 2.2.** Homogeneous Fredholm Integral Equation of the second kind.

$$\phi(x) = \lambda \int_{-1}^{1} xy\phi(y)\,dy, \qquad -1 \leq x \leq 1. \tag{2.1.6}$$

Solution. As in Example 2.1, define

$$A = \int_{-1}^{1} y\phi(y)\,dy. \tag{2.1.7}$$

Then

$$\phi(x) = \lambda A x. \tag{2.1.8}$$

Substituting Eq. (2.1.8) into Eq. (2.1.7), we obtain

$$A = \int_{-1}^{1} \lambda A y^2 dy = \frac{2}{3}\lambda A. \tag{2.1.9}$$

The solution exists only when $\lambda = \frac{3}{2}$. Thus the nontrivial homogeneous solution exists only for $\lambda = \frac{3}{2}$, whence $\phi(x)$ is given by $\phi(x) = \alpha x$ with α arbitrary. If $\lambda \neq \frac{3}{2}$, no nontrivial homogeneous solution exists.

We observe the following correspondence in Examples 2.1 and 2.2:

$$\begin{array}{lll} & \textbf{Inhomogeneous case.} & \textbf{Homogeneous case.} \\ \lambda \neq 3/2 & \text{Unique solution.} & \text{Trivial solution.} \\ \lambda = 3/2 & \text{No solution.} & \text{Infinitely many solutions.} \end{array} \tag{2.1.10}$$

We further note the analogy of an integral equation to a *system of inhomogeneous linear algebraic equations (matrix equations)*:

$$(K - \mu I)\vec{U} = \vec{F} \tag{2.1.11}$$

where K is an $n \times n$ matrix, I is the $n \times n$ identity matrix, $\vec{U}$ and $\vec{F}$ are n-dimensional vectors, and μ is a number. Equation (2.1.11) has the unique solution,

$$\vec{U} = (K - \mu I)^{-1}\vec{F}, \tag{2.1.12}$$

provided that $(K - \mu I)^{-1}$ exists, or equivalently that

$$\det(K - \mu I) \neq 0. \tag{2.1.13}$$

The homogeneous equation corresponding to Eq. (2.1.11) is

$$(K - \mu I)\vec{U} = 0 \quad \text{or} \quad K\vec{U} = \mu\vec{U} \tag{2.1.14}$$

which is the eigenvalue equation for the matrix K. A solution to the homogeneous equation (2.1.14) exists for certain values of $\mu = \mu_n$, which are called the eigenvalues. If μ is equal to an eigenvalue μ_n, $(K - \mu I)^{-1}$ fails to exist and Eq. (2.1.11) has generally no finite solution.

❑ **Example 2.3.** Change the inhomogeneous term x of Example 2.1 to 1.

$$\phi(x) = 1 + \lambda \int_{-1}^{1} xy\phi(y)\,dy, \quad -1 \leq x \leq 1. \tag{2.1.15}$$

Solution. As before, define

$$A = \int_{-1}^{1} y\phi(y)\,dy. \tag{2.1.16}$$

Then

$$\phi(x) = 1 + \lambda Ax. \tag{2.1.17}$$

Substituting Eq. (2.1.17) into Eq. (2.1.16), we obtain $A = \int_{-1}^{1} y(1 + \lambda Ay)\,dy = \frac{2}{3}\lambda A$. Thus, for $\lambda \neq \frac{3}{2}$, the unique solution exists with $A = 0$, and $\phi\,(x) = 1$, while for $\lambda = \frac{3}{2}$, infinitely many solutions exist with A arbitrary and $\phi\,(x) = 1 + \frac{3}{2}Ax$.

The above three examples illustrate the *Fredholm Alternative*:

- For $\lambda = \frac{3}{2}$, the homogeneous problem has a solution, given by

$$\phi_H(x) = \alpha x \quad \text{for any } \alpha.$$

- For $\lambda \neq \frac{3}{2}$, the inhomogeneous problem has a unique solution, given by

$$\phi(x) = \begin{cases} \dfrac{x}{(1 - \frac{2}{3}\lambda)} & \text{when} \quad F(x) = x, \\ 1 & \text{when} \quad F(x) = 1. \end{cases}$$

- For $\lambda = \frac{3}{2}$, the inhomogeneous problem has no solution when $F(x) = x$, while it has infinitely many solutions when $F(x) = 1$. In the former case, $(\phi_H, F) = \int_{-1}^{1} \alpha x \cdot x \, dx \neq 0$, while in the latter case, $(\phi_H, F) = \int_{-1}^{1} \alpha x \cdot 1 \, dx = 0$.

It is, of course, not surprising that Eq. (2.1.15) has infinitely many solutions when $\lambda = 3/2$. Generally, if ϕ_0 is a solution of an inhomogeneous equation, and ϕ_1 is a solution of the corresponding homogeneous equation, then $\phi_0 + a\phi_1$ is also a solution of the inhomogeneous equation, where a is any constant. Thus, if λ is equal to an eigenvalue, an inhomogeneous equation has infinitely many solutions as long as it has one solution. The nontrivial question is: Under what condition can we expect the latter to happen? In the present example, the relevant condition is $\int_{-1}^{1} y \, dy = 0$, which means that the inhomogeneous term (which is 1) multiplied by y and integrated from -1 to 1, is zero. There is a counterpart of this condition for matrix equations. It is well known that, under certain circumstances, the inhomogeneous matrix equation (2.1.11) has solutions even if μ is equal to an eigenvalue. Specifically this happens if the inhomogeneous term $\vec{F}$ is a linear superposition of the vectors each of which forms a column of $(K - \mu I)$. There is another way to phrase this. Consider all vectors $\vec{V}$ satisfying

$$(K^{\mathrm{T}} - \mu I)\vec{V} = 0, \tag{2.1.18}$$

where K^{T} is the transpose of K. The equation above says that $\vec{V}$ is an eigenvector of K^{T} with the eigenvalue μ. It also says that $\vec{V}$ is perpendicular to all row vectors of $(K^{\mathrm{T}} - \mu I)$. If $\vec{F}$ is a linear superposition of the column vectors of $(K - \mu I)$ (which are the row vectors of $(K^{\mathrm{T}} - \mu I)$), then $\vec{F}$ is perpendicular to $\vec{V}$. Therefore, the inhomogeneous equation (2.1.11) has solutions when μ is an eigenvalue, if and only if $\vec{F}$ is perpendicular to all eigenvectors of K^{T} with eigenvalue μ. Similarly an inhomogeneous integral equation has solutions even when λ is equal to an eigenvalue, as long as the inhomogeneous term is perpendicular to all of the eigenfunctions of the transposed kernel (the kernel with $x \leftrightarrow y$) of that particular eigenvalue.

As we have seen in Chapter 1, just as a matrix, a kernel and its transpose have the same eigenvalues. Consequently the homogeneous integral equation with the transposed kernel has no solution if λ is not equal to an eigenvalue of the kernel. Therefore, if λ is not an eigenvalue, any inhomogeneous term is trivially perpendicular to all solutions of the homogeneous integral equation with the transposed kernel, since all of them are trivial. Together with the result in the preceding paragraph, we have arrived at the necessary and sufficient condition for an inhomogeneous integral equation to have a solution: the inhomogeneous term must be perpendicular to all solutions of the homogeneous integral equation with the transposed kernel.

There exists another kind of integral equation in which the unknown appears only in the integrals. Consider one more example of a Fredholm Integral Equation.

❑ **Example 2.4.** Fredholm Integral Equation of the first kind.

Case(A)

$$1 = \int_0^1 xy\phi(y)\,dy, \qquad 0 \le x \le 1. \tag{2.1.19}$$

Case(B)

$$x = \int_0^1 xy\phi(y)\,dy, \qquad 0 \le x \le 1. \tag{2.1.20}$$

Solution. In both cases, divide both sides of the equations by x to obtain

Case(A)

$$\frac{1}{x} = \int_0^1 y\phi(y)\,dy. \tag{2.1.21}$$

Case(B)

$$1 = \int_0^1 y\phi(y)\,dy. \tag{2.1.22}$$

In the case of Eq. (2.1.21), no solution is possible, while in the case of Eq. (2.1.22), infinitely many $\phi(x)$ are possible. Essentially any function $\psi(x)$ which satisfies

$$\int_0^1 y\psi(y)\,dy \ne 0 \quad \text{or} \quad \infty,$$

can be made a solution to Eq. (2.1.20). Indeed,

$$\phi(x) = \frac{\psi(x)}{\displaystyle\int_0^1 y\psi(y)\,dy} \tag{2.1.23}$$

will do.

Therefore, for the kind of integral equations considered in Example 2.4, no solution exists for some inhomogeneous terms, while infinitely many solutions exist for some other inhomogeneous terms.

Next, we shall consider an example of a Volterra Integral Equation of the second kind with the transformation of an integral equation into an ordinary differential equation.

❑ **Example 2.5.** Volterra Integral Equation of the second kind.

$$\phi(x) = ax + \lambda x \int_0^x \phi(x')\, dx'. \tag{2.1.24}$$

Solution. Divide both sides of Eq. (2.1.24) by x to obtain

$$\frac{\phi(x)}{x} = a + \lambda \int_0^x \phi(x')\, dx'. \tag{2.1.25}$$

Differentiate both sides of Eq. (2.1.25) with respect to x to obtain

$$\frac{d}{dx}\left(\frac{\phi(x)}{x}\right) = \lambda\phi(x). \tag{2.1.26}$$

By setting

$$u(x) = \frac{\phi(x)}{x},$$

the following differential equation results,

$$\frac{du(x)}{u(x)} = \lambda x\; dx. \tag{2.1.27}$$

By integrating both sides,

$$\ln u(x) = \frac{1}{2}\lambda x^2 + \text{constant}.$$

Hence the solution is given by

$$u(x) = Ae^{\frac{1}{2}\lambda x^2}, \quad \text{or} \quad \phi(x) = Axe^{\frac{1}{2}\lambda x^2}. \tag{2.1.28}$$

To determine the integration constant A in Eq. (2.1.28), note that as $x \to 0$, based on the integral equation (2.1.24), $\phi(x)$ above behaves as

$$\phi(x) \to ax + O(x^3) \tag{2.1.29}$$

while our solution (2.1.28) behaves as

$$\phi(x) \to Ax + O(x^3). \tag{2.1.30}$$

Hence, from Eqs. (2.1.29) and (2.1.30), we identify

$$A = a.$$

Thus the final form of the solution is

$$\phi(x) = axe^{\frac{1}{2}\lambda x^2}, \tag{2.1.31}$$

which is the unique solution for all λ.

We observe three points:

1. The integral equation (2.1.24) has a unique solution for all values of λ. It follows that the corresponding homogeneous integral equation, obtained from Eq. (2.1.24) by setting $a = 0$, does not have a nontrivial solution. Indeed, this can be directly verified by setting $a = 0$ in Eq. (2.1.31). This means that the kernel for Eq. (2.1.24) has no eigenvalues. This is true for all *square-integrable* kernels of the Volterra type.

2. While the solution to the differential equation (2.1.26) or (2.1.27) contains an arbitrary constant, the solution to the corresponding integral equation (2.1.24) does not. More precisely, Eq. (2.1.24) is equivalent to Eq. (2.1.26) or Eq. (2.1.27) plus an initial condition.

3. The transformation of the Volterra Integral Equation of the second kind to an ordinary differential equation is possible whenever the kernel of the Volterra integral equation is a sum of the factored terms.

In the above example, we solved the integral equation by transforming it into a differential equation. This is not often possible. On the other hand, it is, in general, easy to transform a differential equation into an integral equation. However, to avoid any misunderstanding, let me state that we never solve a differential equation by such a transformation. Indeed, an integral equation is much more difficult to solve than a differential equation in a closed form. Only very rarely can this be done. Therefore, whenever it is possible to transform an integral equation into a differential equation, it is a good idea to do so. On the other hand, there are advantages in transforming a differential equation into an integral equation. This transformation may facilitate the discussion of the existence and uniqueness of the solution, the spectrum of the eigenvalue and the analyticity of the solution. It also enables us to obtain the perturbative solution of the equation.

2.2 Relationship of Integral Equations with Differential Equations and Green's Functions

To provide the reader with a sense of bearing, we shall discuss the transformation of a differential equation to an integral equation. This transformation is accomplished by the use of *Green's functions*.

As an example, consider the one-dimensional Schrödinger equation with potential $U(x)$:

$$\left(\frac{d^2}{dx^2} + k^2\right)\phi(x) = U(x)\phi(x). \tag{2.2.1}$$

It is assumed that $U(x)$ vanishes rapidly as $|x| \to \infty$. Although Eq. (2.2.1) is most usually thought of as an initial value problem, let us suppose that we are given

$$\phi(0) = a, \quad \text{and} \quad \phi'(0) = b, \tag{2.2.2}$$

and we are interested in the solution for $x > 0$.

Green's function: First treat the right-hand side of Eq. (2.2.1) as an inhomogeneous term $f(x)$. Namely, consider the following inhomogeneous problem:

$$L\phi(x) = f(x) \quad \text{with} \quad L = \frac{d^2}{dx^2} + k^2, \tag{2.2.3}$$

and the boundary conditions specified by Eq. (2.2.2). Multiply both sides of Eq. (2.2.3) by $g(x, x')$ and integrate with respect to x from 0 to ∞. Then

$$\int_0^\infty g(x, x')L\phi(x) = \int_0^\infty g(x, x')f(x)\, dx. \tag{2.2.4}$$

Integrate by parts twice on the left-hand side of Eq. (2.2.4) to obtain

$$\int_0^\infty (Lg(x, x'))\phi(x)\, dx + g(x, x')\phi'(x)\Big|_{x=0}^{x=\infty} - \frac{dg(x, x')}{dx}\phi(x)\Big|_{x=0}^{x=\infty}$$

$$= \int_0^\infty g(x, x')f(x)\, dx. \tag{2.2.5}$$

In the boundary terms, $\phi'(0)$ and $\phi(0)$ are known. To eliminate unknown terms, we require

$$g(\infty, x') = 0 \quad \text{and} \quad \frac{dg}{dx}(\infty, x') = 0. \tag{2.2.6}$$

Also, choose $g(x, x')$ to satisfy

$$Lg(x, x') = \delta(x - x'). \tag{2.2.7}$$

Then we find from Eq. (2.2.5):

$$\phi(x') = bg(0, x') - a\frac{dg}{dx}(0, x') + \int_0^\infty g(x, x')f(x)\, dx. \tag{2.2.8}$$

Solution for $g(x, x')$: The governing equation and the boundary conditions are given by

$$\left(\frac{d^2}{dx^2} + k^2\right) g(x, x') = \delta(x - x') \quad \text{on} \quad x \in (0, \infty) \quad \text{with} \quad x' \in (0, \infty), \tag{2.2.9}$$

Boundary condition 1,

$$g(\infty, x') = 0, \tag{2.2.10}$$

Boundary condition 2,

$$\frac{dg}{dx}(\infty, x') = 0. \tag{2.2.11}$$

For $x < x'$,

$$g(x, x') = A \sin kx + B \cos kx. \tag{2.2.12}$$

For $x > x'$,

$$g(x, x') = C \sin kx + D \cos kx. \tag{2.2.13}$$

Applying the boundary conditions, (2.2.10) and (2.2.11) above, results in $C = D = 0$. Thus

$$g(x, x') = 0 \quad \text{for} \quad x > x'. \tag{2.2.14}$$

Now, integrate the differential equation (2.2.9) across x' to obtain

$$\frac{dg}{dx}(x' + \varepsilon, x') - \frac{dg}{dx}(x' - \varepsilon, x') = 1, \tag{2.2.15}$$

$$g(x' + \varepsilon, x') = g(x' - \varepsilon, x'). \tag{2.2.16}$$

Letting $\varepsilon \to 0$, we obtain the equations for A and B.

$$\begin{cases} A \sin kx' + B \cos kx' &= 0, \\ -A \cos kx' + B \sin kx' &= \frac{1}{k}. \end{cases}$$

Thus A and B are determined to be

$$A = \frac{-\cos kx'}{k}, \quad B = \frac{\sin kx'}{k}, \tag{2.2.17}$$

and the Green's function is found to be

$$g(x, x') = \begin{cases} \dfrac{\sin k(x' - x)}{k} & \text{for} \quad x < x', \\ 0 & \text{for} \quad x > x'. \end{cases} \tag{2.2.18}$$

Equation (2.2.8) becomes

$$\phi(x') = b\frac{\sin kx'}{k} + a \cos kx' + \int_0^{x'} \frac{\sin k(x' - x)}{k} f(x)\, dx.$$

Changing x to ξ and x' to x, and recalling that $f(x) = U(x)\phi(x)$, we find

$$\phi(x) = a \cos kx + b\frac{\sin kx}{k} + \int_0^{x} \frac{\sin k(x - \xi)}{k} U(\xi)\phi(\xi)\, d\xi, \tag{2.2.19}$$

which is a Volterra Integral Equation of the second kind.

Next consider the very important *scattering problem* for the Schrödinger equation:

$$\left(\frac{d^2}{dx^2} + k^2\right)\phi(x) = U(x)\phi(x) \quad \text{on} \quad -\infty < x < \infty, \tag{2.2.20}$$

where the potential $U(x) \to 0$ as $|x| \to \infty$. We might expect that

$$\begin{cases} \phi(x) & \to & Ae^{ikx} + Be^{-ikx} & \text{as} \quad x \to -\infty, \\ \phi(x) & \to & Ce^{ikx} + De^{-ikx} & \text{as} \quad x \to +\infty. \end{cases}$$

Now (with an $e^{-i\omega t}$ implicitly multiplying $\phi(x)$), the term e^{ikx} represents a wave going to the right while e^{-ikx} is a wave going to the left. In the scattering problem, we suppose that there is an incident wave with amplitude 1 (i.e., $A = 1$), a reflected wave with amplitude R, (i.e., $B = R$) and a transmitted wave with amplitude T (i.e., $C = T$). Both R and T are unknown still. Also as $x \to +\infty$, there is no left-going wave (i.e., $D = 0$). Thus the problem is to solve

$$\left(\frac{d^2}{dx^2} + k^2\right)\phi(x) = U(x)\phi(x), \tag{2.2.21}$$

with boundary conditions,

$$\begin{cases} \phi(x \to -\infty) &= e^{ikx} + Re^{-ikx}, \\ \phi(x \to +\infty) &= Te^{ikx}. \end{cases} \tag{2.2.22}$$

Green's function: Multiply both sides of Eq. (2.2.21) by $g(x, x')$, integrate with respect to x from $-\infty$ to $+\infty$, and integrate by parts twice. The result is

$$\int_{-\infty}^{+\infty} \phi(x)\left(\frac{d^2}{dx^2} + k^2\right) g(x, x')\, dx + g(\infty, x')\frac{d\phi}{dx}(\infty) - g(-\infty, x')\frac{d\phi}{dx}(-\infty)$$

$$-\frac{dg}{dx}(\infty, x')\phi(\infty) + \frac{dg}{dx}(-\infty, x')\phi(-\infty) = \int_{-\infty}^{+\infty} g(x, x')U(x)\phi(x)\, dx. \tag{2.2.23}$$

We require that Green's function satisfies

$$\left(\frac{d^2}{dx^2} + k^2\right) g(x, x') = \delta(x - x'). \tag{2.2.24}$$

Then Eq. (2.2.23) becomes

$$\phi(x') + g(\infty, x')Tike^{ikx} - g(-\infty, x')\left[ike^{ikx} - Rike^{-ikx}\right]$$

$$-\frac{dg}{dx}(\infty, x')Te^{ikx} + \frac{dg}{dx}(-\infty, x')\left[e^{ikx} + Re^{-ikx}\right] = \int_{-\infty}^{+\infty} g(x, x')U(x)\phi(x)\, dx. \tag{2.2.25}$$

We require that terms involving the unknowns T and R vanish in Eq. (2.2.25), i.e.,

$$\begin{cases} \frac{dg}{dx}(\infty, x') &= ikg(\infty, x'), \\ \frac{dg}{dx}(-\infty, x') &= -ikg(-\infty, x'). \end{cases} \tag{2.2.26}$$

These conditions, (2.2.26), are the appropriate boundary conditions for $g(x, x')$. Hence we obtain

$$\phi(x') = \left[ikg(-\infty, x') - \frac{dg}{dx}(-\infty, x')\right] e^{ikx} + \int_{-\infty}^{+\infty} g(x, x')U(x)\phi(x)\, dx. \tag{2.2.27}$$

Solution for $g(x, x')$: The governing equation for $g(x, x')$ is

$$\left(\frac{d^2}{dx^2} + k^2\right) g(x, x') = \delta(x - x'),$$

and the boundary conditions are Eq. (2.2.26). The solution to this problem is found to be

$$g(x, x') = \begin{cases} A'e^{ikx} & \text{for} \quad x > x', \\ B'e^{-ikx} & \text{for} \quad x < x'. \end{cases}$$

At $x = x'$, there exists a discontinuity in the first derivative $\frac{dg}{dx}$ of g with respect to x.

$$\frac{dg}{dx}(x'_+, x') - \frac{dg}{dx}(x'_-, x') = 1, \quad g(x'_+, x') = g(x'_-, x').$$

From these two conditions, A' and B' are determined to be $A' = e^{-ikx'}/2ik$ and $B' = e^{ikx'}/2ik$. Thus the Green's function $g(x, x')$ for this problem is given by

$$g(x, x') = \frac{1}{2ik} e^{ik|x-x'|}.$$

Now, the first term on the right-hand side of Eq. (2.2.27) assumes the following form:

$$ikg(-\infty, x') - \frac{dg}{dx}(-\infty, x') = 2ikB'e^{-ikx} = e^{ik(x'-x)}.$$

Hence Eq. (2.2.27) becomes

$$\phi(x') = e^{ikx'} + \int_{-\infty}^{+\infty} \frac{e^{ik|x-x'|}}{2ik} U(x)\phi(x)\, dx. \tag{2.2.28}$$

Changing x to ξ and x' to x in Eq. (2.2.28), we have

$$\phi(x) = e^{ikx} + \int_{-\infty}^{+\infty} \frac{e^{ik|\xi-x|}}{2ik} U(\xi)\phi(\xi)\, d\xi. \tag{2.2.29}$$

This is the Fredholm Integral Equation of the second kind.

Reflection: As $x \to -\infty$, $|\xi - x| = \xi - x$ so that

$$\phi(x) \to e^{ikx} + e^{-ikx} \int_{-\infty}^{+\infty} \frac{e^{ik\xi}}{2ik} U(\xi)\phi(\xi)\, d\xi.$$

From this, the *reflection coefficient* R is found.

$$R = \int_{-\infty}^{+\infty} \frac{e^{ik\xi}}{2ik} U(\xi)\phi(\xi)\, d\xi.$$

Transmission: As $x \to +\infty$, $|\xi - x| = x - \xi$ so that

$$\phi(x) \to e^{ikx} \left[1 + \int_{-\infty}^{+\infty} \frac{e^{-ik\xi}}{2ik} U(\xi)\phi(\xi)\, d\xi \right].$$

From this, the *transmission coefficient* T is found.

$$T = 1 + \int_{-\infty}^{+\infty} \frac{e^{-ik\xi}}{2ik} U(\xi)\phi(\xi)\, d\xi.$$

These R and T are still unknowns since $\phi(\xi)$ is not known, but for $|U(\xi)| \ll 1$ (weak potential), we can approximate $\phi(x)$ by e^{ikx}. Then the approximate equations for R and T are given by

$$R \simeq \int_{-\infty}^{+\infty} \frac{e^{2ik\xi}}{2ik} U(\xi)\, d\xi, \quad \text{and} \quad T \simeq 1 + \int_{-\infty}^{+\infty} \frac{1}{2ik} U(\xi)\, d\xi.$$

Also, by approximating $\phi(\xi)$ by $e^{ik\xi}$ in the integrand of Eq. (2.2.29) on the right-hand side, we have, as the first approximation,

$$\phi(x) \simeq e^{ikx} + \int_{-\infty}^{+\infty} \frac{e^{ik|\xi - x|}}{2ik} U(\xi) e^{ik\xi}\, d\xi.$$

By continuing the iteration, we can generate the **Born series** for $\phi(x)$.

We shall discuss the Born approximation thoroughly in Section 2.5.

2.3 Sturm–Liouville System

Consider the linear differential operator

$$L = \frac{1}{r(x)} \left[\frac{d}{dx} \left(p(x) \frac{d}{dx} \right) - q(x) \right] \tag{2.3.1}$$

where

$$r(x),\ p(x) > 0 \quad \text{on} \quad 0 < x < 1, \tag{2.3.2}$$

together with the inner product defined with $r(x)$ as the weight,

$$(f, g) = \int_0^1 f(x) g(x) \cdot r(x)\, dx. \tag{2.3.3}$$

Examine the inner product (g, Lf) by integral by parts twice, to obtain,

$$\begin{aligned}(g, Lf) &= \int_0^1 dx \cdot r(x) \cdot g(x) \frac{1}{r(x)} \left[\frac{d}{dx} \left(p(x) \frac{df(x)}{dx} \right) - q(x) f(x) \right] \\ &= p(1)\left[f'(1)g(1) - f(1)g'(1)\right] - p(0)\left[f'(0)g(0) - f(0)g'(0)\right] + (Lg, f).\end{aligned} \tag{2.3.4}$$

Suppose that the boundary conditions on $f(x)$ are

$$f(0) = 0 \quad \text{and} \quad f(1) = 0, \tag{2.3.5}$$

and the adjoint boundary conditions on $g(x)$ are

$$g(0) = 0 \quad \text{and} \quad g(1) = 0. \tag{2.3.6}$$

(Many other boundary conditions of the type

$$\alpha f(0) + \beta f'(0) = 0 \tag{2.3.7}$$

also work.) Then the boundary terms in Eq. (2.3.4) disappear and we have

$$(g, Lf) = (Lg, f), \tag{2.3.8}$$

i.e., L is *self-adjoint* with the given weighted inner product.

Now examine the eigenvalue problem. The governing equation and boundary conditions are given by

$$L\phi(x) = \lambda\phi(x), \quad \text{with} \quad \phi(0) = 0, \quad \text{and} \quad \phi(1) = 0,$$

i.e.,

$$\frac{d}{dx}\left[p(x)\frac{d}{dx}\phi(x)\right] - q(x)\phi(x) = \lambda r(x)\phi(x), \tag{2.3.9}$$

with the boundary conditions

$$\phi(0) = 0, \quad \text{and} \quad \phi(1) = 0. \tag{2.3.10}$$

Suppose that $\lambda = 0$ is not an eigenvalue (i.e., the homogeneous problem has no nontrivial solutions) so that the Green's function exists. (Otherwise we have to define the modified Green's function.) Suppose that the second-order ordinary differential equation

$$\frac{d}{dx}\left[p(x)\frac{d}{dx}y(x)\right] - q(x)y(x) = 0 \tag{2.3.11}$$

has two independent solutions $y_1(x)$ and $y_2(x)$ such that

$$y_1(0) = 0 \quad \text{and} \quad y_2(1) = 0. \tag{2.3.12}$$

In order for $\lambda = 0$ not to be an eigenvalue, we must make sure that the only C_1 and C_2 for which $C_1y_1(0) + C_2y_2(0) = 0$ and $C_1y_1(1) + C_2y_2(1) = 0$ are not nontrivial. This requires

$$y_1(1) \neq 0 \quad \text{and} \quad y_2(0) \neq 0. \tag{2.3.13}$$

Now, to find the Green's function, multiply the eigenvalue equation (2.3.9) by $G(x, x')$ and integrate from 0 to 1. Using the boundary conditions

$$G(0, x') = 0 \quad \text{and} \quad G(1, x') = 0, \tag{2.3.14}$$

we obtain, after integrating by parts twice,

$$\int_0^1 \phi(x)\left[\frac{d}{dx}\left(p(x)\frac{dG(x,x')}{dx}\right) - q(x)G(x,x')\right] dx$$
$$= \lambda\int_0^1 G(x,x')r(x)\phi(x)\,dx. \quad (2.3.15)$$

Requiring that the Green's function $G(x,x')$ should satisfy

$$\frac{d}{dx}\left(p(x)\frac{dG(x,x')}{dx}\right) - q(x)G(x,x') = \delta(x-x') \quad (2.3.16)$$

with the boundary conditions (2.3.14), we arrive at the following equation

$$\phi(x') = \lambda\int_0^1 G(x,x')r(x)\phi(x)\,dx. \quad (2.3.17)$$

This is an homogeneous Fredholm integral equation of the second kind, once $G(x,x')$ is known.

Solution for $G(x,x')$: Recalling Eqs. (2.3.12), (2.3.13) and (2.3.14), we have

$$G(x,x') = \begin{cases} Ay_1(x) + By_2(x) & \text{for} \quad x < x', \\ Cy_1(x) + Dy_2(x) & \text{for} \quad x > x'. \end{cases}$$

From the boundary conditions (2.3.14) of $G(x,x')$, and (2.3.12) and (2.3.13) of $y_1(x)$ and $y_2(x)$, we immediately have

$$B = 0, \quad \text{and} \quad C = 0.$$

Thus we have

$$G(x,x') = \begin{cases} Ay_1\,(x) & \text{for} \quad x < x', \\ Dy_2\,(x) & \text{for} \quad x > x'. \end{cases}$$

In order to determine A and D, integrate Eq. (2.3.16) across x' with respect to x, and make use of the continuity of $G(x,x')$ at $x = x'$ which results in

$$p(x')\left[\frac{dG}{dx}(x'_+,x') - \frac{dG}{dx}(x'_-,x')\right] = 1,$$

$$G(x'_+,x') = G(x'_-,x'),$$

or,

$$Ay_1(x') = Dy_2(x'),$$

$$Dy'_2(x') - Ay'_1(x') = 1/p(x').$$

Noting that

$$W\left(y_1(x), y_2(x)\right) \equiv y_1(x)y_2'(x) - y_2(x)y_1'(x) \tag{2.3.18}$$

is the *Wronskian* of the differential equation (2.3.11), we obtain A and D as

$$\begin{cases} A = \dfrac{y_2(x')}{p(x')W\left(y_1(x'), y_2(x')\right)}, \\ D = \dfrac{y_1(x')}{p(x')W\left(y_1(x'), y_2(x')\right)}. \end{cases}$$

Now, it can easily be proved that

$$p(x)W\left(y_1(x), y_2(x)\right) = \text{constant}, \tag{2.3.19}$$

for the differential equation (2.3.11). Denoting this constant by

$$p(x)W\left(y_1(x), y_2(x)\right) = C',$$

we simplify A and D as

$$\begin{cases} A = \dfrac{y_2(x')}{C'}, \\ D = \dfrac{y_1(x')}{C'}. \end{cases}$$

Thus the Green's function $G(x, x')$ for the Sturm–Liouville system is given by

$$G(x, x') = \begin{cases} \dfrac{y_1(x)\, y_2(x')}{C'} & \text{for} \quad x < x', \\ \dfrac{y_1(x')\, y_2(x)}{C'} & \text{for} \quad x > x', \end{cases} \tag{2.3.20}$$

$$= \frac{y_1(x_<)y_2(x_>)}{C'} \quad \text{for} \quad \begin{cases} x_< = \dfrac{(x + x')}{2} - \dfrac{|x - x'|}{2}, \\ x_> = \dfrac{(x + x')}{2} + \dfrac{|x - x'|}{2}. \end{cases} \tag{2.3.21}$$

Thus the Sturm–Liouville eigenvalue problem is equivalent to the homogeneous Fredholm integral equation of the second kind,

$$\phi(x) = \lambda \int_0^1 G(\xi, x) r(\xi)\phi(\xi)\, d\xi. \tag{2.3.22}$$

We remark that the Sturm–Liouville eigenvalue problem turns out to have a complete set of eigenfunctions in the space $\mathbb{L}_2(0, 1)$ as long as $p(x)$ and $r(x)$ are analytic and positive on $(0, 1)$.

The kernel of Eq. (2.3.22) is

$$K(\xi, x) = r(\xi)G(\xi, x).$$

This kernel can be symmetrized by defining

$$\psi(x) = \sqrt{r(x)}\phi(x).$$

Then the integral equation (2.3.22) becomes

$$\psi(x) = \lambda \int_0^1 \sqrt{r(\xi)}G(\xi, x)\sqrt{r(x)}\psi(\xi)\, d\xi. \tag{2.3.23}$$

Now, the kernel of Eq. (2.3.23),

$$\sqrt{r(\xi)}G(\xi, x)\sqrt{r(x)}$$

is symmetric since $G(\xi, x)$ is symmetric.

Symmetry of the Green's function, called *reciprocity*, is true in general for any *self-adjoint operator*. The proof of this fact is as follows: Consider

$$L_x G(x, x') = \delta(x - x'), \tag{2.3.24}$$

$$L_x G(x, x'') = \delta(x - x''). \tag{2.3.25}$$

Take the inner product of Eq. (2.3.24) with $G(x, x'')$ from the left and Eq. (2.3.25) with $G(x, x')$ from the right.

$$(G(x, x''), L_x G(x, x')) = (G(x, x''), \delta(x - x')),$$

$$(L_x G(x, x''), G(x, x')) = (\delta(x - x''), G(x, x')).$$

Since L_x is assumed to be self-adjoint, subtracting the two equations above results in

$$G^*(x', x'') = G(x'', x'). \tag{2.3.26}$$

If G is real, we have

$$G(x', x'') = G(x'', x'),$$

i.e., $G(x', x'')$ is symmetric.

2.4 Green's Function for Time-Dependent Scattering Problem

The time-dependent Schrödinger equation takes the following form after setting $\hbar = 1$ and $2m = 1$,

$$\left(i\frac{\partial}{\partial t} + \frac{\partial^2}{\partial x^2}\right)\psi(x, t) = V(x, t)\psi(x, t). \tag{2.4.1}$$

Assume

$$\begin{cases} \lim_{|t|\to\infty} V(x,t) = 0, \\ \lim_{t\to-\infty} \exp[i\omega_0 t]\psi(x,t) = \exp[ik_0 x], \end{cases} \tag{2.4.2}$$

from which, we find

$$\omega_0 = k_0^2. \tag{2.4.3}$$

Define the Green's function $G(x,t;x',t')$ by requiring

$$\psi(x,t) = \exp\left[i(k_0 x - k_0^2 t)\right] + \int_{-\infty}^{+\infty} dt' \int_{-\infty}^{+\infty} dx' G(x,t;x',t')V(x',t')\psi(x',t'). \tag{2.4.4}$$

In order to satisfy partial differential equation (2.4.1), we require

$$\left(i\frac{\partial}{\partial t} + \frac{\partial^2}{\partial x^2}\right) G(x,t;x',t') = \delta(t-t')\delta(x-x'). \tag{2.4.5}$$

We also require that

$$G(x,t;x',t') = 0 \quad \text{for} \quad t < t'. \tag{2.4.6}$$

Note that the initial condition at $t = -\infty$ is satisfied as well as *causality*. Note also that the set of equations could be obtained by the methods we used in the previous two examples. To solve the above equations, Eqs. (2.4.5) and (2.4.6), we Fourier transform in time and space, i.e., we write

$$\begin{cases} \tilde{G}(k,\omega;x',t') &= \displaystyle\int_{-\infty}^{+\infty} dx \int_{-\infty}^{+\infty} dt\, e^{-ikx} e^{-i\omega t} G(x,t;x',t'), \\ G(x,t;x',t') &= \displaystyle\int_{-\infty}^{+\infty} \frac{dk}{2\pi} \int_{-\infty}^{+\infty} \frac{d\omega}{2\pi} e^{+ikx} e^{+i\omega t} \tilde{G}(k,\omega;x',t'). \end{cases} \tag{2.4.7}$$

Taking the Fourier transform of the original equation (2.4.5), we find

$$G(x,t;x',t') = \int_{-\infty}^{+\infty} \frac{dk}{2\pi} \int_{-\infty}^{+\infty} \frac{d\omega}{2\pi} \left(\frac{-1}{\omega + k^2}\right) e^{ik(x-x')} e^{i\omega(t-t')}. \tag{2.4.8}$$

Where do we use the condition that $G(x,t;x',t') = 0 \quad \text{for} \quad t < t'$? Consider the ω integration in the complex ω plane as in Figure 2.1,

$$\int_{-\infty}^{+\infty} d\omega \frac{1}{\omega + k^2} e^{i\omega(t-t')}. \tag{2.4.9}$$

We find that there is a singularity right on the path of integration at $\omega = -k^2$. We either have to go above or below it. Upon writing ω as $\omega = \omega_1 + i\omega_2$, we have the following bound,

$$\left| e^{i\omega(t-t')} \right| = \left| e^{i\omega_1(t-t')} \right| \left| e^{-\omega_2(t-t')} \right| = e^{-\omega_2(t-t')}. \tag{2.4.10}$$

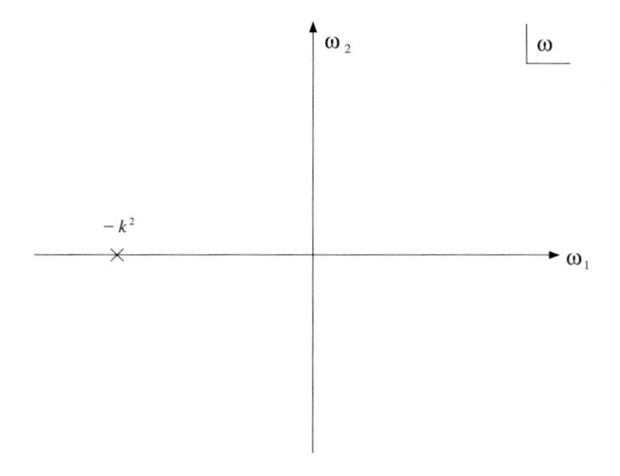

Fig. 2.1: The location of the singularity of the integrand of Eq. (2.4.9) in the complex ω plane.

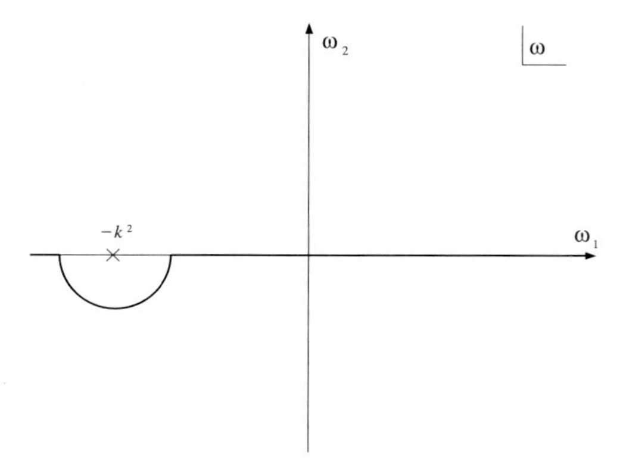

Fig. 2.2: The contour of the complex ω integration of Eq. (2.4.9).

For $t < t'$, we close the contour in the lower half-plane to carry out the contour integral. Since we want G to be zero in this case, we want no singularities inside the contour in that case. This prompts us to take the contour to be as in Figure 2.2. For $t > t'$, when we close the contour in the upper half-plane, we get the contribution from the pole at $\omega = -k^2$.

The result of the calculation is given by

$$\int_{-\infty}^{+\infty} d\omega \frac{1}{\omega + k^2} e^{i\omega(t-t')} = \begin{cases} 2\pi i \cdot e^{ik(x-x')-ik^2(t-t')}, & t > t', \\ 0, & t < t'. \end{cases} \tag{2.4.11}$$

We remark that the idea of the deformation of the contour satisfying causality is often expressed by taking the singularity to be at $-k^2 + i\varepsilon$ ($\varepsilon > 0$) as in Figure 2.3, whence we replace the denominator $\omega + k^2$ with $\omega + k^2 - i\varepsilon$,

$$G(x, t; x', t') = \int_{-\infty}^{+\infty} \frac{dk}{2\pi} \int_{-\infty}^{+\infty} \frac{d\omega}{2\pi} \left(\frac{-1}{\omega + k^2 - i\varepsilon} \right) e^{ik(x-x')+i\omega(t-t')}. \tag{2.4.12}$$

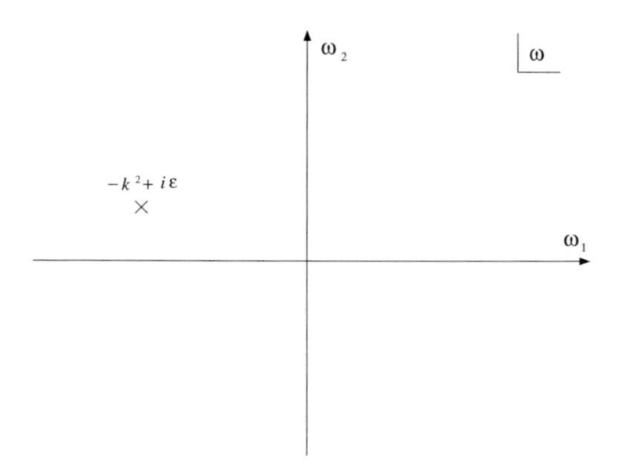

Fig. 2.3: The singularity of the integrand of Eq. (2.4.9) at $\omega = -k^2$ gets shifted to $\omega = -k^2+i\varepsilon$ ($\varepsilon > 0$) in the upper half-plane of the complex ω plane.

This shifts the singularity above the real axis and is equivalent, as $\varepsilon \to 0^+$, to our previous solution. After the ω integration in the complex ω plane is performed, the k integration can be done by completing the square in the exponent of Eq. (2.4.12), but the resulting Gaussian integration is a bit more complicated than the diffusion equation.

The result is given by

$$G(x,t;x',t') = \begin{cases} \sqrt{\frac{i}{4\pi(t-t')}} e^{i(x-x')^2/4(t-t')} & \text{for} \quad t > t', \\ 0 & \text{for} \quad t < t', \end{cases} \tag{2.4.13}$$

where $\hbar = 1$ and $2m = 1$.

In the case of the diffusion equation,

$$\left(-\frac{1}{D}\frac{\partial}{\partial t} + \frac{\partial^2}{\partial x^2} \right) \psi(x,t) = 0, \tag{2.4.14}$$

Equation (2.4.13) reduces to the Green's function for the diffusion equation,

$$G(x,t;x',t') = \begin{cases} \sqrt{\frac{1}{4\pi\kappa(t-t')}} e^{-(x-x')^2/4D(t-t')} & \text{for} \quad t > t', \\ 0 & \text{for} \quad t < t', \end{cases} \tag{2.4.15}$$

which satisfies the following equation,

$$\left(-\frac{1}{D}\frac{\partial}{\partial t} + \frac{\partial^2}{\partial x^2} \right) G(x,t;x',t') = \delta(t-t')\delta(x-x'), \tag{2.4.16}$$
$$G(x,t;x',t') = 0 \quad \text{for} \quad t < t',$$

where the diffusion constant D is given by

$$D = \frac{K}{C\rho} = \frac{\text{(thermal conductivity)}}{\text{(specific heat)} \times \text{(density)}}.$$

These two expressions, Eqs. (2.4.13) and (2.4.15), are related by the analytic continuation, $t \to -it$. The diffusion constant D plays the role of the inverse of the Planck constant $\hbar$.

We shall devote the next section for the more formal discussion of the scattering problem, namely the orthonormality of the outgoing and incoming wave and the unitarity of the S matrix.

2.5 Lippmann–Schwinger Equation

In a nonrelativistic scattering problem of quantum mechanics, we have the

Macroscopic causality of Stueckelberg. *When we regard the potential $V(t, \mathbf{r})$ as a function of t, we have no scattered wave, $\psi_{scatt}(t, \mathbf{r}) = 0$, for $t < T$, if $V(t, \mathbf{r}) = 0$ for $t < T$.*

We employ the

Adiabatic switching hypothesis. *We can take the limit $T \to -\infty$ after the computation of the scattered wave, $\psi_{scatt}(t, \mathbf{r})$.*

We derive the Lippmann–Schwinger equation, and prove the orthonormality of the outgoing wave and the incoming wave and the unitarity of the S matrix.

Lippmann–Schwinger Equation We shall begin with the time-dependent Schrödinger equation with the time-dependent potential $V(t, \mathbf{r})$,

$$i\hbar\frac{\partial}{\partial t}\psi(t, \mathbf{r}) = [H_0 + V(t, \mathbf{r})]\psi(t, \mathbf{r}).$$

In order to use the macroscopic causality, we assume

$$V(t, \mathbf{r}) = \begin{cases} V(\mathbf{r}) & \text{for} \quad t \geq T, \\ 0 & \text{for} \quad t < T. \end{cases}$$

For $t < T$, the particle obeys the free equation,

$$i\hbar\frac{\partial}{\partial t}\psi(t, \mathbf{r}) = H_0\psi(t, \mathbf{r}). \tag{2.5.1}$$

We write the solution of Eq. (2.5.1) as $\psi_{\text{inc}}(t, \mathbf{r})$. The wave function for the general time t is written as

$$\psi(t, \mathbf{r}) = \psi_{\text{inc}}(t, \mathbf{r}) + \psi_{\text{scatt}}(t, \mathbf{r}),$$

where we have

$$\left(i\hbar\frac{\partial}{\partial t} - H_0\right)\psi_{\text{scatt}}(t, \mathbf{r}) = V(t, \mathbf{r})\psi(t, \mathbf{r}). \tag{2.5.2}$$

We introduce the retarded Green's function for Eq. (2.5.2) as

$$\begin{cases} (i\hbar(\partial/\partial t) - H_0)K_{\text{ret}}(t, \mathbf{r}; t', \mathbf{r}') & = \quad \delta(t - t')\delta^3(\mathbf{r} - \mathbf{r}'), \\ K_{\text{ret}}(t, \mathbf{r}; t', \mathbf{r}') & = \quad 0 \quad \text{for} \quad t < t'. \end{cases} \tag{2.5.3}$$

The formal solution to Eq. (2.5.2) is given by

$$\psi_{\text{scatt}}(t,\mathbf{r}) = \int_{-\infty}^{\infty}\int K_{\text{ret}}(t,\mathbf{r};t',\mathbf{r}')V(t',\mathbf{r}')\psi(t',\mathbf{r}')\,dt'd\mathbf{r}'. \tag{2.5.4}$$

We note that the integrand of Eq. (2.5.4) survives only for $t \geq t' \geq T$. We now take the limit $T \to -\infty$, thus losing the t-dependence of $V(t,\mathbf{r})$,

$$\psi_{\text{scatt}}(t,\mathbf{r}) = \int_{-\infty}^{\infty}\int K_{\text{ret}}(t,\mathbf{r};t',\mathbf{r}')V(\mathbf{r}')\psi(t',\mathbf{r}')\,dt'd\mathbf{r}'.$$

When H_0 has no explicit space–time dependence, we have from the translation invariance that

$$K_{\text{ret}}(t,\mathbf{r};t',\mathbf{r}') = K_{\text{ret}}(t-t';\mathbf{r}-\mathbf{r}').$$

Adding $\psi_{\text{inc}}(t,\mathbf{r})$ to $\psi_{\text{scatt}}(t,\mathbf{r})$, Eq. (2.5.4), we have

$$\begin{aligned}\psi(t,\mathbf{r}) &= \psi_{\text{inc}}(t,\mathbf{r}) + \psi_{\text{scatt}}(t,\mathbf{r}) \\ &= \psi_{\text{inc}}(t,\mathbf{r}) + \int_{-\infty}^{\infty}\int K_{\text{ret}}(t-t';\mathbf{r}-\mathbf{r}')V(\mathbf{r}')\psi(t',\mathbf{r}')\,dt'd\mathbf{r}'.\end{aligned} \tag{2.5.5}$$

This equation is the integral equation determining the total wave function, given the incident wave. We rewrite this equation in a time-independent form. For this purpose, we set

$$\begin{aligned}\psi_{\text{inc}}(t,\mathbf{r}) &= \exp\left[\frac{-iEt}{\hbar}\right]\psi_{\text{inc}}(\mathbf{r}), \\ \psi(t,\mathbf{r}) &= \exp\left[\frac{-iEt}{\hbar}\right]\psi(\mathbf{r}).\end{aligned}$$

Then, from Eq. (2.5.5), we obtain

$$\psi(\mathbf{r}) = \psi_{\text{inc}}(\mathbf{r}) + \int G(\mathbf{r}-\mathbf{r}';E)V(\mathbf{r}')\psi(\mathbf{r}')\,d\mathbf{r}'. \tag{2.5.6}$$

Here $G(\mathbf{r}-\mathbf{r}';E)$ is given by

$$G(\mathbf{r}-\mathbf{r}';E) = \int_{-\infty}^{\infty} dt' \exp\left[\frac{iE(t-t')}{\hbar}\right] K_{\text{ret}}(t-t';\mathbf{r}-\mathbf{r}'). \tag{2.5.7}$$

Setting

$$K_{\text{ret}}(t-t';\mathbf{r}-\mathbf{r}') = \int \frac{dEd^3p}{(2\pi)^4}\exp\left[\frac{\{i\mathbf{p}(\mathbf{r}-\mathbf{r}') - iE(t-t')\}}{\hbar}\right]K(E,\mathbf{p}),$$

$$\delta(t-t')\delta^3(\mathbf{r}-\mathbf{r}') = \int \frac{dEd^3p}{(2\pi)^4}\exp\left[\frac{\{i\mathbf{p}(\mathbf{r}-\mathbf{r}') - iE(t-t')\}}{\hbar}\right],$$

substituting into Eq. (2.5.3), and writing $H_0 = \mathbf{p}^2/2m$, we obtain

$$\left(E - \frac{\mathbf{p}^2}{2m}\right)K(E,\mathbf{p}) = 1.$$

The solution consistent with the *retarded boundary condition* is

$$K(E,\mathbf{p}) = \frac{1}{E - (\mathbf{p}^2/2m) + i\varepsilon}, \qquad \text{with} \quad \varepsilon \text{ positive infinitesimal.}$$

Namely

$$K_{\text{ret}}(t-t';\mathbf{r}-\mathbf{r}') = \frac{1}{(2\pi)^4}\int dE d^3p \frac{\exp\left[\frac{\{i\mathbf{p}(\mathbf{r}-\mathbf{r}')-iE(t-t')\}}{\hbar}\right]}{E-(\mathbf{p}^2/2m)+i\varepsilon}.$$

Substituting this into Eq. (2.5.7) and setting $E = (\hbar\mathbf{k})^2/2m$, we obtain

$$G(\mathbf{r}-\mathbf{r}';E) = \frac{1}{(2\pi)^3}\int d^3p \frac{\exp[i\mathbf{p}(\mathbf{r}-\mathbf{r}')/\hbar]}{E-(\mathbf{p}^2/2m)+i\varepsilon} = -\frac{m}{2\pi}\frac{\exp[ik\,|\mathbf{r}-\mathbf{r}'|]}{|\mathbf{r}-\mathbf{r}'|}.$$

In Eq. (2.5.6), since the Fourier transform of $G(\mathbf{r}-\mathbf{r}';E)$ is written as

$$\frac{1}{E - H_0 + i\varepsilon},$$

the equation (2.5.6) can be written formally as

$$\Psi = \Phi + \frac{1}{E - H_0 + i\varepsilon} V\Psi, \qquad E > 0, \tag{2.5.8}$$

where we wrote $\Psi = \psi(\mathbf{r}), \Phi = \psi_{\text{inc}}(\mathbf{r})$, and the incident wave Φ satisfies the free particle equation,

$$(E - H_0)\Phi = 0.$$

Operating $(E - H_0)$ on Eq. (2.5.8) from the left, we obtain the Schrödinger equation,

$$(E - H_0)\Psi = V\Psi. \tag{2.5.9}$$

We note that equation (2.5.9) is the *differential equation* whereas equation (2.5.8) is the *integral equation*, which embodies the boundary condition.

For the bound state problem ($E < 0$), since the operator $(E - H_0)$ is negative definite and has a unique inverse, we have

$$\Psi = \frac{1}{E - H_0} V\Psi, \qquad E < 0. \tag{2.5.10}$$

We call Eqs. (2.5.8) and (2.5.10) the Lippmann–Schwinger equation (the L-S equation in short). The $+i\varepsilon$ in the denominator of Eq. (2.5.8) makes the scattered wave the outgoing spherical wave. The presence of the $+i\varepsilon$ in Eq. (2.5.8) enforces the *outgoing wave condition.* It is mathematically convenient also to introduce the $-i\varepsilon$ into Eq. (2.5.8), which makes the scattered wave the incoming spherical wave and thus enforces the *incoming wave condition.*

We construct two kinds of wave functions:

$$\Psi_a^{(+)} = \Phi_a + \frac{1}{E_a - H_0 + i\varepsilon} V\Psi_a^{(+)}, \quad \text{outgoing wave condition,} \tag{2.5.11.+}$$

$$\Psi_a^{(-)} = \Phi_a + \frac{1}{E_a - H_0 - i\varepsilon} V\Psi_a^{(-)}, \quad \text{incoming wave condition.} \tag{2.5.11.−}$$

The formal solution to the L-S equation was obtained by G. Chew and M. Goldberger. By iteration of Eq. (2.5.11.+), we have

$$\Psi_a^{(+)} = \Phi_a + \frac{1}{E_a - H_0 + i\varepsilon}\left(1 + V\frac{1}{E_a - H_0 + i\varepsilon} + \cdots\right)V\Phi_a$$

$$= \Phi_a + \frac{1}{E_a - H_0 + i\varepsilon}\left(1 - V\frac{1}{E_a - H_0 + i\varepsilon}\right)^{-1} V\Phi_a = \Phi_a + \frac{1}{E_a - H + i\varepsilon}V\Phi_a.$$

Here we have used the operator identity $A^{-1}B^{-1} = (BA)^{-1}$ and $H = H_0 + V$ represents the total Hamiltonian.

We write the formal solution for $\Psi_a^{(+)}$ and $\Psi_a^{(-)}$ together as:

$$\Psi_a^{(+)} = \Phi_a + \frac{1}{E_a - H + i\varepsilon}V\Phi_a, \tag{2.5.12.+}$$

$$\Psi_a^{(-)} = \Phi_a + \frac{1}{E_a - H - i\varepsilon}V\Phi_a. \tag{2.5.12.−}$$

Orthonormality of $\Psi_a^{(+)}$: We will prove the orthonormality only for $\Psi_a^{(+)}$:

$$\begin{aligned}
(\Psi_b^{(+)}, \Psi_a^{(+)}) &= (\Phi_b, \Psi_a^{(+)}) + \left(\frac{1}{E_b - H + i\varepsilon}V\Phi_b, \Psi_a^{(+)}\right) \\
&= (\Phi_b, \Psi_a^{(+)}) + \left(\Phi_b, V\frac{1}{E_b - H - i\varepsilon}\Psi_a^{(+)}\right) \\
&= (\Phi_b, \Psi_a^{(+)}) + \frac{1}{E_b - E_a - i\varepsilon}\left(\Phi_b, V\Psi_a^{(+)}\right) \\
&= (\Phi_b, \Phi_a) + \left(\Phi_b, \frac{1}{E_a - H_0 + i\varepsilon}V\Psi_a^{(+)}\right) \\
&\quad + \frac{1}{E_b - E_a - i\varepsilon}\left(\Phi_b, V\Psi_a^{(+)}\right) \\
&= \delta_{ba} + \left(\frac{1}{E_a - E_b + i\varepsilon} + \frac{1}{E_b - E_a - i\varepsilon}\right)\left(\Phi_b, V\Psi_a^{(+)}\right) \\
&= \delta_{ba}.
\end{aligned} \tag{2.5.13}$$

Thus $\Psi_a^{(+)}$ forms a complete and orthonormal basis. The same proof applies for $\Psi_a^{(-)}$ also. Frequently, the orthonormality of $\Psi_a^{(\pm)}$ is assumed on the outset. We have proved the orthonormality of $\Psi_a^{(\pm)}$ using the L-S equation and the formal solution due to G. Chew and M. Goldberger.

In passing, we state that, in relativistic quantum field theory in the L.S.Z. formalism, the outgoing wave $\Psi_a^{(+)}$ is called the *in-state* and is written as $\Psi_a^{(\mathrm{in})}$, and the incoming wave $\Psi_a^{(-)}$ is called the *out-state* and is written as $\Psi_a^{(\mathrm{out})}$. There exists some confusion on this matter.

Unitarity of the S matrix: We define the S matrix by

$$S_{ba} = (\Psi_b^{(-)}, \Psi_a^{(+)}) = (\Psi_b^{(\mathrm{out})}, \Psi_a^{(\mathrm{in})}). \tag{2.5.14}$$

This definition states that the S matrix transforms one complete set to another complete set. By first making use of the formal solution of G. Chew and M. Goldberger first and then using the $L - S$ equation as before, we obtain

$$\begin{aligned} S_{ba} &= \delta_{ba} + \left(\frac{1}{E_a - E_b + i\varepsilon} + \frac{1}{E_b - E_a + i\varepsilon} \right) \left(\Phi_b, V\Psi_a^{(+)} \right) \\ &= \delta_{ba} - 2\pi i \delta(E_b - E_a) \left(\Phi_b, V\Psi_a^{(+)} \right). \end{aligned} \tag{2.5.15}$$

We define the T matrix by

$$T_{ba} = \left(\Phi_b, V\Psi_a^{(+)} \right). \tag{2.5.16}$$

Then we have

$$S_{ba} = \delta_{ba} - 2\pi i \delta(E_b - E_a) T_{ba}. \tag{2.5.17}$$

If the S matrix is unitary, it satisfies

$$\hat{S}^\dagger \hat{S} = \hat{S}\hat{S}^\dagger = 1. \tag{2.5.18}$$

These unitarity conditions are equivalent to the following conditions in terms of the T matrix:

$$T_{ba}^\dagger - T_{ba} = \begin{cases} 2\pi i \sum_n T_{bn}^\dagger \delta(E_b - E_n) T_{na}, \\ 2\pi i \sum_n T_{bn} \delta(E_b - E_n) T_{na}^\dagger, \end{cases} \quad \text{with} \quad E_b = E_a. \tag{2.5.19, 20}$$

In order to prove the unitarity of the S matrix, Eq. (2.5.18), it suffices to prove Eqs. (2.5.19) and (2.5.20), which are expressed in terms of the T matrix.

We first note

$$T_{ba}^\dagger = T_{ab}^* = (\Phi_a, V\Psi_b^{(+)})^* = (V\Psi_b^{(+)}, \Phi_a) = (\Psi_b^{(+)}, V\Phi_a).$$

Then

$$T_{ba}^\dagger - T_{ba} = (\Psi_b^{(+)}, V\Phi_a) - (\Phi_b, V\Psi_a^{(+)}).$$

Inserting the formal solution of G. Chew and M. Goldberger to $\Psi_a^{(+)}$ and $\Psi_b^{(+)}$ above, we have

$$\begin{aligned} T_{ba}^\dagger - T_{ba} &= (\Phi_b, V\Phi_a) + \left(\frac{1}{E_b - H + i\varepsilon} V\Phi_b, V\Phi_a \right) \\ &\quad - (\Phi_b, V\Phi_a) - \left(\Phi_b, V \frac{1}{E_a - H + i\varepsilon} V\Phi_a \right) \\ &= \left(V\Phi_b, \left(\frac{1}{E_b - H - i\varepsilon} - \frac{1}{E_b - H + i\varepsilon} \right) V\Phi_a \right) \\ &= (V\Phi_b, 2\pi i \delta(E_b - H) V\Phi_a), \end{aligned} \tag{2.5.21}$$

where, in the one line above the last line of Eq. (2.5.21), we used the fact that $E_b = E_a$. Inserting the complete orthonormal basis, $\Psi^{(-)}$, between the product of the operators in Eq. (2.5.21), we have

$$\begin{aligned} T^{\dagger}_{ba} - T_{ba} &= \sum_n (V\Phi_b, \Psi^{(-)}_n) 2\pi i \delta(E_b - E_n)(\Psi^{(-)}_n, V\Phi_a) \\ &= 2\pi i \sum_n T^{\dagger}_{bn} \delta(E_b - E_n) T_{na}. \end{aligned}$$

This is Eq. (2.5.19).

If we insert the complete orthonormal basis, $\Psi^{(+)}$, between the product of the operators in Eq. (2.5.21), we obtain

$$\begin{aligned} T^{\dagger}_{ba} - T_{ba} &= \sum_n (V\Phi_b, \Psi^{(+)}_n) 2\pi i \delta(E_b - E_n)(\Psi^{(+)}_n, V\Phi_a) \\ &= 2\pi i \sum_n T_{bn} \delta(E_b - E_n) T^{\dagger}_{na}. \end{aligned}$$

This is Eq. (2.5.20). Thus the S matrix defined by Eq. (2.5.14) is unitary. The unitarity of the S matrix is *equivalent* to the fact that the outgoing wave set $\{\Psi^{(+)}_a\}$ and the incoming wave set $\{\Psi^{(-)}_b\}$ form the complete orthonormal basis, respectively.

In fact the S matrix has to be unitary since the S matrix transforms one complete orthonormal set to another complete orthonormal set.

2.6 Problems for Chapter 2

2.1. (Due to H. C.) Solve

$$\phi(x) = 1 + \lambda \int_0^1 (xy + x^2 y^2)\phi(y)\, dy.$$

a) Show that this is equivalent to a 2×2 matrix equation.

b) Find the eigenvalues and the corresponding eigenvectors of the kernel.

c) Find the solution of the inhomogeneous equation if $\lambda \neq$ eigenvalues.

d) Solve the corresponding Fredholm integral equation of the first kind.

2.2. (Due to H. C.) Solve

$$\phi(x) = 1 + \lambda \int_0^x xy\phi(y)\, dy.$$

Discuss the solution of the homogeneous equation.

2.3. (Due to H. C.) Consider the integral equation,

$$\phi(x) = f(x) + \lambda \int_{-\infty}^{+\infty} e^{-(x^2+y^2)}\phi(y)\, dy, \quad -\infty < x < \infty.$$

a) Solve this equation for

$$f(x) = 0.$$

For what values of λ, does it have non-trivial solutions?

b) Solve this equation for

$$f(x) = x^m, \quad \text{with} \quad m = 0, 1, 2, \ldots.$$

Does this inhomogeneous equation have any solutions when λ is equal to an eigenvalue of the kernel?

Hint: You may express your results in terms of the Gamma function,

$$\Gamma(z) = \int_0^\infty t^{z-1} e^{-t} dt, \quad \operatorname{Re} z > 0.$$

2.4. (Due to H. C.) Solve the following integral equation,

$$u(\theta) = 1 + \lambda \int_0^{2\pi} \sin(\phi - \theta) u(\phi) d\phi, \quad 0 \leq \theta < 2\pi,$$

where $u(\theta)$ is periodic with period 2π. Does the kernel of this equation have any real eigenvalues?

Hint: Notice

$$\sin(\phi - \theta) = \sin\phi \cos\theta - \cos\phi \sin\theta.$$

2.5. (Due to D. M.) Consider the integral equation,

$$\phi(x) = 1 + \lambda \int_0^1 \frac{x^n - y^n}{x - y} \phi(y)\, dy, \quad 0 \leq x \leq 1.$$

a) Solve this equation for $n = 2$. For what values of λ, does the equation have no solutions?

b) Discuss how you would solve this integral equation for arbitrary positive integer n.

2.6. (Due to D. M.) Solve the integral equation,

$$\phi(x) = 1 + \int_0^1 (1 + x + y + xy)^\nu \phi(y)\, dy, \quad 0 \leq x \leq 1, \quad \nu : \text{real}.$$

Hint: Notice that the kernel,

$$(1 + x + y + xy)^{\nu},$$

can be factorized.

2.7. In the Fredholm integral equation of the second kind, if the kernel is given by

$$K(x, y) = \sum_{n=1}^{N} g_n(x) h_n(y),$$

show that the integral equation is equivalent to an $N \times N$ matrix equation.

2.8. (Due to H. C.) Consider the motion of an harmonic oscillator with a time-dependent spring constant,

$$\frac{d^2}{dt^2} x + \omega^2 x = -A(t)x,$$

where ω is a constant and $A(t)$ is a complicated function of t. Transform this differential equation together with the boundary conditions,

$$x(T_i) = x_i \quad \text{and} \quad x(T_f) = x_f,$$

to an integral equation.

Hint: Construct a Green's function $G(t, t')$ satisfying

$$G(T_i, t') = G(T_f, t') = 0.$$

2.9. Generalize the discussion of Section 2.4 to the case of three spatial dimensions and transform the Schrödinger equation with the initial condition,

$$\lim_{t \to -\infty} e^{i\omega t} \psi(\vec{x}, t) = e^{ikz},$$

to an integral equation.

Hint: Construct a Green's function $G(t, t')$ satisfying

$$\left(i \frac{\partial}{\partial t} + \vec{\nabla}^2 \right) G(\vec{x}, t; \vec{x}', t') = \delta(t - t')\delta^3(\vec{x} - \vec{x}'), \quad G(\vec{x}, t; \vec{x}', t') = 0 \quad \text{for} \quad t < t'.$$

2.10. (Due to H. C.) Consider the equation

$$\left[-\frac{\partial^2}{\partial t^2} + \frac{\partial^2}{\partial x^2} - m^2 \right] \phi(x, t) = U(x, t)\phi(x, t).$$

If the initial and final conditions are

$$\phi(x, -T) = f(x) \quad \text{and} \quad \phi(x, T) = g(x),$$

transform the equation to an integral equation.

Hint: Consider the Green's function

$$\left[-\frac{\partial^2}{\partial t^2}+\frac{\partial^2}{\partial x^2}-m^2\right]G(x,t;x',t')=\delta(x-x')\delta(t-t').$$

2.11. (Due to D. M.) The time-independent Schrödinger equation with the periodic potential, $V(x)=-(a^2+k^2\cos^2 x)$, reads as

$$\frac{d^2}{dx^2}\psi(x)+(a^2+k^2\cos^2 x)\psi(x)=0.$$

Show directly that even periodic solutions of this equation, which are even Mathieu functions, satisfy the homogeneous integral equation,

$$\psi(x)=\lambda\int_{-\pi}^{\pi}\exp[k\cos x\cos y]\psi(y)\,dy.$$

Hint: Show that $\phi(x)$ defined by

$$\phi(x)\equiv\int_{-\pi}^{\pi}\exp[k\cos x\cos y]\psi(y)\,dy$$

is even and periodic, and satisfies the above time-independent Schrödinger equation. Thus, $\psi(x)$ is a constant multiple of $\phi(x)$,

$$\psi(x)=\lambda\phi(x).$$

2.12. (Due to H. C.) Consider the differential equation,

$$\frac{d^2}{dt^2}\phi(t)=\lambda e^{-t}\phi(t),\quad 0\le t<\infty,\quad \lambda=\text{constant},$$

together with the initial conditions,

$$\phi(0)=0\quad\text{and}\quad\phi'(0)=1.$$

a) Find the partial differential equation for the Green's function $G(t,t')$. Determine the form of $G(t,t')$ when $t\neq t'$.

b) Transform the differential equation for $\phi(t)$, together with the initial conditions, to an integral equation. Determine the conditions on $G(t,t')$.

c) Determine $G(t,t')$.

d) Substitute your answer for $G(t,t')$ into the integral equation and verify explicitly that the integral equation is equivalent to the differential equation together with the initial conditions.

e) Does the initial value problem have a solution for all λ? If so, is the solution unique?

2.13. In the Volterra integral equation of the second kind, if the kernel is given by

$$K(x, y) = \sum_{n=1}^{N} g_n(x) h_n(y),$$

show that the integral equation can be reduced to an ordinary differential equation of Nth order.

2.14. Consider the partial differential equation of the form,

$$\left(\frac{\partial^2}{\partial t^2} - \frac{\partial^2}{\partial x^2} \right) \phi(x, t) = p(x, t) - \lambda \frac{\partial^2}{\partial x^2} \phi(x, t) \cdot \frac{\partial}{\partial x} \phi(x, t),$$

where

$$-\infty < x < \infty, \quad t \geq 0,$$

and λ is a constant, with the initial conditions specified by

$$\phi(x, 0) = a(x), \quad \text{and} \quad \frac{\partial}{\partial t} \phi(x, 0) = b(x).$$

This partial differential equation describes the displacement of a vibrating string under the distributed load $p(x, t)$.

a) Find the Green's function for this partial differential equation.

b) Express this initial value problem in terms of an integral equation using the Green's function found in a). Explain how you would find an approximate solution if λ were small.

Hint: By applying the Fourier transform in x, find a function $\phi_0(x, t)$ which satisfies the wave equation,

$$\left(\frac{\partial^2}{\partial t^2} - \frac{\partial^2}{\partial x^2} \right) \phi_0(x, t) = 0,$$

and the given initial conditions.

3 Integral Equations of Volterra Type

3.1 Iterative Solution to Volterra Integral Equation of the Second Kind

Consider the *inhomogeneous Volterra integral equation of the second kind*,

$$\phi(x) = f(x) + \lambda \int_0^x K(x,y)\phi(y)\,dy, \qquad 0 \le x, y \le h, \tag{3.1.1}$$

with $f(x)$ and $K(x,y)$ *square-integrable*,

$$\|f\|^2 = \int_0^h |f(x)|^2 \;dx < \infty, \tag{3.1.2}$$

$$\|K\|^2 = \int_0^h dx \int_0^x dy\,|K(x,y)|^2 < \infty. \tag{3.1.3}$$

Also, define

$$A(x) = \int_0^x |K(x,y)|^2 \;dy. \tag{3.1.4}$$

Note that the upper limit of y integration is x. Note also that the Volterra integral equation is a special case of the Fredholm integral equation with the kernel

$$K(x,y) = 0 \quad \text{for} \quad x < y < h. \tag{3.1.5}$$

We will prove in the following that, for Eq. (3.1.1),

(1) A solution *exists* for all values of λ,

(2) The solution is *unique* for all values of λ,

(3) The iterative solution is *convergent* for all values of λ.

We start our discussion with the construction of an *iterative solution*. Consider a series solution of the usual form,

$$\phi(x) = \phi_0(x) + \lambda\phi_1(x) + \lambda^2\phi_2(x) + \cdots + \lambda^n\phi_n(x) + \cdots. \tag{3.1.6}$$

Applied Mathematics in Theoretical Physics. Michio Masujima

ISBN: 3-527-40534-8

Substituting the series solution (3.1.6) into Eq. (3.1.1), we have

$$\sum_{k=0}^{\infty}\lambda^k\phi_k\left(x\right)=f(x)+\lambda\int_0^x K(x,y)\sum_{j=0}^{\infty}\lambda^j\phi_j(y)\,dy.$$

Collecting like powers of λ, we have

$$\phi_0(x)=f(x),$$

$$\phi_n(x)=\int_0^x K(x,y)\phi_{n-1}(y)\,dy,\quad n=1,2,3,\cdots. \tag{3.1.7}$$

We now examine convergence of the series solution (3.1.6). Applying the *Schwarz inequality* to Eq. (3.1.7), we have

$$\phi_1\left(x\right)=\int_0^x K(x,y)f(y)\,dy$$

$$\Rightarrow\left|\phi_1\left(x\right)\right|^2\leq A(x)\int_0^x\left|f(y)\right|^2\,dy\leq A(x)\left\|f\right\|^2,$$

$$\phi_2\left(x\right)=\int_0^x K(x,y)\phi_1(y)\,dy$$

$$\Rightarrow\left|\phi_2\left(x\right)\right|^2\leq A(x)\int_0^x\left|\phi_1(y)\right|^2\,dy\leq A(x)\int_0^x dx_1A(x_1)\left\|f\right\|^2,$$

and

$$\phi_3(x)=\int_0^x K(x,y)\phi_2(y)\,dy$$

$$\Rightarrow\left|\phi_3(x)\right|^2\leq A(x)\int_0^x dx_2\left|\phi_2(x_2)\right|^2\leq\frac{1}{2}A(x)\left[\int_0^x dx_1A(x_1)\right]^2\left\|f\right\|^2.$$

In general, we have

$$\left|\phi_n\left(x\right)\right|^2\leq\frac{1}{(n-1)!}A(x)\left[\int_0^x dy\,A(y)\right]^{n-1}\left\|f\right\|^2,\quad n=1,2,3,\cdots. \tag{3.1.8}$$

Define

$$B(x)\equiv\int_0^x dy\,A(y). \tag{3.1.9}$$

Thus, from Eqs. (3.1.8) and (3.1.9), we obtain the following bound on $\phi_n\left(x\right)$,

$$\left|\phi_n\left(x\right)\right|\leq\frac{1}{\sqrt{(n-1)!}}\sqrt{A(x)}\left[B(x)\right]^{(n-1)/2}\left\|f\right\|,\quad n=1,2,3,\cdots. \tag{3.1.10}$$

We now examine the convergence of the series solution (3.1.6).

$$|\phi(x) - f(x)| = \left|\sum_{n=1}^{\infty} \lambda^n \phi_n(x)\right| \leq \sum_{n=1}^{\infty} |\lambda|^n \frac{1}{\sqrt{(n-1)!}} \sqrt{A(x)}\,[B(x)]^{(n-1)/2}\,\|f\|$$
$$= \sqrt{A(x)}\,\|f\| \sum_{n=1}^{\infty} |\lambda|^n\,[B(x)]^{(n-1)/2}\,\frac{1}{\sqrt{(n-1)!}}. \tag{3.1.11}$$

Letting

$$a_n = |\lambda|^n\,[B(x)]^{(n-1)/2}\,\frac{1}{\sqrt{(n-1)!}},$$

and applying the ratio test on the right-hand side of Eq. (3.1.11), we have

$$\frac{a_{n+1}}{a_n} = \frac{\left\{|\lambda|^{n+1}\,[B(x)]^{n/2}\,\sqrt{(n-1)!}\right\}}{\left\{|\lambda|^n\,[B(x)]^{(n-1)/2}\,\sqrt{n!}\right\}}$$
$$= |\lambda|\,\frac{\sqrt{B(x)}}{\sqrt{n}},$$

whence we have,

$$\lim_{n\to\infty} \frac{a_{n+1}}{a_n} = 0 \quad \text{for} \quad \text{all} \quad \lambda. \tag{3.1.12}$$

Thus the series solution (3.1.6) converges for all λ, provided that $\|f\|$, $A(x)$ and $B(x)$ exist and are finite.

We have proved statements (1) and (3) under the condition that the kernel $K(x, y)$ and the inhomogeneous term $f(x)$ are *square-integrable*, Eqs. (3.1.2) and (3.1.3).

To show that Eq. (3.1.1) has a *unique solution* which is *square-integrable* ($\|\phi\| < \infty$), we shall prove that

$$R_n(x) \to 0 \quad \text{as} \quad n \to \infty, \tag{3.1.13}$$

where

$$\lambda^{n+1} R_{n+1}(x) \equiv \phi(x) - \sum_{k=0}^{k=n} \lambda^k \phi_k(x), \tag{3.1.14}$$

and

$$R_n(x) = \int_0^x K(x, y) R_{n-1}(y)\,dy, \qquad R_0(x) = \phi(x).$$

Repeating the same procedure as above, we can establish

$$[R_n(x)]^2 \leq \frac{\|\phi\|^2\,A(x)[B(x)]^{n-1}}{(n-1)!}. \tag{3.1.15}$$

Thus $R_{n+1}(x)/R_n(x)$ vanishes as $n \to \infty$. Returning to Eq. (3.1.14), we see that the iterative solution $\phi(x)$, Eq. (3.1.6), is *unique*.

The *uniqueness* of the solution of the *inhomogeneous Volterra integral equation* implies that the *square-integrable solution of the homogeneous Volterra integral equation*

$$\psi_{\mathrm{H}}(x) = \lambda \int_0^x K(x,y)\psi_{\mathrm{H}}(y)\,dy \tag{3.1.16}$$

is trivial,

$$\psi_{\mathrm{H}}(x) \equiv 0.$$

Otherwise, $\phi(x) + c\psi_{\mathrm{H}}(x)$ would also be a solution of the inhomogeneous Volterra integral equation (3.1.1), in contradiction to our conclusion that the solution of Eq. (3.1.1) is unique.

The result above can be used to prove the existence and uniqueness of the solution of a differential equation. As noted in Chapter 2, a differential equation can always be transformed into an integral equation. Consider the following second-order ordinary differential equation:

$$\left[a_0(x)\frac{d^2}{dx^2} + a_1(x)\frac{d}{dx} + a_2(x)\right] u(x) = f(x). \tag{3.1.17}$$

We shall transform this equation together with some initial conditions into a Volterra integral equation. We first perform a transformation to $\phi(x)$ by setting

$$u(x) = \exp\left[-\frac{1}{2}\int^x \frac{a_1(y)}{a_0(y)}\,dy\right]\phi(x),$$

to reduce Eq. (3.1.17) into the form

$$\left[\frac{d^2}{dx^2} + q(x)\right]\phi(x) = F(x).$$

Moving the term $q(x)\phi(x)$ to the right-hand side, we can immediately convert this differential equation into a Volterra integral equation of the second kind. Hence the existence and the uniqueness of the solution for the differential equation (3.1.17) can be deduced.

When the coefficient functions $a_i(x)$ $(i = 0, 1, 2)$ of Eq. (3.1.17) depend on some parameter, say β, and are analytic in β, what can we say about the *analyticity* of the solution as a function of β? All the terms in the iteration series (3.1.6) are analytic in β by construction. The iteration series is absolutely convergent so that the solution to Eq. (3.1.17) is analytic in β.

3.2 Solvable cases of Volterra Integral Equation

We list a few solvable cases of the Volterra integral equation.

Case (1): The kernel is equal to a sum of n *factorized terms*.

We demonstrated the reduction of such an integral equations into an nth order ordinary differential equation in Problem 6 in Chapter 2.

Case (2): The kernel is *translational.*

$$K(x, y) = K(x - y). \tag{3.2.1}$$

Consider

$$\phi(x) = f(x) + \lambda \int_0^x K(x - y)\phi(y)\, dy \quad \text{on} \quad 0 \leq x < \infty. \tag{3.2.2}$$

We note that the second term on the right-hand side of Eq. (3.2.2) is a *convolution integral.* We shall use the *Laplace transform,*

$$L\{F(x)\} \equiv \bar{F}(s) \equiv \int_0^\infty dx F(x) e^{-sx}. \tag{3.2.3}$$

Taking the Laplace transform of the integral equation (3.2.2), we obtain

$$\bar{\phi}(s) = \bar{f}(s) + \lambda \bar{K}(s)\bar{\phi}(s). \tag{3.2.4}$$

Solving Eq. (3.2.4) for $\bar{\phi}(s)$, we obtain

$$\bar{\phi}(s) = \frac{\bar{f}(s)}{1 - \lambda \bar{K}(s)}. \tag{3.2.5}$$

Applying the inverse Laplace transform to Eq. (3.2.5), we obtain

$$\phi(x) = \int_{\gamma - i\infty}^{\gamma + i\infty} \frac{ds}{2\pi i} e^{sx} \frac{\bar{f}(s)}{1 - \lambda \bar{K}(s)}, \tag{3.2.6}$$

where the inversion path $(\gamma \pm i\infty)$ in the complex s plane lies to the right of all singularities of the integrand as indicated in Figure 3.1.

Before we solve an example, we first recall the Laplace transform and some of its properties.

Definition: Suppose $F(t)$ is defined on $[0, \infty)$ with $F(t) = 0$ for $t < 0$. Then the Laplace transform of $F(t)$ is defined by

$$\bar{F}(s) \equiv L\{F(t)\} \equiv \int_0^\infty F(t) e^{-st} dt. \tag{3.2.7}$$

The inversion is given by

$$F(t) = \frac{1}{2\pi i} \int_{\gamma - i\infty}^{\gamma + i\infty} \bar{F}(s) e^{st}\, ds = L^{-1}\{\bar{F}(s)\}. \tag{3.2.8}$$

Properties:

$$L\left\{\frac{d}{dt} F(t)\right\} = s\bar{F}(s) - \bar{F}(0). \tag{3.2.9}$$

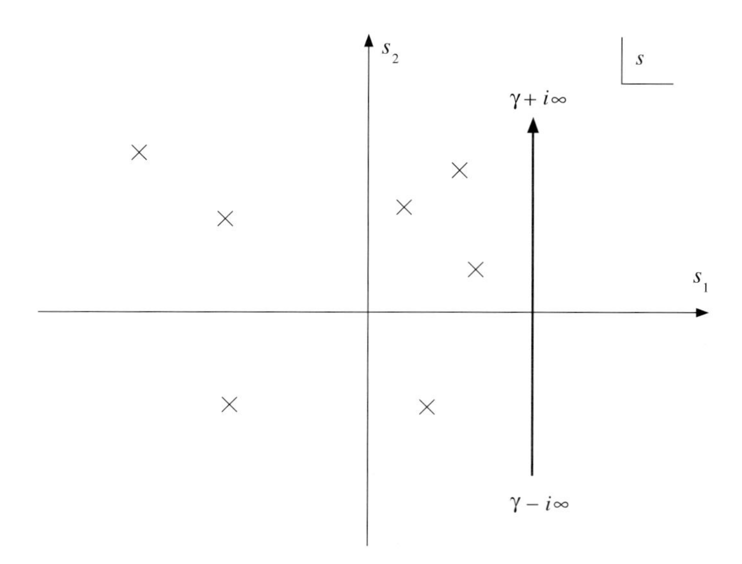

Fig. 3.1: The inversion path of the Laplace transform $\bar{\phi}(s)$ in the complex s plane from $s = \gamma - i\infty$ to $s = \gamma + i\infty$, which lies to the right of all singularities of the integrand.

The Laplace transform of *convolution*

$$H(t) = \int_0^t G(t - t') F(t') \, dt' = \int_0^t G(t')F(t - t') \, dt' \tag{3.2.10}$$

is given by

$$\bar{H}(s) = \bar{G}(s) \bar{F}(s), \tag{3.2.11}$$

which is already used in deriving Eq. (3.2.4).
The Laplace transforms of 1 and t^n are respectively given by

$$L\{1\} = \frac{1}{s}, \tag{3.2.12}$$

and

$$L\{t^n\} = \frac{\Gamma(n+1)}{s^{n+1}}. \tag{3.2.13}$$

The *Gamma Function*: $\Gamma(z)$ is defined by

$$\Gamma(z) = \int_0^\infty t^{z-1} e^{-t} dt. \tag{3.2.14}$$

Properties of the Gamma Function:

$$\Gamma(1) = 1, \tag{3.2.15a}$$

$$\Gamma\left(\tfrac{1}{2}\right) = \sqrt{\pi}, \tag{3.2.15b}$$

$$\Gamma(z+1) = z\Gamma(z), \tag{3.2.15c}$$

$$\Gamma(n+1) = n!, \tag{3.2.15d}$$

$$\Gamma(z)\Gamma(1-z) = \pi/\sin(\pi z), \tag{3.2.15e}$$

$$\Gamma(z) \text{ is singular at } z = 0, -1, -2, \ldots. \tag{3.2.15f}$$

Derivation of the Abel integral equation: The descent time of a frictionless ball on the side of a hill is known as a function of its initial height x. Let us find the shape of the hill. Starting with initial velocity zero, the speed of the ball at height y is obtained by solving

$$\frac{1}{2}mv^2 = mg(x-y),$$

from which $v = \sqrt{2g(x-y)}$. Let the shape of the hill to be given by $\xi = f(y)$. Then the arc length is given by

$$ds = \sqrt{(dy)^2 + (d\xi)^2} = \sqrt{1 + (f'(y))^2}\,|dy|\,.$$

The descent time to height $y = 0$ is given by

$$T(x) = \int dt = \int \frac{dt}{ds}\,ds = \int \frac{ds}{ds/dt} = \int \frac{ds}{v} = \int_{y=x}^{y=0} \frac{\sqrt{1+(f'(y))^2}}{\sqrt{2g(x-y)}}\,|dy|\,.$$

Since y is decreasing, dy is negative so that $|dy| = -dy$. Thus the descent time is given by

$$T(x) = \int_0^x \frac{\phi(y)}{\sqrt{x-y}}\,dy \tag{3.2.16}$$

with

$$\phi(y) = \frac{1}{\sqrt{2g}}\sqrt{1+(f'(y))^2}. \tag{3.2.17}$$

So, given the descent time $T(x)$ as a function of the initial height x, we solve the Abel integral equation (3.2.16) for $\phi(x)$, and then solve (3.2.17) for $f'(y)$ which gives the shape of the curve $\xi = f(y)$.

We solve an example of the Volterra integral equation with the translational kernel derived above, Eq. (3.2.16).

❑ **Example 3.1. Abel Integral Equation.**

$$\int_0^x \frac{\phi(x')}{\sqrt{x-x'}}\,dx' = f(x), \quad \text{with} \quad f(0) = 0. \tag{3.2.18}$$

Solution. Take the Laplace transform of $x^{-\frac{1}{2}}$.

$$L\left\{x^{-\frac{1}{2}}\right\} = \frac{\Gamma\left(\frac{1}{2}\right)}{s^{\frac{1}{2}}} = \sqrt{\frac{\pi}{s}}.$$

Then the Laplace transform of Eq. (3.2.18) is

$$\sqrt{\frac{\pi}{s}}\bar{\phi}(s) = \bar{f}(s).$$

Solving for $\bar{\phi}(s)$, we obtain

$$\bar{\phi}(s) = \sqrt{\frac{s}{\pi}}\bar{f}(s).$$

Unfortunately $\sqrt{s}$ is not the Laplace transform of anything since, for any function $g(t)$, we must have $\bar{g}(s) \to 0$ as $s \to \infty$. (Recall the definition (3.2.7) of the Laplace transform.)

Thus we rewrite

$$\begin{aligned}\phi(x) &= \frac{1}{2\pi i}\int_{\gamma-i\infty}^{\gamma+i\infty} e^{sx}\left(\bar{f}(s)\sqrt{s}/\sqrt{\pi}\right) ds \\ &= \frac{1}{2\pi i}\frac{d}{dx}\int_{\gamma-i\infty}^{\gamma+i\infty} \frac{1}{\pi}\sqrt{\frac{\pi}{s}}\bar{f}(s)\, e^{sx}\, ds \\ &= \frac{1}{\pi}\frac{d}{dx}L^{-1}\left\{\sqrt{\frac{\pi}{s}}\bar{f}(s)\right\} \\ &= \frac{1}{\pi}\frac{d}{dx}\int_0^x \frac{f(x')}{\sqrt{x-x'}}\, dx',\end{aligned} \tag{3.2.19}$$

where the convolution theorem, Eqs. (3.2.10) and (3.2.11), has been applied.

As an extension of Case(2), we can solve a system of Volterra integral equations of the second kind with the translational kernels,

$$\phi_i(x) = f_i(x) + \sum_{j=1}^{n}\int_0^x K_{ij}(x-y)\phi_j(y)\, dy, \qquad i = 1, \cdots, n, \tag{3.2.20}$$

where $K_{ij}(x)$ and $f_i(x)$ are known functions with Laplace transforms $\bar{K}_{ij}(s)$ and $\bar{f}(s)$. Taking the Laplace transform of (3.2.20), we obtain

$$\bar{\phi}_i(s) = \bar{f}_i(s) + \sum_{j=1}^{n}\bar{K}_{ij}(s)\bar{\phi}_j(s), \qquad i = 1, \cdots, n. \tag{3.2.21}$$

Equation (3.2.21) is a system of linear algebraic equations for $\bar{\phi}_i(s)$. We can solve (3.2.21) for $\bar{\phi}_i(s)$ easily and apply the inverse Laplace transform to $\bar{\phi}_i(s)$ to obtain $\phi_i(x)$.

3.3 Problems for Chapter 3

3.1. Consider the Volterra integral equation of the first kind,

$$f(x) = \int_0^x K(x,y)\phi(y)\,dy, \qquad 0 \le x \le h.$$

Show that, by differentiating the above equation with respect to x, we can transform this integral equation to a Volterra integral equation of the second kind as long as

$$K(x,x) \neq 0.$$

3.2. Transform the radial Schrödinger equation

$$\left[\frac{d^2}{dr^2} - \frac{l(l+1)}{r^2} + k^2 - V(r)\right]\psi(r) = 0, \quad \text{with} \quad \psi(r) \sim r^{l+1} \quad \text{as} \quad r \to 0,$$

to a Volterra integral equation of the second kind.

Hint: There are two ways to define the homogeneous equation.

(i)

$$\left[\frac{d^2}{dr^2} - \frac{l(l+1)}{r^2} + k^2\right]\psi_{\mathrm{H}}(r) = 0 \quad \Rightarrow \quad \psi_{\mathrm{H}}(r) = \begin{cases} krj_l(kr) \\ krh_l(kr) \end{cases},$$

where $j_l(kr)$ is the lth-order spherical Bessel function and $h_l(kr)$ is the lth-order spherical Hankel function.

(ii)

$$\left[\frac{d^2}{dr^2} - \frac{l(l+1)}{r^2}\right]\psi_{\mathrm{H}}(r) = 0 \quad \Rightarrow \quad \psi_{\mathrm{H}}(r) = r^{l+1}, \text{ and } r^{-l}.$$

There exist two equivalent Volterra integral equations of the second kind for this problem.

3.3. Solve the generalized Abel equation,

$$\int_0^x \frac{\phi(y)}{(x-y)^\alpha}\,dy = f(x), \qquad 0 < \alpha < 1.$$

3.4. Solve

$$\int_0^x \phi(y)\ln(x-y)\,dy = f(x), \quad \text{with} \quad f(0) = 0.$$

3.5. Solve

$$\phi(x) = 1 + \int_x^\infty e^{\alpha(x-y)}\phi(y)\,dy, \qquad \alpha > 0.$$

Hint: Reduce the integral equation to the ordinary differential equation.

3.6. Solve

$$\phi(x) = 1 + \lambda \int_0^x e^{-(x-y)} \phi(y)\, dy.$$

3.7. (Due to H. C.) Solve

$$\phi(x) = 1 + \int_1^x \frac{1}{x+y} \phi(y)\, dy, \quad x \geq 1.$$

Find the asymptotic behavior of $\phi(x)$ as $x \to \infty$.

3.8. (Due to H. C.) Solve the integro-differential equation,

$$\frac{\partial}{\partial t}\phi(x,t) = -ix\phi(x,t) + \lambda \int_{-\infty}^{+\infty} g(y)\phi(y,t)\, dy, \quad \text{with} \quad \phi(x,0) = f(x),$$

where $f(x)$ and $g(x)$ are given. Find the asymptotic form of $\phi(x,t)$ as $t \to \infty$.

3.9. Solve

$$\int_0^x \frac{1}{\sqrt{x-y}} \phi(y)\, dy + \int_0^1 2xy\phi(y)\, dy = 1.$$

3.10. Solve

$$\lambda \int_0^x \frac{1}{(x-y)^{1/3}} \phi(y)\, dy + \lambda \int_0^1 \phi(y)\, dy = 1.$$

3.11. Solve

$$\lambda \int_0^1 K(x,y)\phi(y)\, dy = 1,$$

where

$$K(x,y) = \begin{cases} (x-y)^{-1/4} + xy & \text{for} \quad 0 \leq y \leq x \leq 1, \\ xy & \text{for} \quad 0 \leq x < y \leq 1. \end{cases}$$

3.12. (Due to H. C.) Solve

$$\phi(x) = \lambda \int_0^x J_0(xy)\phi(y)\, dy,$$

where $J_0(x)$ is the zeroth-order Bessel function of the first kind.

3.13. Solve

$$x^2\phi_\mu(x) - \int_0^x K_{(\mu)}(x,y)\phi_\mu(y)\, dy = \begin{cases} 0 & \text{for} \quad \mu \geq 1, \\ -x^2 & \text{for} \quad \mu = 0, \end{cases}$$

where

$$K_{(\mu)}(x,y) = -x - (x^2 - \mu^2)(x-y).$$

Hint: Setting

$$\phi_\mu(x) = \sum_{n=0}^{\infty} a_n^{(\mu)} x^n,$$

find the recursive relation of $a_n^{(\mu)}$ and solve for $a_n^{(\mu)}$.

3.14. Solve a system of the integral equations,

$$\phi_1(x) = 1 - 2\int_0^x \exp[2(x-y)]\phi_1(y)\,dy + \int_0^x \phi_2(y)\,dy,$$

$$\phi_2(x) = 4x - \int_0^x \phi_1(y)\,dy + 4\int_0^x (x-y)\phi_2(y)\,dy.$$

3.15. Solve a system of the integral equations,

$$\phi_1(x) + \phi_2(x) - \int_0^x (x-y)\phi_1(y)\,dy = ax,$$

$$\phi_1(x) - \phi_2(x) - \int_0^x (x-y)^2\phi_2(y)\,dy = bx^2.$$

3.16. (Due to D. M.) Consider the Volterra integral equation of the second kind,

$$\phi(x) = f(x) + \lambda\int_0^x \exp[x^2 - y^2]\phi(y)\,dy, \quad x > 0.$$

a) Sum up the iteration series exactly and find the general solution to this equation. Verify that the solution is analytic in λ.

b) Solve this integral equation by converting it into a differential equation.

Hint: Multiply both sides by $\exp[-x^2]$ and differentiate.

3.17. (Due to H. C.) The convolution of $f_1(x)$, $f_2(x)$, $\cdots$, $f_n(x)$ is defined as

$$C(x) \equiv \int_0^\infty dx_n \cdots \int_0^\infty dx_1 \prod_{i=1}^n f_i(x_i)\,\delta\left(x - \sum_{i=1}^n x_i\right).$$

a) Verify that for $n = 2$, this is the convolution defined in the text, and that

$$\tilde{C}(s) = \prod_{i=1}^n \tilde{f}_i(s).$$

b) If $f_1(x) = f(x)$, and $f_2(x) = f_3(x) = \cdots = f_n(x) = 1$, show that $C(x)$ is the nth integral of $f(x)$. Show also that

$$\tilde{C}(s) = \tilde{f}(s)s^{1-n},$$

and hence

$$C(x) = \int_0^x \frac{(x-y)^{n-1}}{(n-1)!} f(y)\,dy.$$

c) With the result in b), can you define the "one-third integral" of $f(x)$?

4 Integral Equations of the Fredholm Type

4.1 Iterative Solution to the Fredholm Integral Equation of the Second Kind

Consider the *inhomogeneous Fredholm Integral Equation of the second kind*,

$$\phi(x) = f(x) + \lambda \int_0^h dx' K(x, x')\phi(x'), \qquad 0 \le x \le h, \tag{4.1.1}$$

and assume that $f(x)$ and $K(x, x')$ are both *square-integrable*,

$$\|f\|^2 < \infty, \tag{4.1.2}$$

and

$$\|K\|^2 \equiv \int_0^h dx \int_0^h dx' \, |K(x, x')|^2 < \infty. \tag{4.1.3}$$

Suppose that we look for an *iterative solution* in λ,

$$\phi(x) = \phi_0(x) + \lambda\phi_1(x) + \lambda^2\phi_2(x) + \cdots + \lambda^n\phi_n(x) + \cdots, \tag{4.1.4}$$

which, when substituted into Eq. (4.1.1), yields

$$\begin{aligned}
\phi_0(x) &= f(x), \\
\phi_1(x) &= \int_0^h dy_1 \, K(x, y_1)\phi_0(y_1) = \int_0^h dy_1 \, K(x, y_1) f(y_1), \\
\phi_2(x) &= \int_0^h dy_2 \, K(x, y_2)\phi_1(y_2) = \int_0^h dy_2 \int_0^h dy_1 \, K(x, y_2) K(y_2, y_1) f(y_1).
\end{aligned}$$

In general, we have

$$\begin{aligned}
\phi_n(x) &= \int_0^h dy_n \, K(x, y_n)\phi_{n-1}(y_n) = \int_0^h dy_n \int_0^h dy_{n-1} \cdots \int_0^h dy_1 \\
&\quad \times K(x, y_n) K(y_n, y_{n-1}) \cdots K(y_2, y_1) f(y_1).
\end{aligned} \tag{4.1.5}$$

Applied Mathematics in Theoretical Physics. Michio Masujima

ISBN: 3-527-40534-8

Bounds: First, in order to establish the bound on $|\phi_n(x)|$, define

$$A(x) \equiv \int_0^h dy \, |K(x,y)|^2 . \tag{4.1.6}$$

Then the square of the *norm* of the kernel $K(x,y)$ is given by

$$\|K\|^2 = \int_0^h dx \, A(x). \tag{4.1.7}$$

Examine the iteration series (4.1.4) and apply the Schwarz inequality. Each term in the iteration series (4.1.4) is bounded as follows:

$$\phi_1(x) = \int_0^h dy_1 \, K(x,y_1) f(y_1)$$

$$\Rightarrow |\phi_1(x)|^2 \leq A(x) \, \|f\|^2 ,$$

$$\phi_2(x) = \int_0^h dy_2 \, K(x,y_2) \phi_1(y_2)$$

$$\Rightarrow |\phi_2(x)|^2 \leq A(x) \, \|\phi_1\|^2 \leq A(x) \, \|f\|^2 \, \|K\|^2 ,$$

and

$$\phi_3(x) = \int_0^h dy_3 \, K(x,y_3) \phi_2(y_3)$$

$$\Rightarrow |\phi_3(x)|^2 \leq A(x) \, \|\phi_2\|^2 \leq A(x) \, \|f\|^2 \, \|K\|^4 .$$

In general, we have

$$|\phi_n(x)|^2 \leq A(x) \, \|f\|^2 \, \|K\|^{2(n-1)} .$$

Thus the bound on $|\phi_n(x)|$ is established as

$$|\phi_n(x)| \leq \sqrt{A(x)} \, \|f\| \, \|K\|^{n-1} , \qquad n = 1, 2, 3, \cdots . \tag{4.1.8}$$

Now examine the whole series (4.1.4).

$$\phi(x) - f(x) = \lambda \phi_1(x) + \lambda^2 \phi_2(x) + \lambda^3 \phi_3(x) + \cdots .$$

Taking the absolute value of both sides, applying the triangular inequality on the right-hand side, and using the bound on $|\phi_n(x)|$ established as in Eq. (4.1.8), we have

$$\begin{aligned} |\phi(x) - f(x)| &\leq |\lambda| \, |\phi_1(x)| + |\lambda|^2 \, |\phi_2(x)| + \cdots \\ &= \sum_{n=1}^{\infty} |\lambda|^n \, |\phi_n(x)| \\ &\leq \sum_{n=1}^{\infty} |\lambda|^n \sqrt{A(x)} \cdot \|f\| \cdot \|K\|^{n-1} \\ &= |\lambda| \sqrt{A(x)} \cdot \|f\| \cdot \sum_{n=0}^{\infty} |\lambda|^n \cdot \|K\|^n . \end{aligned} \tag{4.1.9}$$

Now, the series on the right-hand side of the inequality (4.1.9)

$$\sum_{n=0}^{\infty} |\lambda|^n \cdot \|K\|^n$$

converges as long as

$$|\lambda| < \frac{1}{\|K\|}, \tag{4.1.10}$$

converging to

$$\frac{1}{1 - |\lambda| \cdot \|K\|}.$$

Therefore, in that case, (i.e., for Eq. (4.1.10)), the assumed series (4.1.4) is a convergent series, giving us a solution $\phi(x)$ to the integral equation (4.1.1), which is analytic inside the disk (4.1.10).

Symbolically the integral equation (4.1.1) can be written as if it is an algebraic equation,

$$\phi = f + \lambda K\phi \Rightarrow (1 - \lambda K)\phi = f$$

$$\Rightarrow \phi = \frac{f}{1 - \lambda K} = (1 + \lambda K + \lambda^2 K^2 + \cdots)f,$$

which converges only for $|\lambda K| < 1$.

Uniqueness: Inside the disk, Eq. (4.1.10), we can establish the uniqueness of the solution by showing that the corresponding homogeneous problem has no nontrivial solutions. Consider the homogeneous problem,

$$\phi_H(x) = \lambda \int_0^h K(x, y)\phi_H(y). \tag{4.1.11}$$

Applying the Schwarz inequality to Eq. (4.1.11),

$$|\phi_H(x)|^2 \leq |\lambda|^2 A(x) \|\phi_H\|^2 .$$

Integrating both sides with respect to x from 0 to h,

$$\|\phi_H\|^2 \leq |\lambda|^2 \cdot \|\phi_H\|^2 \int_0^h A(x)\, dx = |\lambda|^2 \cdot \|K\|^2 \cdot \|\phi_H\|^2 ,$$

i.e.,

$$\|\phi_H\|^2 \cdot (1 - |\lambda|^2 \cdot \|K\|^2) \leq 0. \tag{4.1.12}$$

Since $|\lambda| \cdot \|K\| < 1$, the inequality (4.1.12) can only be satisfied if and only if

$$\|\phi_H\| = 0$$

or

$$\phi_H \equiv 0. \tag{4.1.13}$$

Thus the homogeneous problem has only a trivial solution, so the inhomogeneous problem has a unique solution.

4.2 Resolvent Kernel

Returning to the series solution (4.1.4), we find that after defining the *iterated kernels* as follows,

$$K_1(x, y) = K(x, y), \tag{4.2.1}$$

$$K_2(x, y) = \int_0^h dy_2 \, K(x, y_2) K(y_2, y), \tag{4.2.2}$$

$$K_3(x, y) = \int_0^h dy_3 \int_0^h dy_2 \, K(x, y_3) K(y_3, y_2) K(y_2, y), \tag{4.2.3}$$

and generally

$$K_n(x, y) = \int_0^h dy_n \int_0^h dy_{n-1} \cdots \int_0^h dy_2 \, K(x, y_n) K(y_n, y_{n-1}) \cdots K(y_2, y), \tag{4.2.4}$$

we may write each term in the series (4.1.4) as

$$\phi_1(x) = \int_0^h dy \, K_1(x, y) f(y), \tag{4.2.5}$$

$$\phi_2(x) = \int_0^h dy \, K_2(x, y) f(y), \tag{4.2.6}$$

$$\phi_3(x) = \int_0^h dy \, K_3(x, y) f(y), \tag{4.2.7}$$

and generally

$$\phi_n(x) = \int_0^h dy \, K_n(x, y) f(y). \tag{4.2.8}$$

Therefore, we have

$$\phi(x) = f(x) + \sum_{n=1}^{\infty} \lambda^n \int_0^h dy \, K_n(x, y) f(y). \tag{4.2.9}$$

Now, define the *resolvent kernel* $H(x, y; \lambda)$ to be

$$\begin{aligned} -H(x, y; \lambda) &\equiv K_1(x, y) + \lambda K_2(x, y) + \lambda^2 K_3(x, y) + \cdots \\ &= \sum_{n=1}^{\infty} \lambda^{n-1} K_n(x, y). \end{aligned} \tag{4.2.10}$$

Then the solution (4.2.9) can be expressed compactly as,

$$\phi(x) = f(x) - \lambda \int_0^h dy \, H(x, y; \lambda) f(y), \tag{4.2.11}$$

with the resolvent $H(x, y; \lambda)$ defined by Eq. (4.2.10). We have in effect shown that $H(x, y; \lambda)$ exists and is analytic for

$$|\lambda| < \frac{1}{\|K\|}. \tag{4.2.12}$$

❑ **Example 4.1.** Solve the Fredholm Integral Equation of the second kind,

$$\phi(x) = f(x) + \lambda \int_0^1 e^{x-y}\phi(y)\, dy. \tag{4.2.13}$$

Solution. We have, for the iterated kernels,

$$K_1(x, y) = K(x, y) = e^{x-y},$$

$$K_2(x, y) = \int_0^1 d\xi\, K(x, \xi)K(\xi, y) = e^{x-y},$$

$$K_3(x, y) = \int_0^1 d\xi\, K_1(x, \xi)K_2(\xi, y) = e^{x-y},$$

and hence

$$K_n(x, y) = e^{x-y} \quad \text{for} \quad \text{all} \quad n. \tag{4.2.14}$$

Then we have as the resolvent kernel of this problem,

$$-H(x, y; \lambda) = \sum_{n=1}^{\infty} \lambda^{n-1} e^{x-y} = e^{x-y}(1 + \lambda + \lambda^2 + \lambda^3 + \cdots). \tag{4.2.15}$$

For

$$|\lambda| < 1, \tag{4.2.16}$$

we have

$$H(x, y; \lambda) = -\frac{e^{x-y}}{1-\lambda}. \tag{4.2.17}$$

Thus the solution to this problem is given by

$$\phi(x) = f(x) + \frac{\lambda}{1-\lambda}\int_0^1 dy\, e^{x-y} f(y). \tag{4.2.18}$$

We note that, in this case, not only do we know the radius of convergence for the series solution (or for the resolvent kernel) as in Eq. (4.2.16), but we also know the nature of the singularity (a simple pole at $\lambda = 1$). In fact, our solution (4.2.18) is valid for all values of λ, even those which have $|\lambda| > 1$, with the exception of $\lambda = 1$. The question we now wish to address is whether, in general, we can say more about the nature of the singularity in $H(x, y; \lambda)$, than simply knowing the radius of the disk in which we have a convergence.

Properties of the resolvent: We now derive some properties of the resolvent $H(x, y; \lambda)$ which will be useful to us later. Consider the original integral operator written as

$$\tilde{K} = \int_0^h dy\, K(x, y), \tag{4.2.19}$$

and the operator corresponding to the resolvent as

$$\tilde{H} = \int_0^h dy\, H(x, y; \lambda). \tag{4.2.20}$$

Note the integrals are with respect to the second argument in Eqs. (4.2.19) and (4.2.20). Now, the operators $\tilde{K}^2$, or $\tilde{H}^2$, or $\tilde{K}\tilde{H}$, or $\tilde{H}\tilde{K}$ are defined in the usual way:

$$\begin{aligned}\tilde{K}^2\phi = \tilde{K}(\tilde{K}\phi) &= \tilde{K}\left(\int_0^h dy_1\, K(x, y_1)\phi(y_1)\right) \\ &= \int_0^h dy_2\, K(x, y_2) \int_0^h dy_1\, K(y_2, y_1)\phi(y_1),\end{aligned}$$

i.e.,

$$\tilde{K}^2 = \int_0^h dy_2 \int_0^h dy_1\, K(x, y_2)K(y_2, y_1). \tag{4.2.21}$$

We wish to show that the two operators $\tilde{K}$ and $\tilde{H}$ commute, i.e.,

$$\tilde{K}\tilde{H} = \tilde{H}\tilde{K}. \tag{4.2.22}$$

The original integral equation can be written as

$$\phi = f + \lambda\tilde{K}\phi \tag{4.2.23}$$

while the solution obtained by the resolvent takes the form

$$\phi = f - \lambda\tilde{H}f. \tag{4.2.24}$$

Defining $\tilde{I}$ to be the identity operator

$$\tilde{I} = \int_0^h dy\, \delta(x - y), \tag{4.2.25}$$

Eqs. (4.2.23) and (4.2.24) can be written as

$$f = (\tilde{I} - \lambda\tilde{K})\phi, \tag{4.2.26}$$

$$\phi = (\tilde{I} - \lambda\tilde{H})f. \tag{4.2.27}$$

Then, combining Eqs. (4.2.26) and (4.2.27), we obtain

$$f = (\tilde{I} - \lambda\tilde{K})(\tilde{I} - \lambda\tilde{H})f \quad \text{and} \quad \phi = (\tilde{I} - \lambda\tilde{H})(\tilde{I} - \lambda\tilde{K})\phi.$$

In other words, we have

$$(\tilde{I} - \lambda\tilde{K})(\tilde{I} - \lambda\tilde{H}) = \tilde{I} \quad \text{and} \quad (\tilde{I} - \lambda\tilde{H})(\tilde{I} - \lambda\tilde{K}) = \tilde{I}.$$

Thus we obtain

$$\tilde{K} + \tilde{H} = \lambda\tilde{K}\tilde{H} \quad \text{and} \quad \tilde{K} + \tilde{H} = \lambda\tilde{H}\tilde{K}.$$

Hence we have established the identity

$$\tilde{K}\tilde{H} = \tilde{H}\tilde{K}, \tag{4.2.28}$$

i.e., $\tilde{K}$ and $\tilde{H}$ commute. This can be written explicitly as

$$\int_0^h dy_2 \int_0^h dy_1\, K(x,y_2)H(y_2,y_1) = \int_0^h dy_2 \int_0^h dy_1\, H(x,y_2)K(y_2,y_1).$$

Let both of these operators act on the function $\delta(y_1 - y)$. Then we find

$$\int_0^h d\xi\, K(x,\xi)H(\xi,y) = \int_0^h d\xi\, H(x,\xi)K(\xi,y). \tag{4.2.29}$$

Similarly the operator equation

$$\tilde{K} + \tilde{H} = \lambda\tilde{K}\tilde{H} \tag{4.2.30}$$

may be written as

$$H(x,y;\lambda) = -K(x,y) + \lambda \int_0^h K(x,\xi)H(\xi,y;\lambda)\, d\xi. \tag{4.2.31}$$

We will find this to be a useful relation later.

4.3 Pincherle–Goursat Kernel

Let us now examine the problem of determining a more explicit formula for the resolvent which points out more clearly the nature of the singularities of $H(x,y;\lambda)$ in the complex λ plane. We do this for two different cases. First, we look at the case of a kernel which is given by a finite sum of separable terms (the so-called *Pincherle–Goursat kernel*). Secondly we examine the case of a general kernel which we decompose into a sum of a Pincherle–Goursat kernel and a remainder, which can be made as small as possible.

Pincherle–Goursat kernel: Suppose that we are given the kernel which is a finite sum of separable terms,

$$K(x,y) = \sum_{n=1}^{N} g_n(x)h_n(y), \tag{4.3.1}$$

i.e., we are given the following integral equation,

$$\phi(x) = f(x) + \lambda \int_0^h \sum_{n=1}^N g_n(x) h_n(y) \phi(y)\, dy. \tag{4.3.2}$$

Define β_n to be

$$\beta_n \equiv \int_0^h h_n(y)\phi(y)\, dy. \tag{4.3.3}$$

Then the integral equation (4.3.2) takes the form,

$$\phi(x) = f(x) + \lambda \sum_{k=1}^N g_k(x)\beta_k. \tag{4.3.4}$$

Substituting Eq. (4.3.4) into the expression (4.3.3) for β_n, we have

$$\beta_n = \int_0^h dy\, h_n(y) f(y) + \int_0^h dy\, h_n(y) \cdot \lambda \sum_{k=1}^N g_k(y)\beta_k. \tag{4.3.5}$$

Hence, upon letting

$$A_{nk} = \int_0^h dy\, h_n(y) g_k(y), \tag{4.3.6}$$

and

$$\alpha_n = \int_0^h dy\, h_n(y) f(y), \tag{4.3.7}$$

Equation (4.3.5) takes the form

$$\beta_n = \alpha_n + \lambda \sum_{k=1}^N A_{nk}\beta_k,$$

or

$$\sum_{k=1}^N (\delta_{nk} - \lambda A_{nk})\beta_k = \alpha_n, \tag{4.3.8}$$

which is equivalent to the $N \times N$ matrix equation

$$(I - \lambda A)\vec{\beta} = \vec{\alpha}, \tag{4.3.9}$$

where the αs are known and the βs are unknown. If the determinant of the matrix $(I - \lambda A)$ is denoted by $\tilde{D}(\lambda)$

$$\tilde{D}(\lambda) = \det(I - \lambda A),$$

the inverse of $(I - \lambda A)$ can be written as

$$(I - \lambda A)^{-1} = \frac{1}{\tilde{D}(\lambda)} \cdot D,$$

where D is a matrix whose ij-th element is the cofactor of the ji-th element of $I - \lambda A$. (We recall that the cofactor of a_{ij} is given by $(-1)^{i+j} \det M_{ij}$, where $\det M_{ij}$ is the minor determinant obtained by deleting the row and column to which a_{ij} belongs.) Therefore,

$$\beta_n = \frac{1}{\tilde{D}(\lambda)} \cdot \sum_{k=1}^{N} D_{nk}\alpha_k. \tag{4.3.10}$$

From Eqs. (4.3.4) and (4.3.10), we obtain the solution $\phi(x)$ as

$$\phi(x) = f(x) + \frac{\lambda}{\tilde{D}(\lambda)} \sum_{n=1}^{N}\sum_{k=1}^{N} g_n(x) D_{nk}\alpha_k. \tag{4.3.11}$$

Writing out α_k explicitly, we have

$$\phi(x) = f(x) + \frac{\lambda}{\tilde{D}(\lambda)} \sum_{n=1}^{N}\sum_{k=1}^{N} g_n(x) D_{nk} \int_0^h dy\, h_k(y) f(y). \tag{4.3.12}$$

Comparing Eq. (4.3.12) with the definition of the resolvent $H(x, y; \lambda)$

$$\phi(x) = f(y) - \lambda \int_0^h dy\, H(x, y; \lambda) f(y), \tag{4.3.13}$$

we obtain the resolvent for the case of the Pincherle–Goursat kernel as

$$-H(x, y; \lambda) = \frac{1}{\tilde{D}(\lambda)} \sum_{n=1}^{N}\sum_{k=1}^{N} g_n(x) D_{nk} h_k(y). \tag{4.3.14}$$

Note that this is a ratio of two polynomials in λ.

We note that the cofactors of the matrix $(I - \lambda A)$ are polynomials in λ and hence have no singularities in λ. Thus the numerator of $H(x, y; \lambda)$ has no singularities. Then the only singularities of $H(x, y; \lambda)$ occur at the zeros of the denominator

$$\tilde{D}(\lambda) = \det(I - \lambda A),$$

which is a polynomial of degree N in λ. Therefore, $H(x, y; \lambda)$ in this case has, at most, N singularities which are poles in the complex λ plane. At the poles of $H(x, y; \lambda)$ in λ, the homogeneous problem has nontrivial solutions.

General kernel: By approximating a general kernel as a sum of a Pincherle–Goursat kernel (plus a small remainder term), we can now prove that in any finite region of the complex λ plane, there can be, at most, a finit number of singularities. Consider the integral equation

$$\phi(x) = f(x) + \lambda \int_0^h K(x,y)\phi(y)\,dy, \tag{4.3.15}$$

with a general *square-integrable* kernel $K(x,y)$. Suppose we are interested in examining the singularities of $H(x,y;\lambda)$ in the region $|\lambda| < 1/\varepsilon$ in the complex λ plane (with ε possibly quite small). For this purpose, we can always find an approximation to the kernel $K(x,y)$ in the form (with N sufficiently large)

$$K(x,y) = \sum_{n=1}^{N} g_n(x)h_n(y) + R(x,y) \tag{4.3.16}$$

with

$$\|R\| < \varepsilon. \tag{4.3.17}$$

The integral equation (4.3.15) then becomes

$$\phi(x) = f(x) + \lambda \int_0^h \sum_{n=1}^{N} g_n(x)h_n(y)\phi(y)\,dy + \lambda \int_0^h R(x,y)\phi(y)\,dy.$$

Define

$$F(x) = f(x) + \lambda \int_0^h \sum_{n=1}^{N} g_n(x)h_n(y)\phi(y)\,dy. \tag{4.3.18}$$

Then

$$\phi(x) = F(x) + \lambda \int_0^h R(x,y)\phi(y)\,dy.$$

Let $H_R(x,y;\lambda)$ be the resolvent kernel corresponding to $R(x,y)$,

$$-H_R(x,y;\lambda) = R(x,y) + \lambda R_2(x,y) + \lambda^2 R_3(x,y) + \cdots,$$

whence we have

$$\phi(x) = F(x) - \lambda \int_0^h H_R(x,y;\lambda)F(y)\,dy. \tag{4.3.19}$$

Substituting the given expression (4.3.18) for $F(x)$ into Eq. (4.3.19), we have

$$\begin{aligned}\phi(x) = f(x) &+ \lambda \int_0^h \sum_{n=1}^{N} g_n(x)h_n(y)\phi(y)\,dy \\ &- \lambda \int_0^h H_R(x,y;\lambda)\left[f(y) + \lambda \int_0^h \sum_{n=1}^{N} g_n(y)h_n(z)\phi(z)\,dz\right] dy.\end{aligned}$$

Define

$$\tilde{F}(x) \equiv f(x) - \lambda \int_0^h H_R(x, y; \lambda) f(y)\, dy.$$

Then

$$\begin{aligned}\phi(x) = \tilde{F}(x) + \lambda \int_0^h dy \sum_{n=1}^N g_n(x) h_n(y) \phi(y) \\ - \lambda^2 \int_0^h dy \int_0^h dz\, H_R(x, y; \lambda) \left(\sum_{n=1}^N g_n(y) h_n(z) \right) \phi(z).\end{aligned} \tag{4.3.20}$$

In the above expression (4.3.20), interchange y and z in the last term on the right-hand side,

$$\begin{aligned}&\phi(x) \\ &= \tilde{F}(x) + \lambda \int_0^h dy \left[\sum_{n=1}^N g_n(x) h_n(y) - \lambda \int_0^h dz\, H_R(x, z; \lambda) \sum_{n=1}^N g_n(z) h_n(y) \right] \phi(y).\end{aligned}$$

Then we have

$$\phi(x) = \tilde{F}(x) + \lambda \int_0^h dy \left(\sum_{n=1}^N G_n(x; \lambda) h_n(y) \right) \phi(y)$$

where

$$G_n(x; \lambda) = g_n(x) - \lambda \int_0^h dz\, H_R(x, z; \lambda) g_n(z).$$

We have thus reduced the integral equation with the general kernel to one with a *Pincherle–Goursat type kernel*. The only difference is that the entries in the new kernel

$$\sum_{n=1}^N G_n(x; \lambda) h_n(y)$$

also depend on λ through the dependence of $G_n(x; \lambda)$ on λ. We know, however, that for

$$|\lambda| \cdot \|R\| < 1,$$

the resolvent $H_R(x, y; \lambda)$ is analytic in λ. Hence $G_n(x; \lambda)$ is also analytic in λ. Therefore, the singularities in the complex λ plane are still found by setting

$$\det(I - \lambda A) = 0,$$

where the kn-th element of A is given by

$$A_{kn} = \int_0^h dy\, h_k(y) G_n(y; \lambda) = A_{kn}(\lambda).$$

Since the entries A_{kn} depend on λ analytically, the function $\det(I - \lambda A)$ is an analytic function of λ (but not necessarily a polynomial of degree N) and hence it has finitely many zeros in the region $|\lambda| \cdot \|R\| < 1$, or

$$|\lambda| < \frac{1}{\varepsilon} < \frac{1}{\|R\|}.$$

This concludes the proof that in any disk $|\lambda| < 1/\varepsilon$, there are a finit number of singularities of λ for the integral equation (4.3.15).

4.4 Fredholm Theory for a Bounded Kernel

We now consider the case of a general kernel as approached by Fredholm. We shall show that the resolvent kernel can be written as a ratio of the entire functions of λ, whence the singularities in λ occur when the function in the denominator is zero.

Consider

$$\phi(x) = f(x) + \lambda \int_0^h K(x, y)\phi(y)\,dy, \quad 0 \leq x \leq h. \tag{4.4.1}$$

Discretize the above equation by letting

$$\varepsilon = \frac{h}{N}, \qquad x_i = i\varepsilon, \qquad y_j = j\varepsilon,$$

$$i, j = 0, 1, 2, \ldots, N.$$

Also let

$$\phi_i = \phi(x_i), \qquad f_i = f(x_i), \qquad K_{ij} = K(x_i, y_j).$$

The discrete version of the integral equation (4.4.1) takes the form

$$\phi_i = f_i + \lambda \sum_{j=1}^{N} K_{ij}\phi_j\varepsilon,$$

i.e.,

$$\sum_{j=1}^{N} (\delta_{ij} - \lambda\varepsilon K_{ij})\phi_j = f_i. \tag{4.4.2}$$

Define $\tilde{D}(\lambda)$ to be

$$\tilde{D}(\lambda) = \det(I - \lambda\varepsilon K).$$

Writing out $\tilde{D}(\lambda)$ explicitly, we have

$$\tilde{D}(\lambda) = \begin{vmatrix} 1-\lambda\varepsilon K_{11}, & -\lambda\varepsilon K_{12}, & \cdot\;\cdot\;\cdot & -\lambda\varepsilon K_{1N} \\ -\lambda\varepsilon K_{21}, & 1-\lambda\varepsilon K_{22}, & \cdot\;\cdot\;\cdot & -\lambda\varepsilon K_{2N} \\ \cdot & \cdot & & \cdot \\ \cdot & \cdot & & \cdot \\ \cdot & \cdot & & \cdot \\ -\lambda\varepsilon K_{N1}, & -\lambda\varepsilon K_{N2}, & \cdot\;\cdot\;\cdot & 1-\lambda\varepsilon K_{NN} \end{vmatrix}.$$

This determinant can be expanded as

$$\tilde{D}(\lambda) = \tilde{D}(0) + \lambda\tilde{D}'(0) + \frac{\lambda^2}{2!}\tilde{D}''(0) + \cdots + \frac{\lambda^N}{N!}\tilde{D}^{(N)}(0).$$

Using the fact that

$$\frac{d}{d\lambda}\left|\vec{a}_1, \vec{a}_2, \cdots, \vec{a}_N\right| = \left|\frac{d}{d\lambda}\vec{a}_1, \vec{a}_2, \cdots, \vec{a}_N\right| + \left|\vec{a}_1, \frac{d}{d\lambda}\vec{a}_2, \cdots, \vec{a}_N\right| + \cdots + \left|\vec{a}_1, \vec{a}_2, \cdots, \frac{d}{d\lambda}\vec{a}_N\right|,$$

we finally obtain (after considerable algebra),

$$\tilde{D}(\lambda) = 1 - \lambda\varepsilon\sum_{i=1}^{N} K_{ii} + \frac{\lambda^2\varepsilon^2}{2!}\sum_{i=1}^{N}\sum_{j=1}^{N}\begin{vmatrix} K_{ii}, & K_{ij} \\ K_{ji}, & K_{jj} \end{vmatrix} - \frac{\lambda^3\varepsilon^3}{3!}\sum_{i=1}^{N}\sum_{j=1}^{N}\sum_{k=1}^{N}\begin{vmatrix} K_{ii}, & K_{ij}, & K_{ik} \\ K_{ji}, & K_{jj}, & K_{jk} \\ K_{ki}, & K_{kj}, & K_{kk} \end{vmatrix} + \cdots.$$

In the limit as $n \to \infty$, each sum, when multiplied by ε, is approximates a corresponding integral, i.e.,

$$\sum_{i=1}^{N}\varepsilon K_{ii} \to \int_0^h K(x,x)\,dx,$$

$$\sum_{i=1}^{N}\sum_{j=1}^{N}\varepsilon^2\begin{vmatrix} K_{ii}, & K_{ij} \\ K_{ji}, & K_{jj} \end{vmatrix} \to \int_0^h dx\int_0^h dy\begin{vmatrix} K(x,x), & K(x,y) \\ K(y,x), & K(y,y) \end{vmatrix}.$$

Define

$$K\begin{pmatrix} x_1, & x_2, & \dots, & x_n \\ y_1, & y_2, & \dots, & y_n \end{pmatrix} \equiv \begin{vmatrix} K(x_1,y_1), & K(x_1,y_2), & \cdot \ \cdot \ \cdot & K(x_1,y_n) \\ K(x_2,y_1), & \cdot & \cdot \ \cdot \ \cdot & K(x_2,y_n) \\ \cdot & \cdot & & \cdot \\ \cdot & \cdot & & \cdot \\ \cdot & \cdot & & \cdot \\ K(x_n,y_1), & \cdot & & K(x_n,y_n) \end{vmatrix}.$$

Then, in the limit as $N \to \infty$, we find (on renaming $\tilde{D}$ as D)

$$D(\lambda) = 1 + \sum_{n=1}^{\infty} \frac{(-1)^n \lambda^n}{n!} D_n$$

with

$$D_n = \int_0^h dx_1 \int_0^h dx_2 \cdots \int_0^h dx_n \, K\begin{pmatrix} x_1, & x_2, & \dots, & x_n \\ x_1, & x_2, & \dots, & x_n \end{pmatrix}.$$

We expect singularities in the resolvent $H(x, y; \lambda)$ to occur only when the determinant $D(\lambda)$ vanishes. Thus we hope to show that $H(x, y; \lambda)$ can be expressed as the ratio,

$$H(x,y;\lambda) \equiv \frac{D(x,y;\lambda)}{D(\lambda)}. \tag{4.4.3}$$

So we need to obtain the numerator $D(x, y; \lambda)$ and show that it is entire. We also show that the power series given above for $D(\lambda)$ has an infinite radius of convergence and thus represents an analytic function.

To this end, we make use of the fact that $K(x, y)$ is *bounded* (by assumption) and also invoke the *Hadamard inequality* which says

$$|\det [\vec{v}_1, \vec{v}_2, \cdots, \vec{v}_n]| \le \|\vec{v}_1\| \, \|\vec{v}_2\| \cdots \|\vec{v}_n\| \, .$$

This has the interpretation that the volume of the parallelepiped whose edges are $\vec{v}_1$ through $\vec{v}_n$ is less than the product of the lengths of those edges. Suppose that $|K(x, y)|$ is bounded by A on $x, y \in [0, h]$. Then

$$K\begin{pmatrix} x_1, & \dots, & x_n \\ x_1, & \dots, & x_n \end{pmatrix} = \begin{vmatrix} K(x_1,x_1), & \dots, & K(x_1,x_n) \\ \cdot & & \cdot \\ K(x_n,x_1), & \dots, & K(x_n,x_n) \end{vmatrix}$$

is bounded by

$$\left| K\begin{pmatrix} x_1, & \dots, & x_n \\ x_1, & \dots, & x_n \end{pmatrix} \right| \le (\sqrt{n}A)^n,$$

since the norm of each column is less than $\sqrt{n}A$. This implies

$$|D_n| = \left| \int_0^h dx_1 \cdots \int_0^h dx_n \, K \begin{pmatrix} x_1, & \cdots, & x_n \\ x_1, & \cdots, & x_n \end{pmatrix} \right| \leq h^n n^{n/2} A^n.$$

Thus

$$\left| \sum_{n=1}^{\infty} \frac{(-1)^n \lambda^n}{n!} D_n \right| \leq \sum_{n=1}^{\infty} \frac{|\lambda|^n \, h^n n^{n/2} A^n}{n!}. \tag{4.4.4}$$

Letting

$$a_n = \frac{|\lambda|^n \, h^n n^{n/2} A^n}{n!},$$

and applying the ratio test to the right-hand side of the inequality (4.4.4), we have

$$\begin{aligned} \lim_{n\to\infty} \frac{a_{n+1}}{a_n} &= \lim_{n\to\infty} \frac{(|\lambda|^{n+1} \, h^{n+1} (n+1)^{(n+1)/2} A^{n+1} \cdot n!)}{(|\lambda|^n \, h^n n^{n/2} A^n \cdot (n+1)!)} \\ &= \lim_{n\to\infty} \left[|\lambda| \, hA \left(1 + \frac{1}{n}\right)^{n/2} \frac{1}{\sqrt{n+1}} \right] = 0. \end{aligned}$$

Hence the series converges for all λ. We conclude that $D(\lambda)$ is an entire function of λ.

The last step we need to take is to find the numerator $D(x, y; \lambda)$ of the resolvent and show that it, too, is an entire function of λ. For this purpose, we recall that the resolvent itself, $H(x, y; \lambda)$, satisfies the integral equation,

$$H(x, y; \lambda) = -K(x, y) + \lambda \int_0^h K(x, z) H(z, y; \lambda) \, dz. \tag{4.4.5}$$

Therefore, upon multiplying the integral equation (4.4.5) by $D(\lambda)$ and using the definition (4.4.3) of $D(x, y; \lambda)$, we have

$$D(x, y; \lambda) = -K(x, y) D(\lambda) + \lambda \int_0^h K(x, z) D(z, y; \lambda) \, dz. \tag{4.4.6}$$

Since $D(\lambda)$ has the expansion

$$D(\lambda) = \sum_{n=0}^{\infty} \frac{(-\lambda)^n}{n!} D_n \quad \text{with} \quad D_0 = 1, \tag{4.4.7}$$

we seek an expansion for $D(x, y; \lambda)$ of the form

$$D(x, y; \lambda) = \sum_{n=0}^{\infty} \frac{(-\lambda)^n}{n!} C_n(x, y). \tag{4.4.8}$$

Substituting Eqs. (4.4.7) and (4.4.8) into the integral equation (4.4.6) for $D(x, y; \lambda)$, we find

$$\sum_{n=0}^{\infty} \frac{(-\lambda)^n}{n!} C_n(x, y) = -\sum_{n=0}^{\infty} \frac{(-\lambda)^n}{n!} D_n K(x, y) - \sum_{n=0}^{\infty} \int_0^h \frac{(-\lambda)^{n+1}}{n!} K(x, z) C_n(z, y)\, dz.$$

Collecting like powers of λ, we get

$$C_0(x, y) = -K(x, y) \quad \text{for} \quad n = 0,$$

$$C_n(x, y) = -D_n K(x, y) - n \int_0^h K(x, z) C_{n-1}(z, y)\, dz \quad \text{for} \quad n = 1, 2, \ldots.$$

Let us calculate the first few of these:

$$C_0(x, y) = -K(x, y).$$

From this, we have

$$\begin{aligned} C_1(x, y) &= -K(x, y) D_1 + \int_0^h K(x, z) K(z, y)\, dz \\ &= \int_0^h dx_1 \left(K(x, x_1) K(x_1, y) - K(x_1, x_1) K(x, y) \right) \\ &= -\int_0^h dx_1 K \begin{pmatrix} x, & x_1 \\ y, & x_1 \end{pmatrix}, \end{aligned}$$

from which, we obtain

$$\begin{aligned} C_2(x, y) &= -\int_0^h dx_1 \int_0^h dx_2 \Bigg[K(x, y) K \begin{pmatrix} x_1, & x_2 \\ x_1, & x_2 \end{pmatrix} \\ &\quad - K(x, x_1) K \begin{pmatrix} x_1, & x_2 \\ y, & x_2 \end{pmatrix} + K(x, x_2) K \begin{pmatrix} x_1, & x_2 \\ y, & x_1 \end{pmatrix} \Bigg] \\ &= -\int_0^h dx_1 \int_0^h dx_2 K \begin{pmatrix} x, & x_1, & x_2 \\ y, & x_1, & x_2 \end{pmatrix}. \end{aligned}$$

In general, we have

$$C_n(x, y) = -\int_0^h dx_1 \int_0^h dx_2 \cdots \int_0^h dx_n\, K \begin{pmatrix} x, & x_1, & x_2, & \cdots, & x_n \\ y, & x_1, & x_2, & \cdots, & x_n \end{pmatrix}.$$

Therefore, we have the numerator $D(x, y; \lambda)$ of $H(x, y; \lambda)$,

$$D(x, y; \lambda) = \sum_{n=0}^{\infty} \frac{(-\lambda)^n}{n!} C_n(x, y), \tag{4.4.9}$$

with

$$C_n(x,y) = -\int_0^h dx_1 \cdots \int_0^h dx_n \, K\begin{pmatrix} x, & x_1, & x_2, & \ldots, & x_n \\ y, & x_1, & x_2, & \ldots, & x_n \end{pmatrix},$$

$$n = 1, 2, \ldots, \quad (4.4.10)$$

and

$$C_0(x,y) = -K(x,y). \quad (4.4.11)$$

We prove that the power series for $D(x,y;\lambda)$ converges for all λ. First, by the Hadamard inequality, we have the following bounds,

$$\left| K\begin{pmatrix} x, & x_1, & \ldots, & x_n \\ y, & x_1, & \ldots, & x_n \end{pmatrix} \right| \leq (\sqrt{n+1}A)^{n+1},$$

since K above is a $(n+1)\times(n+1)$ determinant with each entry less than A, i.e.,

$$|K(x,y)| < A.$$

Then the bound on $C_n(x,y)$ is given by

$$|C_n(x,y)| \leq h^n(\sqrt{n+1}A)^{n+1}.$$

Thus we have the bound on $D(x,y;\lambda)$ as

$$|D(x,y;\lambda)| \leq \sum_{n=0}^{\infty} \frac{|\lambda|^n}{n!} h^n (\sqrt{n+1}A)^{n+1}. \quad (4.4.12)$$

Letting

$$a_n = \frac{|\lambda|^n}{n!} h^n (\sqrt{n+1}A)^{n+1},$$

we apply the ratio test to the right-hand side of inequality (4.4.12).

$$\begin{aligned} \lim_{n\to\infty} \frac{a_n}{a_{n-1}} &= \lim_{n\to\infty} \frac{\left(|\lambda|^n \, h^n (n+1)^{(n+1)/2} A^{n+1} (n-1)!\right)}{\left(|\lambda|^{n-1} \, h^{n-1} n^{n/2} A^n n!\right)} \\ &= \lim_{n\to\infty} |\lambda| \, hA \frac{\sqrt{n+1}}{n} \left(\frac{n+1}{n}\right)^{n/2} \\ &= \lim_{n\to\infty} |\lambda| \, hA \frac{\sqrt{n+1}}{n} \left(1 + \frac{1}{n}\right)^{n/2} \\ &= 0. \end{aligned}$$

Hence the power series expansion for $D(x, y; \lambda)$ converges for all λ, and $D(x, y; \lambda)$ is an entire function of λ.

Finally, we can prove that whenever $H(x, y; \lambda)$ exists (i.e., for all λ such that $D(\lambda) \neq 0$), the solution to the integral equation (4.4.1) is *unique*. This is best done using the operator notation introduced in Section 4.2. Consider the homogeneous problem

$$\phi_{\mathrm{H}} = \lambda \tilde{K} \phi_{\mathrm{H}}. \tag{4.4.13}$$

$\tilde{H}$ acts on both sides to give

$$\tilde{H} \phi_{\mathrm{H}} = \lambda \tilde{H} \tilde{K} \phi_{\mathrm{H}}.$$

Use the identity

$$\lambda \tilde{H} \tilde{K} = \tilde{K} + \tilde{H}$$

to get

$$\tilde{H} \phi_{\mathrm{H}} = \tilde{K} \phi_{\mathrm{H}} + \tilde{H} \phi_{\mathrm{H}}.$$

Hence we get

$$\tilde{K} \phi_{\mathrm{H}} = 0,$$

which implies

$$\phi_{\mathrm{H}} = \lambda \tilde{K} \phi_{\mathrm{H}} = 0. \tag{4.4.14}$$

Thus the homogeneous problem has no nontrivial solutions and the inhomogeneous problem has a unique solution.

Summary of the Fredholm theory for a bounded kernel

The integral equation

$$\phi(x) = f(x) + \lambda \int_0^h K(x, y) \phi(y) \, dy$$

has the solution

$$\phi(x) = f(x) - \lambda \int_0^h H(x, y; \lambda) f(y) \, dy$$

with the resolvent kernel given by

$$H(x, y; \lambda) = \frac{D(x, y; \lambda)}{D(\lambda)}$$

in which

$$D(\lambda) = \sum_{n=0}^{\infty} \frac{(-\lambda)^n}{n!} D_n$$

$$D_n = \int_0^h dx_1 \cdots \int_0^h dx_n \, K \begin{pmatrix} x_1, & \ldots, & x_n \\ x_1, & \ldots, & x_n \end{pmatrix}; \quad D_0 = 1$$

and

$$D(x, y; \lambda) = \sum_{n=0}^{\infty} \frac{(-\lambda)^n}{n!} C_n(x, y),$$

$$C_n(x, y) = -\int_0^h dx_1 \cdots \int_0^h dx_n \, K \begin{pmatrix} x, & x_1, & \ldots, & x_n \\ y, & x_1, & \ldots, & x_n \end{pmatrix};$$

$$C_0(x, y) = -K(x, y)$$

where

$$K \begin{pmatrix} z_1, & z_2, & \ldots, & z_n \\ w_1, & w_2, & \ldots, & w_n \end{pmatrix} \equiv \begin{vmatrix} K(z_1, w_1), & K(z_1, w_2), & \ldots, & K(z_1, w_n) \\ K(z_2, w_1), & K(z_2, w_2), & \ldots, & K(z_2, w_n) \\ \cdot & & & \\ \cdot & & & \\ \cdot & & & \\ K(z_n, w_1), & K(z_n, w_2), & \ldots, & K(z_n, w_n) \end{vmatrix}.$$

4.5 Solvable Example

Consider the following homogeneous integral equation.

❑ **Example 4.2.** Solve

$$\phi(x) = \lambda \int_0^x dy \, e^{-(x-y)} \phi(y) + \lambda \int_x^{\infty} dy \, \phi(y), \qquad 0 \leq x < \infty. \tag{4.5.1}$$

Solution. This is a Fredholm integral equation of the second kind with the kernel

$$K(x, y) = \begin{cases} e^{-(x-y)} & \text{for} \quad 0 \leq y < x < \infty, \\ 1 & \text{for} \quad 0 \leq x \leq y < \infty. \end{cases} \tag{4.5.2}$$

Note that this kernel (4.5.2) is not *square-integrable*. Differentiating both sides of Eq. (4.5.1) once, we find after a little algebra

$$e^x \phi'(x) = -\lambda \int_0^x dy \, e^y \phi(y). \tag{4.5.3}$$

Differentiate both sides of Eq. (4.5.3) once more to obtain the second-order ordinary differential equation of the form

$$\phi''(x) + \phi'(x) + \lambda\phi(x) = 0. \tag{4.5.4}$$

We try a solution of the form

$$\phi(x) = Ce^{\alpha x}. \tag{4.5.5}$$

Substituting Eq. (4.5.5) into Eq. (4.5.4), we obtain

$$\alpha^2 + \alpha + \lambda = 0,$$

or,

$$\alpha = \frac{-1 \pm \sqrt{1 - 4\lambda}}{2}.$$

In general, we obtain

$$\phi(x) = C_1 e^{\alpha_1 x} + C_2 e^{\alpha_2 x}, \tag{4.5.6}$$

with

$$\alpha_1 = \frac{-1 + \sqrt{1 - 4\lambda}}{2}, \quad \alpha_2 = \frac{-1 - \sqrt{1 - 4\lambda}}{2}. \tag{4.5.7}$$

Now, the expression for $\phi'(x)$ given above, Eq. (4.5.3), indicates that

$$\phi'(0) = 0. \tag{4.5.8}$$

This requires

$$\alpha_1 C_1 + \alpha_2 C_2 = 0 \quad \text{or} \quad C_2 = -\frac{\alpha_1}{\alpha_2} C_1.$$

Hence the solution is

$$\phi(x) = C\left[\frac{e^{\alpha_1 x}}{\alpha_1} - \frac{e^{\alpha_2 x}}{\alpha_2}\right].$$

However, in order for the integral equation to make sense, we must require the integral

$$\int_x^\infty dy\, \phi(y)$$

to converge. This requires

$$\mathrm{Re}\, \alpha < 0,$$

which in turn requires

$$\lambda > 0. \tag{4.5.9}$$

Thus, for $\lambda \leq 0$, we have no solution, and for $\lambda > 0$, we have

$$\phi(x) = C\left[\frac{e^{\alpha_1 x}}{\alpha_1} - \frac{e^{\alpha_2 x}}{\alpha_2}\right], \tag{4.5.10}$$

with α_1 and α_2 given by Eq. (4.5.7).

Note that, in this case, we have a continuous spectrum of eigenvalues ($\lambda > 0$) for which the homogeneous problem has a solution. The reason the eigenvalue is not discrete is that $K(x, y)$ is not square-integrable.

4.6 Fredholm Integral Equation with a Translation Kernel

Suppose $x \in (-\infty, +\infty)$ and the kernel is *translation invariant*, i.e.,

$$K(x, y) = K(x - y). \tag{4.6.1}$$

Then the inhomogeneous Fredholm integral equation of the second kind is given by

$$\phi(x) = f(x) + \lambda \int_{-\infty}^{+\infty} K(x - y)\phi(y)\,dy. \tag{4.6.2}$$

Take the *Fourier transform* of both sides of Eq. (4.6.2) to find

$$\hat{\phi}(k) = \hat{f}(k) + \lambda \hat{K}(k)\hat{\phi}(k).$$

Solve for $\hat{\phi}(k)$ to find

$$\hat{\phi}(k) = \frac{\hat{f}(k)}{1 - \lambda \hat{K}(k)}. \tag{4.6.3}$$

Solution $\phi(x)$ is provided by inverting the Fourier transform $\hat{\phi}(k)$ obtained above. It seems very simple, but there are some subtleties involved in the inversion of $\hat{\phi}(k)$. We present some general discussion of the *inversion of the Fourier transform*.

Suppose that the function $F(x)$ has the asymptotic forms

$$F(x) \sim \begin{cases} e^{ax} & \text{as} \quad x \to +\infty \quad (a > 0), \\ e^{bx} & \text{as} \quad x \to -\infty \quad (b > a > 0). \end{cases} \tag{4.6.4}$$

Namely $F(x)$ grows exponentially as $x \to +\infty$, and decays exponentially as $x \to -\infty$. Then the Fourier transform $\hat{F}(k)$

$$\hat{F}(k) = \int_{-\infty}^{+\infty} e^{-ikx} F(x)\,dx \tag{4.6.5}$$

exists as long as

$$-b < \operatorname{Im} k < -a, \tag{4.6.6}$$

since the integrand has magnitude

$$\left|e^{-ikx}F(x)\right| \sim \begin{cases} e^{(k_2+a)x} & \text{as} \quad x \to +\infty, \\ e^{(k_2+b)x} & \text{as} \quad x \to -\infty, \end{cases}$$

where we set

$$k = k_1 + ik_2, \quad \text{with} \quad k_1 \text{ and } k_2 \quad \text{real.}$$

With

$$-b < k_2 < -a,$$

the magnitude of the integrand vanishes exponentially at both ends.

The inverse Fourier transformation then becomes

$$F(x) = \frac{1}{2\pi} \int_{-\infty - i\gamma}^{+\infty - i\gamma} e^{ikx} \hat{F}(k)\, dk \quad \text{with} \quad a < \gamma < b. \tag{4.6.7}$$

Now if $b = a$ such that $F(x) \sim e^{ax}$ for $|x| \to \infty$ $(a > 0)$, then the inversion contour is on $\gamma = a$ and the Fourier transform exists only for $k_2 = -a$.

Similarly, if the function $F(x)$ decays exponentially as $x \to +\infty$ and grows as $x \to -\infty$, we are able to continue defining the Fourier transform and its inverse by proceeding in the upper half-plane.

As an example, a function like

$$F(x) = e^{-\alpha|x|} \tag{4.6.8}$$

which decays exponentially as $x \to \pm\infty$ has a Fourier transform which exists and is analytic in

$$-\alpha < k_2 < \alpha. \tag{4.6.9}$$

With these qualifications, we should be able to invert $\hat{\phi}(k)$ to obtain the solution to the inhomogeneous problem (4.6.2).

Now follows the homogeneous problem,

$$\phi_{\mathrm{H}}(x) = \lambda \int_{-\infty}^{+\infty} K(x-y)\phi_{\mathrm{H}}(y)\, dy. \tag{4.6.10}$$

By Fourier transforming Eq. (4.6.10), we obtain

$$(1 - \lambda\hat{K}(k))\hat{\phi}_{\mathrm{H}}(k) = 0. \tag{4.6.11}$$

If $1 - \lambda\hat{K}(k)$ has no zeros for all k, then we have

$$\hat{\phi}_{\mathrm{H}}(k) = 0 \Rightarrow \phi_{\mathrm{H}}(x) = 0, \tag{4.6.12}$$

i.e., no nontrivial solution exists for the homogeneous problem. If, on the other hand, $1 - \lambda\hat{K}(k)$ has a zero of order n at $k = \alpha$, $\hat{\phi}_{\mathrm{H}}(k)$ can be allowed to be of the form

$$\hat{\phi}_{\mathrm{H}}(k) = C_1\delta(k-\alpha) + C_2\frac{d}{dk}\delta(k-\alpha) + \cdots + C_n\left(\frac{d}{dk}\right)^{n-1}\delta(k-\alpha).$$

On inversion, we find

$$\phi_{\mathrm{H}}(x) = C_1 e^{i\alpha x} + C_2 x e^{i\alpha x} + \cdots + C_n x^{n-1} e^{i\alpha x} = e^{i\alpha x}\sum_{j=1}^{n} C_j x^{j-1}. \quad (4.6.13)$$

For the homogeneous problem, we choose the inversion contour of the Fourier transform based on the asymptotic behavior of the kernel, a point to be discussed in the following example.

❑ **Example 4.3.** Consider the homogeneous integral equation,

$$\phi_{\mathrm{H}}(x) = \lambda\int_{-\infty}^{+\infty} e^{-|x-y|}\phi_{\mathrm{H}}(y)\,dy. \quad (4.6.14)$$

Solution. Since the kernel vanishes exponentially as e^{-y} as $y \to \infty$ and as e^{+y} as $y \to -\infty$, we need not require $\phi_{\mathrm{H}}(y)$ to vanish as $y \to \pm\infty$, rather, more generally we may permit

$$\phi_{\mathrm{H}}(y) \to \begin{cases} e^{(1-\varepsilon)y} & \text{as} \quad y \to \infty, \\ e^{(-1+\varepsilon)y} & \text{as} \quad y \to -\infty, \end{cases}$$

and the integral equation still makes sense. So in the Fourier transform, we may allow

$$-1 < k_2 < 1, \quad (4.6.15)$$

and still have a valid solution.

The Fourier transform of $e^{-\alpha|x|}$ is given by

$$\int_{-\infty}^{+\infty} e^{-ikx}e^{-\alpha|x|}\,dx = \frac{2\alpha}{k^2+\alpha^2}. \quad (4.6.16)$$

Taking the Fourier transform of the homogeneous equation with $\alpha = 1$, we find

$$\hat{\phi}_{\mathrm{H}}(k) = \frac{2\lambda}{k^2+1}\hat{\phi}_H(k),$$

from which, we obtain

$$\frac{k^2+1-2\lambda}{k^2+1}\hat{\phi}_{\mathrm{H}}(k) = 0.$$

So there exists no nontrivial solution unless $k = \pm i\sqrt{1-2\lambda}$. By the inversion formula, $\phi_{\mathrm{H}}(x)$ is a superposition of e^{+ikx} terms with amplitude $\hat{\phi}_{\mathrm{H}}(k)$. But $\hat{\phi}_{\mathrm{H}}(k)$ is zero for all but $k = \pm i\sqrt{1-2\lambda}$. Hence we may conclude tentatively that

$$\phi_{\mathrm{H}}(x) = C_1 e^{-\sqrt{1-2\lambda}x} + C_2 e^{+\sqrt{1-2\lambda}x}.$$

However, we can at most allow $\phi_{\mathrm{H}}(x)$ to grow as fast as e^x as $x \to \infty$ and as e^{-x} as $x \to -\infty$, as we discussed above. Thus further analysis is in order.

Case (1) $1-2\lambda<0, \quad \text{or} \quad \lambda>\frac{1}{2}.$

$\phi_{\mathrm{H}}(x)$ is oscillatory and is given by

$$\phi_{\mathrm{H}}(x)=C_1 e^{-i\sqrt{2\lambda-1}x}+C_2 e^{+i\sqrt{2\lambda-1}x}. \tag{4.6.17}$$

Case (2) $0<1-2\lambda<1, \quad \text{or} \quad 0<\lambda<\frac{1}{2}.$

$\phi_{\mathrm{H}}(x)$ grows less fast than $e^{|x|}$ as $|x|\to\infty$.

$$\phi_{\mathrm{H}}(x)=C_1 e^{-\sqrt{1-2\lambda}x}+C_2 e^{+\sqrt{1-2\lambda}x}. \tag{4.6.18}$$

Case (3) $1-2\lambda\geq 1, \quad \text{or} \quad \lambda\leq 0.$

No acceptable solution for $\phi_{\mathrm{H}}(x)$ exists, since $e^{\pm\sqrt{1-2\lambda}x}$ grows faster than $e^{|x|}$ as $|x|\to\infty$.

Case (4) $\lambda=\frac{1}{2}.$

$$\phi_{\mathrm{H}}(x)=C_1+C_2 x. \tag{4.6.19}$$

Now consider the corresponding inhomogeneous problem.

❑ **Example 4.4.** Consider the inhomogeneous integral equation,

$$\phi(x)=ae^{-\alpha|x|}+\lambda\int_{-\infty}^{+\infty} e^{-|x-y|}\phi(y)\,dy. \tag{4.6.20}$$

Solution. On taking the Fourier Transform of Eq. (4.6.20), we obtain

$$\hat{\phi}(k)=\frac{2a\alpha}{k^2+\alpha^2}+\frac{2\lambda}{k^2+1}\hat{\phi}(k).$$

Solving for $\hat{\phi}(k)$, we obtain

$$\hat{\phi}\,(k)=\frac{2a\alpha(k^2+1)}{(k^2+1-2\lambda)(k^2+\alpha^2)}.$$

To invert the latter transform we note that, depending on whether λ is larger or smaller than $1/2$, the poles $k=\pm\sqrt{2\lambda-1}$ could lie on the real or imaginary axis of the complex k plane. What we can do is to choose any contour for the inversion within the strip

$$-\min(\alpha,1)<k_2<\min(\alpha,1) \tag{4.6.21}$$

to get a *particular solution* to our equation and we may then add any multiple of the homogeneous solution when the latter exists. The reason for choosing the strip (4.6.21) instead of

$$-1<k_2<1$$

in this case is that, in order for the Fourier transform of the inhomogeneous term $e^{-\alpha|x|}$ to exist, we must also restrict our attention to

$$-\alpha<k_2<\alpha.$$

Consider the first three cases given in Example 4.3.

Cases (2) and (3) $\lambda < \frac{1}{2}$.

In these cases, we have $1 - 2\lambda > 0$. Hence $\hat{\phi}(k)$ has simple poles at $k = \pm i\alpha$ and $k = \pm i\sqrt{1-2\lambda}$. To find a particular solution, use the real k axis as the integration contour for the inverse Fourier transformation. Then $\phi_{\text{P}}(x)$ is given by

$$\phi_{\text{P}}(x) = \frac{2a\alpha}{2\pi} \int_{-\infty}^{+\infty} dk\, e^{ikx} \frac{(k^2+1)}{(k^2+1-2\lambda)(k^2+\alpha^2)}.$$

For $x > 0$, we close the contour in the upper half-plane to obtain

$$\begin{aligned}\phi_{\text{P}}(x) &= \frac{2a\alpha}{2\pi} \cdot 2\pi i \left[\text{Res}\,(i\alpha) + \text{Res}\left(i\sqrt{1-2\lambda}\right)\right] \\ &= \frac{a}{1-2\lambda-\alpha^2} \left[\left(1-\alpha^2\right) e^{-\alpha x} - \frac{2\lambda\alpha}{\sqrt{1-2\lambda}} e^{-\sqrt{1-2\lambda}x}\right].\end{aligned}$$

For $x < 0$, we close the contour in the lower half-plane to get an identical result with x replaced by $-x$.

Thus our particular solution $\phi_{\text{P}}(x)$ is given by

$$\phi_{\text{P}}(x) = \frac{a}{1-2\lambda-\alpha^2} \left[\left(1-\alpha^2\right) e^{-\alpha|x|} - \frac{2\lambda\alpha}{\sqrt{1-2\lambda}} e^{-\sqrt{1-2\lambda}|x|}\right]. \tag{4.6.22}$$

For Case (3), this is the unique solution because there exists no acceptable homogeneous solution, while for Case (2) we must also add the homogeneous part given by

$$\phi_{\text{H}}(x) = C_1 e^{-\sqrt{1-2\lambda}x} + C_2 e^{+\sqrt{1-2\lambda}x}.$$

Case (1) $\lambda > \frac{1}{2}$.

In this case, we have $1 - 2\lambda < 0$. Hence $\hat{\phi}(k)$ has simple poles at $k = \pm i\alpha$ and $k = \pm\sqrt{2\lambda-1}$. To do the inversion for the particular solution, we can take any of the contours (1), (2), (3) or (4) as displayed in Figures 4.1 through 4.4, or Principal Value contours which are equivalent to half the sum of the first two contours or half the sum of the latter two contours.

Any of these differs by a multiple of the homogeneous solution. Consider a particular choice (4) for the inversion. For $x > 0$, we close the contour in the upper half-plane. Then our particular solution $\phi_{\text{P}}(x)$ is given by

$$\begin{aligned}\phi_{\text{P}}(x) &= \frac{2a\alpha}{2\pi} \cdot 2\pi i \left[\text{Res}(-\sqrt{2\lambda-1}) + \text{Res}(+i\alpha)\right] \\ &= \frac{a}{1-2\lambda-\alpha^2} \left[(1-\alpha^2)e^{-\alpha x} + \frac{2\lambda\alpha i}{\sqrt{2\lambda-1}} e^{-i\sqrt{2\lambda-1}x}\right].\end{aligned}$$

For $x < 0$, we close the contour in the lower half-plane to get an identical result with x replaced by $-x$.

So, in general, we can write our particular solution with the inversion contour (4) as,

$$\phi_{\text{P}}(x) = \frac{a}{1-2\lambda-\alpha^2} \left[\left(1-\alpha^2\right) e^{-\alpha|x|} + \frac{2\lambda\alpha i}{\sqrt{2\lambda-1}} e^{-i\sqrt{2\lambda-1}|x|}\right], \tag{4.6.23}$$

to which must be added the homogeneous part for Case (1) which reads,

$$\phi_{\text{H}}(x) = C_1 e^{-i\sqrt{2\lambda-1}x} + C_2 e^{+i\sqrt{2\lambda-1}x}.$$

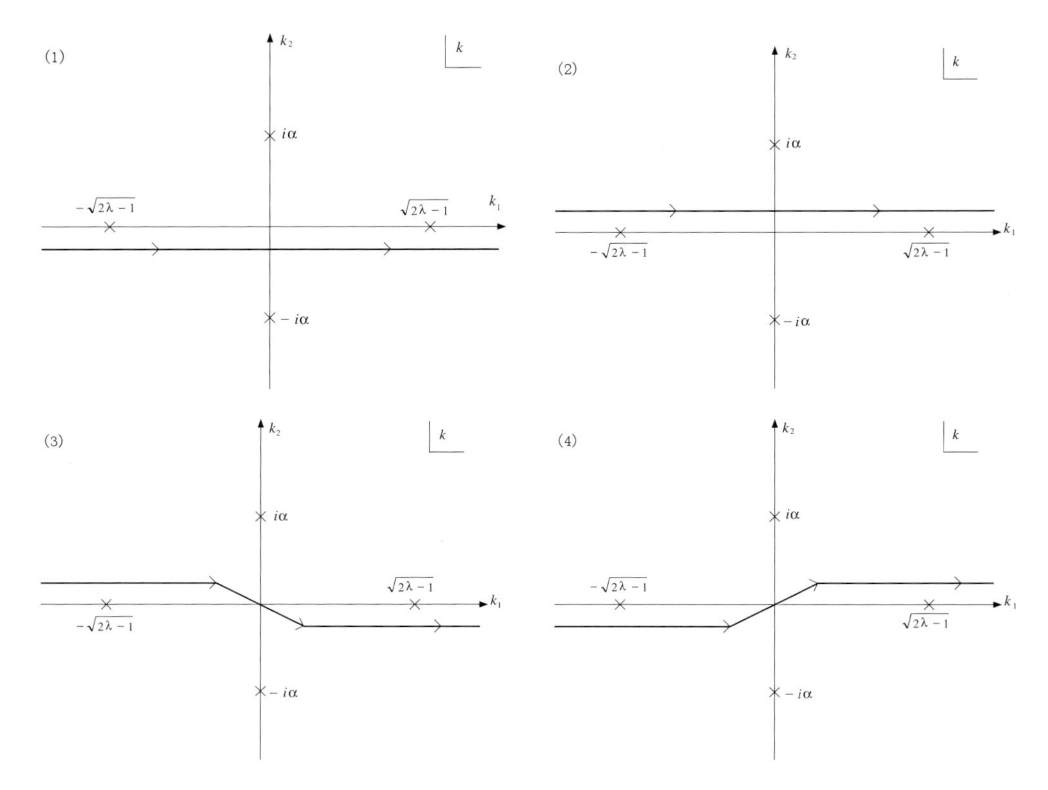

Fig. 4.1: The inversion contour (1)–(4) for **Case (1)–Case (4)**.

4.7 System of Fredholm Integral Equations of the Second Kind

We solve the system of Fredholm integral equations of the second kind,

$$\phi_i(x) - \lambda \int_a^b \sum_{j=1}^{n} K_{ij}(x,y)\phi_j(y)\,dy = f_i(x), \qquad i = 1, 2, \cdots, n, \tag{4.7.1}$$

where the kernels $K_{ij}(x,y)$ are *square-integrable*. We first extend the basic interval from $[a,b]$ to $[a, a+n(b-a)]$, and set

$$\begin{aligned} x + (i-1)(b-a) &= X < a + i(b-a), \\ y + (j-1)(b-a) &= Y < a + j(b-a), \end{aligned} \tag{4.7.2}$$

$$\phi(X) = \phi_i(x), \qquad K(X,Y) = K_{ij}(x,y), \qquad f(X) = f_i(x). \tag{4.7.3}$$

We then obtain the Fredholm integral equation of the second kind,

$$\phi(X) - \lambda \int_a^{a+n(b-a)} K(X,Y)\phi(Y)dY = f(X), \tag{4.7.4}$$

where the kernel $K(X,Y)$ is discontinuous in general but is *square-integrable* on account of the square-integrability of $K_{ij}(x,y)$. The solution $\phi(X)$ to Eq. (4.7.4) provides the solutions $\phi_i(x)$ to Eq. (4.7.1) with Eqs. (4.7.2) and (4.7.3).

4.8 Problems for Chapter 4

4.1. Calculate $D(\lambda)$ for

$$\text{a)} \quad K(x,y) = \begin{cases} xy, & y \le x, \\ 0, & \text{otherwise.} \end{cases}$$

$$\text{b)} \quad K(x,y) = xy, \quad 0 \le x,y \le 1.$$

$$\text{c)} \quad K(x,y) = \begin{cases} g(x)h(y), & y \le x, \\ 0, & \text{otherwise.} \end{cases}$$

$$\text{d)} \quad K(x,y) = g(x)h(y), \quad 0 \le x,y \le 1.$$

Find zero of $D(\lambda)$ for each case.

4.2. (Due to H. C.) Solve

$$\phi(x) = \lambda \left\{ \int_0^x dy \frac{\phi(y)}{(y+1)^2} \left(\frac{y}{x}\right)^a + \int_x^{+\infty} dy \frac{\phi(y)}{(y+1)^2} \right\}, \quad a > 0.$$

Find all eigenvalues and eigenfunctions.

4.3. (Due to H. C.) Solve the Fredholm integral equation of the second kind, given that

$$K(x-y) = e^{-|x-y|}, \qquad f(x) = \begin{cases} x, & x > 0, \\ 0, & x < 0. \end{cases}$$

4.4. (Due to H. C.) Solve the Fredholm integral equation of the second kind, given that

$$K(x-y) = e^{-|x-y|}, \qquad f(x) = x \quad \text{for} \quad -\infty < x < +\infty.$$

4.5. (Due to H. C.) Solve

$$\phi(x) + \lambda \int_{-1}^{+1} K(x,y)\phi(y)\,dy = 1, \quad -1 \le x \le 1, \quad K(x,y) = \sqrt{\frac{1-y^2}{1-x^2}}.$$

Find all eigenvalues of $K(x,y)$. Calculate also $D(\lambda)$ and $D(x,y;\lambda)$.

4.6. (Due to H. C.) Solve the Fredholm integral equation of the second kind,

$$\phi(x) = e^{-\frac{x}{2}} + \lambda \int_{-\infty}^{+\infty} \frac{1}{\cosh(x-y)}\phi(y)\,dy.$$

Hint:

$$\int_{-\infty}^{+\infty} \frac{e^{ikx}}{\cosh x}\, dx = \frac{\pi}{\cosh(\pi k/2)}.$$

4.7. (Due to H. C. and D. M.) Consider the integral equation,

$$\phi(x) = \lambda \int_{-\infty}^{+\infty} \frac{dy}{\sqrt{2\pi}} e^{ixy} \phi(y), \qquad -\infty < x < \infty.$$

a) Show that there are only four eigenvalues of the kernel $(1/\sqrt{2\pi}) \exp[ixy]$. What are these?

b) Show by an explicit calculation that the functions,

$$\phi_n(x) = \exp\left[-\frac{x^2}{2}\right] H_n(x),$$

where

$$H_n(x) \equiv (-1)^n \exp[x^2] \frac{d^n}{dx^n} \exp[-x^2],$$

are Hermite polynomials, are eigenfunctions with the corresponding eigenvalues, $(i)^n$, $(n = 0, 1, 2, \ldots)$. Why should one expect $\phi_n(x)$ to be Fourier transforms of themselves?

Hint: Think of the Schrödinger equation for the harmonic oscillator.

c) Using the result in b) and the fact that $\{\phi_n(x)\}_n$ form a complete set, in some sense, show that any square-integrable solution is of the form,

$$\phi(x) = f(x) + C\tilde{f}(x),$$

where $f(x)$ is an arbitrary even or odd, square-integrable function with Fourier transform $\tilde{f}(k)$, and C is a suitable constant. Evaluate C and relate its values to the eigenvalues found in a).

d) From c), construct a solution by taking $f(x) = \exp[-ax^2/2]$, $a > 0$.

4.8. (Due to H. C.) Find an eigenvalue and the corresponding eigenfunction for

$$K(x, y) = \exp\left[-(ax^2 + 2bxy + cy^2)\right], \qquad -\infty < x, y < \infty, \quad a + c > 0.$$

4.9. Consider the homogeneous integral equation,

$$\phi(x) = \lambda \int_{-\infty}^{\infty} K(x, y)\phi(y)\, dy, \qquad -\infty < x < \infty,$$

where

$$K(x, y) = \frac{1}{\sqrt{1 - t^2}} \exp\left[\frac{x^2 + y^2}{2}\right] \exp\left[-\frac{x^2 + y^2 - 2xyt}{1 - t^2}\right], \qquad t \text{ fixed}, \quad 0 < t < 1.$$

(a) Show directly that $\phi_0(x) = \exp[-x^2/2]$ is an eigenfunction of $K(x, y)$ corresponding to eigenvalue $\lambda = \lambda_0 = 1/\sqrt{\pi}$.

(b) Let

$$\phi_n(x) = \exp\left[-\frac{x^2}{2}\right] H_n(x).$$

Assume that $\phi_n = \lambda_n K\phi_n$. Show that $\phi_{n+1} = \lambda_{n+1} K\phi_{n+1}$ with $\lambda_n = t\lambda_{n+1}$. This means that the original integral equation has eigenvalues $\lambda_n = t^{-n}/\sqrt{\pi}$, with the corresponding eigenfunctions $\phi_n(x)$.

4.10. (Due to H. C.) Find the eigenvalues and eigenfunctions of the integral equation,

$$\phi(x) = \lambda \int_0^{\infty} \exp[-xy]\phi(y)\, dy.$$

Hint: Consider the Mellin transform,

$$\Phi(p) = \int_0^{\infty} x^{ip-\frac{1}{2}}\phi(x)\, dx \quad \text{with} \quad \phi(x) = \frac{1}{2\pi}\int_{-\infty}^{\infty} x^{-ip-\frac{1}{2}}\Phi(p)\, dp.$$

4.11. (Due to H. C.) Solve

$$\psi(x) = e^{bx} + \lambda \int_0^x \psi(y)\, dy + 2\lambda \int_x^1 \psi(y)\, dy.$$

4.12. Solve

$$\phi(x) = \lambda \int_{-1}^{+1} K(x, y)\phi(y)\, dy - \frac{1}{2}\int_{-1}^{+1} \phi(y)\, dy \quad \text{with} \quad \phi(\pm 1) = \text{finite},$$

where

$$K(x, y) = \frac{1}{2}\ln\left(\frac{1 + x_<}{1 - x_>}\right),$$

$$x_< = \frac{1}{2}(x + y) - \frac{1}{2}|x - y| \quad \text{and} \quad x_> = \frac{1}{2}(x + y) + \frac{1}{2}|x - y|.$$

4.13. Solve

$$\phi(x) = \lambda \int_{-\infty}^{+\infty} K(x, y)\phi(y)\, dy \quad \text{with} \quad \phi(\pm\infty) = \text{finite},$$

where

$$K(x, y) = \sqrt{\frac{\alpha}{\pi}}\left\{\exp\left[\frac{\alpha}{2}(x^2 + y^2)\right]\int_{-\infty}^{x_<} \exp[-\alpha\tau^2]d\tau \cdot \int_{x_>}^{+\infty} \exp[-\alpha\tau^2]d\tau\right\},$$

$$x_< = \frac{1}{2}(x + y) - \frac{1}{2}|x - y| \quad \text{and} \quad x_> = \frac{1}{2}(x + y) + \frac{1}{2}|x - y|.$$

4.14. Solve

$$\phi(x) = \lambda \int_0^\infty K(x,y)\phi(y)\,dy$$

with

$$|\phi(x)| < \infty \qquad \text{for} \qquad 0 \le x < \infty,$$

where

$$K(x,y) = \frac{\exp[-\beta\,|x-y|]}{2\beta xy}.$$

4.15. (Due to D. M.) Show that the non-trivial solutions of the homogeneous integral equation,

$$\phi(x) = \lambda \int_{-\pi}^{\pi} \left[\frac{1}{4\pi}(x-y)^2 - \frac{1}{2}\,|x-y|\right] \phi(y)\,dy,$$

are $\cos(mx)$ and $\sin(mx)$, where $\lambda = m^2$ and m is any integer.

Hint for Problems 4.12 through 4.15 : The kernels change their forms continuously as x passes through y. Differentiate the given integral equations with respect to x and reduce them to the ordinary differential equations.

4.16. (Due to D. M.) Solve the inhomogeneous integral equation,

$$\phi(x) = f(x) + \lambda \int_0^\infty \cos(2xy)\phi(y)\,dy, \quad x \ge 0,$$

where $\lambda^2 \ne 4/\pi$.

Hint: Multiply both sides of the integral equation by $\cos(2x\xi)$ and integrate over x. Use the identity,

$$\cos(2xy)\cos(2x\xi) = \frac{1}{2}\{\cos[2x(y+\xi)] + \cos[2x(y-\xi)]\},$$

and observe that

$$\begin{aligned}\int_0^\infty \cos(\alpha x)\,dx &= \frac{1}{2}\int_0^\infty (\exp[i\alpha x] + \exp[-i\alpha x])\,dx \\ &= \frac{1}{2}\int_{-\infty}^\infty \exp[i\alpha x]\,dx \\ &= \frac{1}{2}\cdot 2\pi\delta(\alpha) = \pi\delta(\alpha).\end{aligned}$$

4.17. (Due to D. M.) In the theoretical search for "*supergain antennas*", maximizing the directivity in the far field of axially invariant currents $j(\phi)$ that flow along the surface of infinitely long, circular cylinders of radius a, leads to the following Fredholm integral equation for the current density $j(\phi)$,

$$j(\phi) = \exp[ika \sin\phi] - \alpha \int_0^{2\pi} \frac{d\phi'}{2\pi} J_0\left(2ka \sin\frac{\phi - \phi'}{2}\right) j(\phi'), \quad 0 \le \phi < 2\pi,$$

where ϕ is the polar angle of the circular cross-section, k is a positive wave number, α is a parameter (Lagrange multiplier) which expresses a constraint on the current magnitude, $\alpha \ge 0$, and $J_0(x)$ is the zeroth-order Bessel function of the first kind.

a) Determine the eigenvalues of the homogeneous equation.

b) Solve the given inhomogeneous equation in terms of the Fourier series,

$$j(\phi) = \sum_{n=-\infty}^{\infty} f_n \exp[in\phi].$$

Hint: Use the formulas,

$$\exp[ika \sin\phi] = \sum_{n=-\infty}^{\infty} J_n(ka) \exp[in\phi],$$

and

$$J_0\left(2ka \sin\frac{\phi - \phi'}{2}\right) = \sum_{m=-\infty}^{\infty} J_m(ka)^2 \exp[im(\phi - \phi')],$$

where $J_n(x)$ is the nth-order Bessel function of the first kind. Substitution of the Fourier series for $j(\phi)$,

$$j(\phi) = \sum_{n=-\infty}^{\infty} f_n \exp[in\phi],$$

yields the decoupled equation for f_n,

$$f_n = \frac{J_n(ka)}{1 + \alpha J_n(ka)^2}, \qquad -\infty < n < \infty.$$

4.18. (Due to D. M.) Problem 4.17 corresponds to the circular loop in two-dimensions. For the circular disk in two dimensions, we have the following Fredholm integral equation for the current density $j(\phi)$,

$$j(r, \phi) = \exp[ikr \sin\phi] - \frac{2\alpha}{a^2} \int_0^{2\pi} \frac{d\phi'}{2\pi} \int_0^a r' dr'$$
$$\times J_0\left(k\sqrt{r^2 + r'^2 - 2rr' \cos(\phi - \phi')}\right) j(r', \phi'),$$

with

$$0 \le r \le a, \qquad 0 \le \phi < 2\pi.$$

Solve the given inhomogeneous equation in terms of the Fourier series,

$$j(r, \phi) = \sum_{n=-\infty}^{n=\infty} f_n(r) \exp[in\phi].$$

Hint: By using the addition formula,

$$J_0\left(k\sqrt{r^2 + r'^2 - 2rr'\cos(\phi - \phi')}\right) = \sum_{m=-\infty}^{\infty} J_m(kr) J_m(kr') \exp[im(\phi - \phi')],$$

it is found that $f_n(r)$ satisfy the following integral equation,

$$f_n(r) = \left[1 - \frac{2\alpha}{a^2} \int_0^a r' dr' f_n(r') J_n(kr')\right] J_n(kr), \qquad -\infty < n < \infty.$$

Substitution of

$$f_n(r) = \lambda_n J_n(kr)$$

yields

$$\begin{aligned} \lambda_n &= \left[1 + \frac{2\alpha}{a^2} \int_0^a r' dr' J_n(kr')^2\right]^{-1} \\ &= \left[1 + \alpha[J_n(ka)^2 - J_{n+1}(ka) J_{n-1}(ka)]\right]^{-1}. \end{aligned}$$

4.19. (Due to D. M.) Problem 4.17 corresponds to the circular loop in two dimensions. For the circular loop in three dimensions, we have the following Fredholm integral equation for the current density $j(\phi)$,

$$j(\phi) = \exp[ika\sin\phi] - \alpha \int_0^{2\pi} \frac{d\phi'}{2\pi} \mathcal{K}(\phi - \phi') j(\phi'), \qquad 0 \le \phi < 2\pi,$$

with

$$\begin{aligned} \mathcal{K}(\phi) &= \frac{\sin w}{w} + \frac{\cos w}{w^2} - \frac{\sin w}{w^3} \\ &= \frac{1}{4} \int_{-1}^{1} (1 + \xi^2) \exp[iw\xi] \, d\xi, \end{aligned}$$

and

$$w = w(\phi) = 2ka \sin\frac{\phi}{2}.$$

Solve the given inhomogeneous equation in terms of Fourier series,

$$j(\phi) = \sum_{n=-\infty}^{n=\infty} f_n \exp[in\phi].$$

Hint: Following the step employed in Problem 4.17, substitute the Fourier series into the integral equation. The decoupled equation for f_n,

$$f_n = \frac{J_n(ka)}{1+\alpha U_n(ka)}, \qquad -\infty < n < \infty,$$

results, where

$$\begin{aligned} U_n(ka) &= \int_{-\pi}^{\pi} \frac{d\phi}{2\pi} \mathcal{K}(\phi) \exp[-in\phi] \\ &= \frac{1}{8\pi} \int_{-1}^{1} d\xi (1+\xi^2) \int_{-\pi}^{\pi} d\phi \exp[iw(\phi)\xi] \cos(n\phi) \\ &= \frac{1}{2} \int_{0}^{1} d\xi (1+\xi^2) J_{2n}(2ka\xi). \end{aligned}$$

The integral for $U_n(ka)$ can be further simplified by the use of Lommel's function $S_{\mu,\nu}(x)$, and Weber's function $\mathbf{E}_\nu(x)$.

Reference for Problems 4.17, 4.18 and 4.19:

We cite the following article for the Fredholm integral equations of the second kind in the theoretical search for "*supergain antennas*".

Margetis, D., Fikioris, G., Myers, J.M., and Wu, T.T.: Phys. Rev. **E58**., 2531, (1998).

We cite the following article for the Fredholm integral equations of the second kind for the two-dimensional, highly directive currents on large circular loops.

Margetis, D. and Fikioris, G.: Jour. Math. Phys. **41**., 6130, (2000).

We can derive the above-stated Fredholm integral equations of the second kind for the localized, monochromatic, and highly directive classical current distributions in two and three dimensions by maximizing the directivity D in the far field while constraining $C = N/T$, where N is the integral of the square of the magnitude of the current density and T is proportional to the total radiated power. This derivation is the application of the calculus of variations. We derive the homogeneous Fredholm integral equations of the second kind and the inhomogeneous Fredholm integral equations of the second kind in their general forms in Section 9.6 of Chapter 9.

5 Hilbert–Schmidt Theory of Symmetric Kernel

5.1 Real and Symmetric Matrix

We would now like to examine the case of a *symmetric kernel* (*self-adjoint integral operator*) which is also *square-integrable*. Recalling from our earlier discussions in Chapter 1 that self-adjoint operators can be *diagonalized*, our principal aim is to accomplish the same goal for the case of symmetric kernels.

For this purpose, let us first examine the corresponding problem for an $n \times n$ *real* and *symmetric* matrix A. Suppose A has eigenvalues λ_k and normalized eigenvectors $\vec{v}_k$, i.e.,

$$A\vec{v}_k = \lambda_k \vec{v}_k, \quad k = 1, 2, \cdots, n, \tag{5.1.1a}$$

$$\vec{v}_k^{\mathrm{T}} \vec{v}_m = \delta_{km}. \tag{5.1.1b}$$

We may thus write

$$\begin{aligned} A\left[\vec{v}_1, \vec{v}_2, \ldots, \vec{v}_n\right] &= \left[\lambda_1 \vec{v}_1, \lambda_2 \vec{v}_2, \ldots, \lambda_n \vec{v}_n\right] \\ &= \left[\vec{v}_1, \ldots, \vec{v}_n\right] \begin{bmatrix} \lambda_1 & 0 & & \cdot & 0 \\ 0 & \lambda_2 & & \cdot & \\ & & \cdot & \cdot & \cdot \\ \cdot & \cdot & \cdot & \cdot & 0 \\ 0 & & \cdot & 0 & \lambda_n \end{bmatrix}. \end{aligned} \tag{5.1.2}$$

Define the matrix S by

$$S = \left[\vec{v}_1, \vec{v}_2, \cdots, \vec{v}_n\right] \tag{5.1.3a}$$

and consider S^{T}

$$S^{\mathrm{T}} = \begin{bmatrix} \vec{v}_1^{\mathrm{T}} \\ \vdots \\ \vec{v}_n^{\mathrm{T}} \end{bmatrix}. \tag{5.1.3b}$$

Applied Mathematics in Theoretical Physics. Michio Masujima

ISBN: 3-527-40534-8

Then we have

$$S^{\mathrm{T}}S = \begin{bmatrix} \vec{v}_1^{\mathrm{T}} \\ \vdots \\ \vec{v}_n^{\mathrm{T}} \end{bmatrix} [\vec{v}_1, \cdots, \vec{v}_n] = \begin{bmatrix} 1 & 0 & & 0 \\ 0 & 1 & & \\ & & \ddots & 0 \\ 0 & & 0 & 1 \end{bmatrix} = I, \tag{5.1.4a}$$

since we have Eq. (5.1.1b). Hence we have

$$S^{\mathrm{T}} = S^{-1}. \tag{5.1.5}$$

Define

$$D = \begin{bmatrix} \lambda_1 & 0 & & 0 \\ 0 & \lambda_2 & & \\ & & \ddots & 0 \\ 0 & & 0 & \lambda_n \end{bmatrix}. \tag{5.1.6}$$

From Eq. (5.1.2), we have

$$AS = SD. \tag{5.1.7a}$$

Hence we have

$$A = SDS^{-1} = SDS^{\mathrm{T}}. \tag{5.1.7b}$$

The above relation can also be written as

$$A = \sum_{k=1}^{n} \lambda_k \vec{v}_k^{\mathrm{T}} \vec{v}_k, \tag{5.1.8}$$

which represents the diagonalization of a symmetric matrix. Eq. (5.1.7b) is really convenient for calculation of functions of A, e.g.,

$$A^2 = \left(SDS^{-1}\right)\left(SDS^{-1}\right) = SD^2S^{-1} = S \begin{bmatrix} \lambda_1^2 & 0 & & 0 \\ 0 & \ddots & & \\ & & \ddots & 0 \\ 0 & & 0 & \lambda_n^2 \end{bmatrix} S^{-1}.$$

$$e^A = I + A + \frac{1}{2}A^2 + \frac{1}{3!}A^3 + \cdots = S \begin{bmatrix} e^{\lambda_1} & 0 & & 0 \\ 0 & \cdot & & \\ & & \cdot & 0 \\ 0 & & 0 & e^{\lambda_n} \end{bmatrix} S^{-1}.$$

$$f(A) = S \begin{bmatrix} f(\lambda_1) & 0 & & 0 \\ 0 & \cdot & & \\ & & \cdot & 0 \\ 0 & & 0 & f(\lambda_n) \end{bmatrix} S^{-1}. \tag{5.1.9}$$

Finally we have

$$\det A = \det SDS^{-1} = \det S \det D \det S^{-1} = \det D = \prod_{k=1}^{n} \lambda_k, \tag{5.1.10a}$$

$$\text{tr}(A) = \text{tr}\left(SDS^{-1}\right) = \text{tr}\left(DS^{-1}S\right) = \text{tr}\,(D) = \sum_{k=1}^{n} \lambda_k. \tag{5.1.10b}$$

5.2 Real and Symmetric Kernel

Symmetric kernels have the property that when transposed they remain the same as the original kernel. We denote the transposed kernel K^{T} by

$$K^{\mathrm{T}}(x, y) = K(y, x), \tag{5.2.1}$$

and note that when the kernel K is symmetric, we have

$$K^{\mathrm{T}}(x, y) = K(y, x) = K(x, y). \tag{5.2.2}$$

The eigenvalues of $K(x, y)$ and eigenvalues of $K^{\mathrm{T}}(x, y)$ are the same. This is because an eigenvalue λ_n is a zero of $D(\lambda)$. From the definition of $D(\lambda)$, we find that, since a determinant remains the same as we exchange its rows and columns,

$$D(\lambda) \text{ for } K(x, y) = D(\lambda) \text{ for } K^{\mathrm{T}}(x, y). \tag{5.2.3}$$

Thus the spectrum of $K(x, y)$ coincides with that of $K^{\mathrm{T}}(x, y)$.

We will need to make use of the orthogonality property held by the eigenfunctions belonging to each eigenvalue. To show this property, we start with the eigenvalue equations,

$$\phi_n(x) = \lambda_n \int_0^h K(x, y) \phi_n(y)\, dy, \tag{5.2.4}$$

$$\psi_n(x) = \lambda_n \int_0^h K^{\mathrm{T}}(x, y) \psi_n(y)\, dy, \tag{5.2.5}$$

and from the definition of $K^{\mathrm{T}}(x, y)$,

$$\psi_n(x) = \lambda_n \int_0^h K(y, x) \psi_n(y)\, dy. \tag{5.2.6}$$

Multiplying Eq. (5.2.4) by $\psi_m(x)$ and integrating over x, we get

$$\int_0^h \psi_m(x)\phi_n(x)\,dx = \lambda_n \int_0^h dx\,\psi_m(x) \int_0^h K(x,y)\phi_n(y)\,dy$$

$$= \frac{\lambda_n}{\lambda_m} \int_0^h \psi_m(y)\phi_n(y)\,dy,$$

so then

$$\left(1 - \frac{\lambda_n}{\lambda_m}\right) \int_0^h \psi_m(x)\phi_n(x)\,dx = 0. \tag{5.2.7}$$

If

$$\lambda_n \neq \lambda_m,$$

then we have

$$\int_0^h \psi_m(x)\phi_n(x)\,dx = 0 \quad \text{for} \quad \lambda_n \neq \lambda_m. \tag{5.2.8}$$

In the case of finite matrices, we know that the eigenvalues of a symmetric matrix are real and that the matrix is diagonalizable. Also, the eigenvectors are orthogonal to each other. We shall show that the same statements hold true in the case of *square-integrable symmetric kernels*.

If K is symmetric, then

$$\psi_n(x) = \phi_n(x).$$

Then, by Eq. (5.2.8), the eigenfunctions of a symmetric kernel are orthogonal to each other,

$$\int_0^h \phi_m(x)\phi_n(x)\,dx = 0 \quad \text{for} \quad \lambda_n \neq \lambda_m. \tag{5.2.9}$$

Furthermore, the eigenvalues of a symmetric kernel must be real. This is seen by supposing that the eigenvalue λ_n is complex. Then we have

$$\lambda_n \neq \lambda_n^*.$$

The complex conjugate of Eq. (5.2.4) is given by

$$\phi_n^*(x) = \lambda_n^* \int_0^h K(x,y)\phi_n^*(y)\,dy, \tag{5.2.10}$$

implying that λ_n^* and $\phi_n^*(x)$ are an eigenvalue and eigenfunction of the kernel $K(x,y)$. But, Eq. (5.2.9) with

$$\lambda_n \neq \lambda_n^*$$

then requires that

$$\int_0^h \phi_n(x)\phi_n^*(x)\,dx = \int_0^h |\phi_n(x)|^2\,dx = 0, \tag{5.2.11}$$

implying then that

$$\phi_n(x) \equiv 0,$$

which is a contradiction. Thus the eigenvalue must be real,

$$\lambda_n = \lambda_n^*,$$

to avoid this contradiction. Therefore the *eigenfunctions of a symmetric kernel are orthogonal to each other* and the *eigenvalues are real.*

We now, rather boldly, expand the symmetric kernel $K(x, y)$ in terms of $\phi_n(x)$,

$$K(x, y) = \sum_n a_n \phi_n(x). \tag{5.2.12}$$

We then normalize the eigenfunctions such that

$$\int_0^h \phi_n(x)\phi_m(x)\,dx = \delta_{nm}, \tag{5.2.13}$$

(even if there is more than one eigenfunction belonging to a certain eigenvalue, we can choose linear combinations of these eigenfunctions to satisfy Eq. (5.2.13)). From the orthogonality (5.2.13), we find

$$a_n = \int_0^h dx\,\phi_n(x)K(x, y) = \frac{1}{\lambda_n}\phi_n(y), \tag{5.2.14}$$

and thus obtain

Hilbert–Schmidt Theorem: $$K(x, y) = \sum_n \frac{\phi_n(y)\phi_n(x)}{\lambda_n}. \tag{5.2.15}$$

There is a problem though. We do not know if the eigenfunctions $\{\phi_n(x)\}_n$ are complete. In fact, we are often sure that the set $\{\phi_n(x)\}_n$ is not complete. An example is the kernel in the form of a finite sum of factorized terms. However, the content of the *Hilbert–Schmidt Theorem* (which will be proved shortly) is to claim that Eq. (5.2.15) for $K(x, y)$ is valid whether or not $\{\phi_n(x)\}_n$ is complete. The *only conditions* are that $K(x, y)$ be *symmetric* and *square-integrable*.

We calculate the iterated kernel,

$$
\begin{aligned}
K_2(x,y) &= \int_0^h K(x,z)K(z,y)\,dz \\
&= \int_0^h \sum_n \frac{\phi_n(x)\phi_n(z)}{\lambda_n} \sum_m \frac{\phi_m(z)\phi_m(y)}{\lambda_m}\,dz \\
&= \sum_n \sum_m \frac{1}{\lambda_n \lambda_m} \phi_n(x)\delta_{nm}\phi_m(y) \\
&= \sum_n \frac{\phi_n(x)\phi_n(y)}{\lambda_n^2},
\end{aligned}
\tag{5.2.16}
$$

and in general, we obtain

$$
K_j(x,y) = \sum_n \frac{\phi_n(x)\phi_n(y)}{\lambda_n^j}, \qquad j = 2, 3, \cdots. \tag{5.2.17}
$$

Now the definition for the resolvent kernel

$$
H(x,y;\lambda) = -K(x,y) - \lambda K_2(x,y) - \cdots - \lambda^j K_{j+1}(x,y) - \cdots
$$

becomes

$$
\begin{aligned}
H(x,y;\lambda) &= -\sum_n \frac{\phi_n(x)\phi_n(y)}{\lambda_n}\left[1 + \frac{\lambda}{\lambda_n} + \frac{\lambda^2}{\lambda_n^2} + \cdots + \frac{\lambda^j}{\lambda_n^j} + \cdots\right] \\
&= -\sum_n \frac{\phi_n(x)\phi_n(y)}{\lambda_n} \frac{1}{1 - \frac{\lambda}{\lambda_n}},
\end{aligned}
$$

i.e.,

$$
H(x,y;\lambda) = \sum_n \frac{\phi_n(x)\phi_n(y)}{\lambda - \lambda_n}. \tag{5.2.18}
$$

This elegant expression explicitly shows the analytic properties of $H(x,y;\lambda)$ in the complex λ plane. We can use this resolvent to solve the *inhomogeneous Fredholm Integral Equation of the second kind with a symmetric and square-integrable kernel.*

$$
\begin{aligned}
\phi(x) &= f(x) + \lambda \int_0^h K(x,y)\phi(y)\,dy \\
&= f(x) - \lambda \int_0^h H(x,y;\lambda) f(y)\,dy \\
&= f(x) - \lambda \sum_n \frac{\phi_n(x)}{\lambda - \lambda_n} \int_0^h \phi_n(y) f(y)\,dy.
\end{aligned}
\tag{5.2.19}
$$

Denoting

$$f_n \equiv \int_0^h \phi_n(y) f(y)\, dy, \tag{5.2.20}$$

we have the solution to the inhomogeneous equation (5.2.19),

$$\phi(x) = f(x) - \lambda \sum_n \frac{f_n \phi_n(x)}{\lambda - \lambda_n}. \tag{5.2.21}$$

At $\lambda = \lambda_n$, the solution does not exist unless $f_n = 0$, as usual.

As an another application of the eigenfunction expansion (5.2.15), we consider the *Fredholm Integral Equation of the first kind with a symmetric and square-integrable kernel*,

$$f(x) = \int_0^h K(x, y) \phi(y)\, dy. \tag{5.2.22}$$

Denoting

$$\phi_n \equiv \int_0^h \phi_n(y) \phi(y)\, dy, \tag{5.2.23}$$

we have

$$f(x) = \sum_n \frac{\phi_n(x)}{\lambda_n} \phi_n. \tag{5.2.24}$$

Immediately we encounter the problem. Equation (5.2.24) states that $f(x)$ is a linear combination of $\phi_n(x)$. In many cases, the set $\{\phi_n(x)\}_n$ is not complete, and thus $f(x)$ is not necessarily representable by a linear superposition of $\{\phi_n(x)\}_n$ and Eq. (5.2.22) has no solution.

If $f(x)$ is representable by a linear superposition of $\{\phi_n(x)\}_n$, it is easy to obtain ϕ_n. From Eqs. (5.2.20) and (5.2.24),

$$f_n = \int_0^h \phi_n(x) f(x)\, dx = \frac{\phi_n}{\lambda_n}, \tag{5.2.25}$$

and so

$$\phi_n = f_n \lambda_n. \tag{5.2.26}$$

A solution to Eq. (5.2.22) is then given by

$$\phi(x) = \sum_n \phi_n \phi_n(x) = \sum_n \lambda_n f_n \phi_n(x). \tag{5.2.27}$$

If the set $\{\phi_n(x)\}_n$ is not complete, the solution (5.2.27) is not unique. We can add to it any linear combination of $\{\psi_i(x)\}_i$ that is orthogonal to $\{\phi_n(x)\}_n$,

$$\phi(x) = \sum_n \lambda_n f_n \phi_n(x) + \sum_i C_i \psi_i(x), \tag{5.2.28}$$

where

$$\int_0^h \psi_i(x)\phi_n(x)\,dx = 0 \quad \text{for} \quad \text{all } i \text{ and } n. \tag{5.2.29}$$

If the set $\{\phi_n(x)\}_n$ is complete, the solution (5.2.27) is the unique solution. It may, however, still diverge since we have λ_n in the numerator, unless f_n vanishes sufficiently rapidly as $n \to \infty$ to ensure the convergence of the series (5.2.27).

We will now prove the Hilbert–Schmidt expansion, (5.2.15), in order to exhibit its efficiency, but, to avoid getting too mathematical, we will not be completely rigorous.

We will outline a plan of the proof. First, note the following lemma.

Lemma. *For a non-zero normed symmetric kernel,*

$$\infty > \|K\| > 0 \quad \textit{and} \quad K(x,y) = K^T(x,y), \tag{5.2.30}$$

there exists at least one eigenvalue λ_1 and one eigenfunction $\phi_1(x)$ (which we normalize to unity).

To prove Eq. (5.2.15), once this lemma has been established, we can construct a new kernel $\bar{K}(x,y)$ by

$$\bar{K}(x,y) \equiv K(x,y) - \frac{\phi_1(x)\phi_1(y)}{\lambda_1}. \tag{5.2.31}$$

Now $\phi_1(x)$ cannot be an eigenfunction of $\bar{K}(x,y)$ because we have

$$\begin{aligned} \int_0^h \bar{K}(x,y)\phi_1(y)\,dy &= \int_0^h \left[K(x,y) - \frac{\phi_1(x)\phi_1(y)}{\lambda_1}\right]\phi_1(y)\,dy \\ &= \frac{1}{\lambda_1}\phi_1(x) - \frac{1}{\lambda_1}\phi_1(x) = 0, \end{aligned} \tag{5.2.32}$$

which leaves us two possibilities,

(A) $\|\bar{K}\| \equiv 0$.

We have an equality

$$K(x,y) = \frac{\phi_1(x)\phi_1(y)}{\lambda_1}, \tag{5.2.33}$$

except over a set of points x whose measure is zero. A proof for this case is shown.

(B) $\|\bar{K}\| \neq 0$.

By the Lemma, there exists at least one eigenvalue λ_2 and one eigenfunction $\phi_2(x)$ of a kernel $\bar{K}(x,y)$.

$$\lambda_2 \int_0^h \bar{K}(x,y)\phi_2(y)\,dy = \phi_2(x), \tag{5.2.34}$$

i.e.,

$$\lambda_2 \int_0^h \left[K(x,y) - \frac{\phi_1(x)\phi_1(y)}{\lambda_1} \right] \phi_2(y)\, dy = \phi_2(x). \tag{5.2.35}$$

We then show that $\phi_2(x)$ and λ_2 are an eigenfunction and eigenvalue of the original kernel $K(x,y)$ orthogonal to $\phi_1(x)$.

To demonstrate the orthogonality of $\phi_2(x)$ to $\phi_1(x)$, multiply Eq. (5.2.35) by $\phi_1(x)$ and integrate over x.

$$\begin{aligned}\int_0^h \phi_1(x)\phi_2(x)\, dx &= \lambda_2 \int_0^h \phi_1(x)\, dx \int_0^h \left[K(x,y) - \frac{\phi_1(x)\phi_1(y)}{\lambda_1} \right] \phi_2(y)\, dy \\ &= \lambda_2 \int_0^h \left[\frac{1}{\lambda_1}\phi_1(y) - \frac{1}{\lambda_1}\phi_1(y) \right] \phi_2(y)\, dy = 0,\end{aligned}$$

i.e.,

$$\int_0^h \phi_1(x)\phi_2(x)\, dx = 0. \tag{5.2.36}$$

From Eq. (5.2.35), we then have

$$\lambda_2 \int_0^h K(x,y)\phi_2(y)\, dy = \phi_2(x). \tag{5.2.37}$$

Once we find $\phi_2(x)$, we construct a new kernel $\tilde{K}(x,y)$ by

$$\tilde{K}(x,y) \equiv \bar{K}(x,y) - \frac{\phi_2(x)\phi_2(y)}{\lambda_2} = K(x,y) - \sum_{n=1}^{2} \frac{\phi_n(x)\phi_n(y)}{\lambda_n}. \tag{5.2.38}$$

We then repeat the argument for $\tilde{K}(x,y)$. Ultimately either we find after N steps,

$$K(x,y) = \sum_{n=1}^{N} \frac{\phi_n(x)\phi_n(y)}{\lambda_n}, \tag{5.2.39}$$

or we find the infinite series,

$$K(x,y) \approx \sum_{n=1}^{\infty} \frac{\phi_n(x)\phi_n(y)}{\lambda_n}, \tag{5.2.40}$$

and can show that the *remainder* $R(x,y)$, which is defined by

$$R(x,y) \equiv K(x,y) - \sum_{n=1}^{\infty} \frac{\phi_n(x)\phi_n(y)}{\lambda_n}, \tag{5.2.41}$$

cannot have any eigenfunction. If $\psi(x)$ is the eigenfunction of $R(x,y)$,

$$\lambda_0 \int_0^h R(x,y)\psi(y)\,dy = \psi(x), \tag{5.2.42}$$

we know that

(1) $\psi(x)$ is distinct from all $\{\phi_n(x)\}_n$,

$$\psi(x) \neq \phi_n(x) \quad \text{for} \quad n = 1, 2, \cdots, \tag{5.2.43}$$

and that

(2) $\psi(x)$ is orthogonal to all $\{\phi_n(x)\}_n$,

$$\int_0^h \psi(x)\phi_n(x)\,dx = 0 \quad \text{for} \quad n = 1, 2, \cdots. \tag{5.2.44}$$

Then, substituting the definition (5.2.41) of $R(x,y)$ into Eq. (5.2.42) and noting the orthogonality (5.2.44), we find

$$\lambda_0 \int_0^h K(x,y)\psi(y)\,dy = \psi(x), \tag{5.2.45}$$

which is a contradiction of Eq. (5.2.43). Thus we must have

$$\|R\|^2 = \int_0^h \int_0^h R^2(x,y)\,dxdy = 0, \tag{5.2.46}$$

which is the meaning of the $\approx$ in Eq. (5.2.40). Therefore, the formula (5.2.15) holds in the sense of the *mean square convergence*.

Proof of **Lemma**: So we only have to prove the **Lemma** stated with the condition (5.2.30) and the proof of the Hilbert–Schmidt Theorem will be complete. To do so, it is necessary to work with the iterated kernel $K_2(x,y)$, which is also symmetric.

$$K_2(x,y) = \int_0^h K(x,z)K(z,y)\,dz = \int_0^h K(x,z)K(y,z)\,dz. \tag{5.2.47}$$

This is because the trace of $K_2(x,y)$ is always positive.

$$\int_0^h K_2(x,x)\,dx = \int_0^h dx \int_0^h dz\, K^2(x,z) = \|K\|^2 > 0. \tag{5.2.48}$$

First, we will prove that if $K_2(x,y)$ has an eigenvalue, then $K(x,y)$ has at least one eigenvalue equalling one of the square-roots of the former.

Recall the definition of the resolvent kernel of $K(x, y)$,

$$H(x, y; \lambda) = -K(x, y) - \lambda K_2(x, y) - \cdots - \lambda^j K_{j+1}(x, y) - \cdots , \tag{5.2.49}$$

$$H(x, y; -\lambda) = -K(x, y) + \lambda K_2(x, y) - \cdots - (-\lambda)^j K_{j+1}(x, y) - \cdots . \tag{5.2.50}$$

Taking the difference of Eqs. (5.2.49) and (5.2.50), we find

$$\begin{aligned} &\frac{1}{2}[H(x, y; \lambda) - H(x, y; -\lambda)] \\ &\qquad = -\lambda \left[K_2(x, y) + \lambda^2 K_4(x, y) + \lambda^4 K_6(x, y) + \cdots\right] \\ &\qquad = \lambda H_2(x, y; \lambda^2), \end{aligned} \tag{5.2.51}$$

which is the resolvent for $K_2(x, y)$. The equality (5.2.51), which is valid for sufficiently small λ where the series expansion in λ is defined, holds for all λ by analytic continuation.

If c is an eigenvalue of $K_2(x, y)$, $H_2(x, y; \lambda^2)$ has a pole at $\lambda^2 = c$. From Eq. (5.2.51), either $H(x, y; \lambda)$ or $H(x, y; -\lambda)$ must have a pole at $\lambda = \pm\sqrt{c}$. This means that at least one of $\pm\sqrt{c}$ is an eigenvalue of $K(x, y)$.

Now we prove that $K_2(x, y)$ has at least one eigenvalue. We have

$$\begin{aligned} \int_0^h D_2(x, x; s)\, dx / D_2(s) &= \int_0^h H_2(x, x; s)\, dx \\ &= -(A_2 + sA_4 + s^2 A_6 + \cdots) \end{aligned} \tag{5.2.52}$$

where

$$A_m = \int_0^h K_m(x, x)\, dx, \qquad m = 2, 3, \cdots . \tag{5.2.53}$$

If $K_2(x, y)$ has no eigenvalues, then $D_2(s)$ has no zeros, and the series (5.2.52) must be convergent for all values of s. To this end, consider

$$\begin{aligned} A_{m+n} = \int_0^h dx\, K_{m+n}(x, x) &= \int_0^h dx \int_0^h dz\, K_m(x, z) K_n(z, x) \\ &= \int_0^h dx \int_0^h dz\, K_m(x, z) K_n(x, z). \end{aligned} \tag{5.2.54}$$

Applying the Schwarz inequality,

$$\begin{aligned} A_{m+n}^2 &\leq \left[\int_0^h dx \int_0^h dz\, K_m^2(x, z)\right] \left[\int_0^h dx \int_0^h dz\, K_n^2(x, z)\right] \\ &= \left[\int_0^h dx\, K_{2m}(x, x)\right] \left[\int_0^h dx\, K_{2n}(x, x)\right] \\ &= A_{2m} A_{2n}, \end{aligned}$$

i.e., we have

$$A_{m+n}^2 \le A_{2m}A_{2n}. \tag{5.2.55}$$

Setting

$$\begin{cases} m & \to & n-1, \\ n & \to & n+1, \end{cases}$$

in the inequality (5.2.55), we have

$$A_{2n}^2 \le A_{2n-2}A_{2n+2}. \tag{5.2.56}$$

Recalling that

$$A_{2m} > 0, \tag{5.2.57}$$

which is precisely the reason we consider $K_2(x, y)$, we have

$$\frac{A_{2n}}{A_{2n-2}} \le \frac{A_{2n+2}}{A_{2n}}. \tag{5.2.58}$$

Successively, we have

$$\frac{A_{2n+2}}{A_{2n}} \ge \frac{A_{2n}}{A_{2n-2}} \ge \frac{A_{2n-2}}{A_{2n-4}} \ge \cdots \ge \frac{A_4}{A_2} \equiv R_1, \tag{5.2.59}$$

so that

$$A_4 = R_1 A_2,$$

$$A_6 \ge R_1^2 A_2,$$

$$A_8 \ge R_1^3 A_2,$$

and generally

$$A_{2n} \ge R_1^{n-1} A_2. \tag{5.2.60}$$

Thus we have

$$A_2 + sA_4 + s^2 A_6 + s^3 A_8 + \cdots \ge A_2(1 + sR_1 + s^2 R_1^2 + s^3 R_1^3 + \cdots). \tag{5.2.61}$$

The right-hand side of the inequality (5.2.61) diverges for those s such that

$$s \ge \frac{1}{R_1} = \frac{A_2}{A_4}. \tag{5.2.62}$$

Thus

$$\int_0^h H_2(x, x; s)\, dx$$

is divergent for those s satisfying the inequality (5.2.62). Then $K_2(x,y)$ has the eigenvalue s satisfying

$$s \le \frac{A_2}{A_4},$$

and $K(x,y)$ has the eigenvalue λ_1 satisfying

$$|\lambda_1| \le \sqrt{\frac{A_2}{A_4}}. \tag{5.2.63}$$

This completes the proof of the **Lemma** and completes the proof of the Hilbert–Schmidt Theorem. □

The Hilbert–Schmidt expansion, (5.2.15), can be helpful in many problems where a symmetric and *square-integrable* kernel is involved.

❑ **Example 5.1.** Solve the integro-differential equation,

$$\frac{\partial}{\partial t}\phi(x,t) = \int_0^h K(x,y)\phi(y,t)\,dy, \tag{5.2.64a}$$

with the initial condition,

$$\phi(x,0) = f(x), \tag{5.2.64b}$$

where $K(x,y)$ is symmetric and *square-integrable.*

Solution. The Hilbert–Schmidt expansion, (5.2.15), can be applied giving

$$\frac{\partial}{\partial t}\phi(x,t) = \sum_n \frac{\phi_n(x)}{\lambda_n}\int_0^h \phi_n(y)\phi(y,t)\,dy. \tag{5.2.65}$$

Defining $A_n(t)$ by

$$A_n(t) \equiv \int_0^h \phi_n(y)\phi(y,t)\,dy, \tag{5.2.66}$$

and changing the dummy index of summation from n to m in Eq. (5.2.65), we then have

$$\frac{\partial}{\partial t}\phi(x,t) = \sum_m \frac{\phi_m(x)}{\lambda_m}A_m(t). \tag{5.2.67}$$

Taking the time derivative of Eq. (5.2.66) yields

$$\frac{d}{dt}A_n(t) = \int_0^h dy\phi_n(y)\frac{\partial}{\partial t}\phi(y,t). \tag{5.2.68}$$

Substituting Eq. (5.2.67) into Eq. (5.2.68) and, observing that the orthogonality of $\{\phi_m(x)\}_m$ means that only the $m = n$ term is left, we get

$$\frac{d}{dt}A_n(t) = \frac{A_n(t)}{\lambda_n}. \tag{5.2.69}$$

The solution to Eq. (5.2.69) is then

$$A_n(t) = A_n(0)\exp\left[\frac{t}{\lambda_n}\right], \tag{5.2.70}$$

with

$$A_n(0) = \int_0^h dx\, \phi_n(x) f(x). \tag{5.2.71}$$

We can now integrate Eq. (5.2.67) from 0 to t, with $A_n(t)$ from Eq. (5.2.70),

$$\int_0^t dt \frac{\partial}{\partial t}\phi(x,t) = \int_0^t dt \sum_n \frac{\phi_n(x)}{\lambda_n} A_n(0)\exp\left[\frac{t}{\lambda_n}\right]. \tag{5.2.72}$$

The left-hand side of Eq. (5.2.72) is now exact, and we obtain

$$\phi(x,t) - \phi(x,0) = \sum_n \frac{\phi_n(x)}{\lambda_n} A_n(0) \frac{\left(\exp\left[\frac{t}{\lambda_n}\right] - 1\right)}{\left(\frac{1}{\lambda_n}\right)}. \tag{5.2.73}$$

From the initial condition (5.2.64b) and Eq. (5.2.73), we get finally

$$\phi(x,t) = f(x) + \sum_n A_n(0)\left(\exp\left[\frac{t}{\lambda_n}\right] - 1\right)\phi_n(x). \tag{5.2.74}$$

As $t \to \infty$, the asymptotic form of $\phi(x,t)$ is given either by

$$\phi(x,t) = f(x) - \sum_n A_n(0)\phi_n(x) \quad \text{if} \quad \text{all } \lambda_n < 0, \tag{5.2.75}$$

or by

$$\phi(x,t) = A_i(0)\phi_i(x)\exp\left[\frac{t}{\lambda_i}\right] \quad \text{if} \quad 0 < \lambda_i < \text{all other } \lambda_n. \tag{5.2.76}$$

5.3 Bounds on the Eigenvalues

In the process of proving our Lemma in the previous section, we managed to obtain the *upper bound* on the lowest eigenvalue,

$$|\lambda_1| \leq \sqrt{\frac{A_2}{A_4}}.$$

A better upper bound can be obtained as follows. If we call

$$R_2 = \frac{A_6}{A_4},$$

we note that

$$R_2 \geq R_1,$$

or

$$\frac{1}{R_2} \leq \frac{1}{R_1}.$$

Furthermore, we find

$$\begin{aligned} A_2 + sA_4 + s^2A_6 + s^3A_8 + s^4A_{10} + \cdots \\ &= A_2 + sA_4\left[1 + s\left(\frac{A_6}{A_4}\right) + s^2\left(\frac{A_8}{A_4}\right) + \cdots\right] \\ &\geq A_2 + sA_4[1 + sR_2 + s^2R_2^2 + \cdots] \end{aligned}$$

which diverges if

$$sR_2 > 1.$$

Hence we have a singularity for

$$s \leq \frac{1}{R_2} \leq \frac{1}{R_1}.$$

We therefore have an improved upper bound on λ_1,

$$|\lambda_1| \leq \sqrt{\frac{A_4}{A_6}} \leq \sqrt{\frac{A_2}{A_4}}.$$

So, we have the successively better upper bounds on $|\lambda_1|$,

$$\sqrt{\frac{A_2}{A_4}},\ \sqrt{\frac{A_4}{A_6}},\ \sqrt{\frac{A_6}{A_8}},\ \cdots,$$

for the lowest eigenvalue, each better than the previous one, i.e.,

$$|\lambda_1| \leq \cdots \leq \sqrt{\frac{A_6}{A_8}} \leq \sqrt{\frac{A_4}{A_6}} \leq \sqrt{\frac{A_2}{A_4}}. \tag{5.3.1a}$$

Recall also that with a symmetric kernel, we have

$$A_{2m} = \|K_m\|^2. \tag{5.3.2}$$

The upper bounds (5.3.1a), in terms of the norm of the iterated kernels, become

$$|\lambda_1| \leq \cdots \leq \frac{\|K_3\|}{\|K_4\|} \leq \frac{\|K_2\|}{\|K_3\|} \leq \frac{\|K\|}{\|K_2\|}. \tag{5.3.1b}$$

Now consider the question of finding the *lower bounds* for the lowest eigenvalue λ_1. Consider the expansion of the symmetric kernel,

$$K(x,y) \approx \sum_n \frac{\phi_n(x)\phi_n(y)}{\lambda_n}. \tag{5.3.3}$$

This expression (5.3.3) is an *equation in the mean*, and hence there is no guarantee that it is true at any point as an exact equation. In particular, on the line $y = x$ which has zero measure in the square $0 \leq x, y \leq h$, it need not be true. The following equality

$$K(x,x) = \sum_n \frac{\phi_n(x)\phi_n(x)}{\lambda_n}$$

therefore need not be true. Hence

$$\int_0^h K(x,x)\,dx = \sum_n \frac{1}{\lambda_n} \tag{5.3.4}$$

need not be true. The right-hand side of Eq. (5.3.4) may not converge.

However, for

$$K_2(x,y) = \int_0^h K(x,z)K(z,y)\,dz = \sum_n \frac{\phi_n(x)\phi_n(y)}{\lambda_n^2},$$

since we know $K(x,y)$ to be *square-integrable*,

$$A_2 = \int_0^h K_2(x,x)\,dx = \sum_n \frac{1}{\lambda_n^2} \tag{5.3.5}$$

must converge, and, in general, for $m \geq 2$, we have

$$A_m = \sum_n \frac{1}{\lambda_n^m}, \qquad m = 2, 3, \cdots, \tag{5.3.6}$$

and we know that the right-hand side of Eq. (5.3.6) converges.

Consider now the expansion for A_2, namely

$$A_2 = \frac{1}{\lambda_1^2} + \frac{1}{\lambda_2^2} + \frac{1}{\lambda_3^2} + \cdots = \frac{1}{\lambda_1^2}\left[1 + \left(\frac{\lambda_1}{\lambda_2}\right)^2 + \left(\frac{\lambda_1}{\lambda_3}\right)^2 + \cdots\right] \geq \frac{1}{\lambda_1^2},$$

i.e.,

$$\lambda_1^2 \geq \frac{1}{A_2}. \tag{5.3.7}$$

Hence we have a lower bound for the eigenvalue λ_1,

$$|\lambda_1| \geq \frac{1}{\sqrt{A_2}}, \quad \text{or} \quad |\lambda_1| \geq \frac{1}{\|K\|}. \tag{5.3.8}$$

This is *consistent* with our early discussion of the series solution to the Fredholm integral equation of the second kind for which we concluded that when

$$|\lambda| < \frac{1}{\|K\|}, \tag{5.3.9}$$

there are no singularities in λ, so that the first eigenvalue $\lambda = \lambda_1$ must satisfy the inequality (5.3.8).

We can obtain better lower bounds for the eigenvalue λ_1. Consider A_4,

$$A_4 = \frac{1}{\lambda_1^4} + \frac{1}{\lambda_2^4} + \frac{1}{\lambda_3^4} + \cdots = \frac{1}{\lambda_1^4}\left[1 + \left(\frac{\lambda_1}{\lambda_2}\right)^4 + \left(\frac{\lambda_1}{\lambda_3}\right)^4 + \cdots\right] \geq \frac{1}{\lambda_1^4},$$

i.e.,

$$|\lambda_1| \geq \frac{1}{(A_4)^{1/4}}. \tag{5.3.10}$$

This is an improvement over the previously established lower bound since we know from Eqs. (5.3.1a) and (5.3.8) that

$$\frac{1}{A_2^{1/2}} \leq |\lambda_1| \leq \left(\frac{A_2}{A_4}\right)^{1/2},$$

so that

$$A_4 \leq A_2^2,$$

i.e.,

$$\frac{1}{(A_4)^{1/4}} \geq \frac{1}{(A_2)^{1/2}}.$$

Thus $1/(A_4)^{1/4}$ is a better lower bound than $1/(A_2)^{1/2}$.

Proceeding in the same way with A_6, A_8, $\cdots$, we get increasingly better lower bounds,

$$\frac{1}{(A_2)^{1/2}} \leq \frac{1}{(A_4)^{1/4}} \leq \frac{1}{(A_6)^{1/6}} \leq \cdots \leq |\lambda_1|. \tag{5.3.11}$$

Putting both the upper bounds (5.3.1a) and lower bounds (5.3.11) together, we have for the smallest eigenvalue λ_1,

$$\frac{1}{(A_2)^{1/2}} \leq \frac{1}{(A_4)^{1/4}} \leq \frac{1}{(A_6)^{1/6}} \leq \cdots$$
$$\cdots \leq |\lambda_1| \leq \cdots \leq \left(\frac{A_6}{A_8}\right)^{1/2} \leq \left(\frac{A_4}{A_6}\right)^{1/2} \leq \left(\frac{A_2}{A_4}\right)^{1/2}. \tag{5.3.12}$$

Strength permitting, we calculate A_6, A_8, $\cdots$, to obtain successively better upper bounds and lower bounds from Eq. (5.3.12).

5.4 Rayleigh Quotient

Another useful technique for finding the upper bounds for eigenvalues of self-adjoint operators is based on the *Rayleigh quotient*. Consider the self-adjoint integral operator,

$$\tilde{K} = \int_0^h dy\, K(x,y) \quad \text{with} \quad K(x,y) = K(y,x), \tag{5.4.1}$$

with eigenvalues λ_n and eigenfunctions $\phi_n(x)$,

$$\begin{cases} \tilde{K}\phi_n = \dfrac{1}{\lambda_n}\phi_n, \\ (\phi_n, \phi_m) = \delta_{nm}, \end{cases} \tag{5.4.2}$$

where the eigenvalues are ordered such that

$$|\lambda_1| \leq |\lambda_2| \leq |\lambda_3| \leq \cdots .$$

Consider any given function $g(x)$ such that

$$\|g\| \neq 0. \tag{5.4.3}$$

Consider the series

$$\sum_n b_n \phi_n(x) \quad \text{with} \quad b_n = (\phi_n, g). \tag{5.4.4}$$

This series expansion is the projection of $g(x)$ onto the space spanned by the set $\{\phi_n(x)\}_n$, which may not be complete.

We can easily verify the *Bessel inequality*, which says

$$\sum_n b_n^2 \leq \|g\|^2 . \tag{5.4.5}$$

Proof of the Bessel inequality:

$$\left\| g - \sum_n b_n \phi_n \right\|^2 \geq 0,$$

which implies

$$(g, g) - \sum_m b_m(g, \phi_m) - \sum_n b_n(\phi_n, g) + \sum_n \sum_m b_n b_m(\phi_n, \phi_m) \geq 0.$$

Thus we have

$$(g, g) - \sum_m b_m^2 \geq 0,$$

which states

$$\sum_n b_n^2 \leq \|g\|^2 ,$$

completing the proof of the Bessel inequality (5.4.5). □

Now, consider the quadratic form $(g, \tilde{K}g)$,

$$(g, \tilde{K}g) = \int_0^h dx\, g(x)\tilde{K}g(x) = \int_0^h dx \int_0^h dy\, g(x)K(x, y)g(y). \tag{5.4.6}$$

Substituting the expansion

$$K(x, y) \approx \sum_n \frac{\phi_n(x)\phi_n(y)}{\lambda_n}$$

into Eq. (5.4.6), we obtain

$$(g, \tilde{K}g) = \int_0^h dx \int_0^h dy\, g(x) \sum_n \left(\frac{\phi_n(x)\phi_n(y)}{\lambda_n} \right) g(y) = \sum_n \frac{b_n^2}{\lambda_n} . \tag{5.4.7}$$

Then, taking the absolute value of the above quadratic form (5.4.7), we obtain

$$\begin{aligned} \left|(g, \tilde{K}g)\right| = \left| \sum_n \frac{b_n^2}{\lambda_n} \right| &\leq \left\{ \frac{b_1^2}{|\lambda_1|} + \frac{b_2^2}{|\lambda_2|} + \frac{b_3^2}{|\lambda_3|} + \cdots \right\} \\ &= \frac{1}{|\lambda_1|} \left\{ b_1^2 + \left| \frac{\lambda_1}{\lambda_2} \right| b_2^2 + \left| \frac{\lambda_1}{\lambda_3} \right| b_3^2 + \cdots \right\} \\ &\leq \frac{1}{|\lambda_1|} \left\{ b_1^2 + b_2^2 + b_3^2 + \cdots \right\} \\ &\leq \frac{1}{|\lambda_1|} \|g\|^2 . \end{aligned} \tag{5.4.8}$$

Hence, from Eq. (5.4.8), the Rayleigh quotient Q, defined by

$$Q \equiv \frac{(g, \tilde{K}g)}{(g, g)}, \tag{5.4.9}$$

is bounded above by

$$|Q| = \frac{\left|(g, \tilde{K}g)\right|}{\|g\|^2} \leq \frac{1}{|\lambda_1|},$$

i.e., the absolute value of the lowest eigenvalue λ_1 is bounded above by

$$|\lambda_1| \leq \frac{1}{|Q|}. \tag{5.4.10}$$

To find a good upper bound on $|\lambda_1|$, choose a trial function $g(x)$ with adjustable parameters and obtain the minimum of $\frac{1}{|Q|}$. Namely, we have

$$|\lambda_1| \leq \min\left(\frac{1}{|Q|}\right) \tag{5.4.11}$$

with Q given by Eq. (5.4.9).

❑ **Example 5.2.** Find an upper bound on the leading eigenvalue of the symmetric kernel

$$K(x,y) = \begin{cases} (1-x)y, & 0 \leq y < x \leq 1, \\ (1-y)x, & 0 \leq x < y \leq 1, \end{cases}$$

using the Rayleigh quotient.

Solution. Consider the trial function $g(x) = ax$ which is probably not very good. We have

$$(g,g) = \int_0^1 dx\, a^2x^2 = \frac{a^2}{3},$$

and

$$\begin{aligned}(g, \tilde{K}g) &= \int_0^1 dx \int_0^1 dy\, g(x)K(x,y)g(y) \\ &= \int_0^1 dx\, ax \left[\int_0^x dy(1-x)yay + \int_x^1 dy(1-y)xay\right] \\ &= a^2 \int_0^1 dx\, x^2(1-x^2)/6 = \frac{a^2}{30}.\end{aligned}$$

So, the Rayleigh quotient Q is given by

$$Q = \frac{(g, \tilde{K}g)}{(g,g)} = \frac{1}{10},$$

and we get

$$\min\left(\frac{1}{|Q|}\right) = 10.$$

Thus, from Eq. (5.4.11), we obtain

$$|\lambda_1| \leq 10,$$

which is a reasonable upper bound. The exact value of λ_1 turns out to be

$$\lambda_1 = \pi^2 \approx 9.8696,$$

so it is not too bad, considering that the eigenfunction for λ_1 turns out to be $A\sin(\pi x)$, not well approximated by ax.

5.5 Completeness of Sturm–Liouville Eigenfunctions

Consider the *Sturm–Liouville eigenvalue problem*,

$$\frac{d}{dx}\left[p(x)\frac{d}{dx}\phi(x)\right] - q(x)\phi(x) = \lambda r(x)\phi(x) \quad \text{on} \quad [0, h], \tag{5.5.1}$$

with

$$\phi(0) = \phi(h) = 0,$$

and

$$p(x) > 0, \quad r(x) > 0, \quad \text{for} \quad x \in [0, h].$$

We proved earlier that, using the Green's function $G(x, y)$ defined by

$$\frac{d}{dx}\left[p(x)\frac{d}{dx}G(x, y)\right] - q(x)G(x, y) = \delta(x - y), \tag{5.5.2}$$

with

$$G(0, y) = 0, \quad G(h, y) = 0,$$

Eq. (5.5.1) is equivalent to the integral equation

$$\phi(x) = \lambda \int_0^h G(y, x)r(y)\phi(y)\, dy. \tag{5.5.3}$$

Further, we have shown that, since the Sturm–Liouville operator is self-adjoint and symmetric, we have a symmetric Green's function,

$$G(x, y) = G(y, x). \tag{5.5.4}$$

Now, define

$$\psi(x) = \sqrt{r(x)}\phi(x), \tag{5.5.5a}$$

and

$$K(x, y) = \sqrt{r(x)} G(x, y) \sqrt{r(y)}. \tag{5.5.5b}$$

Then $\psi(x)$ satisfies

$$\psi(x) = \lambda \int_0^h K(x, y) \psi(y)\, dy, \tag{5.5.6}$$

which has a *symmetric kernel*. Applying the usual Hilbert–Schmidt theorem, we know that $K(x, y)$ defined above is decomposable in the form,

$$K(x, y) \approx \sum_n \frac{\psi_n(x)\psi_n(y)}{\lambda_n} = \sum_n \sqrt{r(x)r(y)} \frac{\phi_n(x)\phi_n(y)}{\lambda_n}, \tag{5.5.7}$$

with λ_n *real* and *discrete*, and the set $\{\psi_n(x)\}_n$ *orthonormal*, i.e.,

$$\int_0^h \psi_n(x)\psi_m(x)\, dx = \int_0^h r(x)\phi_n(x)\phi_m(x)\, dx = \delta_{nm}. \tag{5.5.8}$$

Note the appearance of the *weight function* $r(x)$ in the middle equation of Eq. (5.5.8).

To prove the *completeness*, we will establish that any function $f(x)$ can be expanded in a series of $\{\psi_n(x)\}_n$ or $\{\phi_n(x)\}_n$. Let us do this for the *differentiable* case (which is *stronger than square-integrable*), i.e., assume $f(x)$ is differentiable. Therefore, given any $f(x)$, we can define $g(x)$ by

$$g(x) \equiv L f(x), \tag{5.5.9}$$

where

$$L = \frac{d}{dx}\left[p(x) \frac{d}{dx} \right] - q(x). \tag{5.5.10}$$

Take the inner product of both sides of Eq. (5.5.9) with $G(x, y)$ to get

$$(G(x, y), g(x)) = f(y), \tag{5.5.11a}$$

i.e.,

$$f(x) = \int_0^h G(x, y) g(y)\, dy = \int_0^h \left(\frac{K(x, y)}{\sqrt{r(x)r(y)}} \right) g(y)\, dy. \tag{5.5.11b}$$

Substituting the expression (5.5.7) for $K(x, y)$ into Eq. (5.5.11b), we obtain

$$\begin{aligned} f(x) = \int_0^h dy \sum_n \left(\frac{\phi_n(x)\phi_n(y)}{\lambda_n} \right) g(y) &= \sum_n \left(\frac{\phi_n(x)}{\lambda_n} \right) (\phi_n, g) \\ &= \sum_n \left(\frac{\beta_n}{\lambda_n} \right) \phi_n(x), \end{aligned} \tag{5.5.12}$$

where we set

$$\beta_n \equiv (\phi_n, g) = \int_0^h \phi_n(y) g(y)\, dy. \tag{5.5.13}$$

Since the above expansion of $f(x)$ in terms of $\phi_n(x)$ is true for any $f(x)$, this demonstrates that the set $\{\phi_n(x)\}_n$ is *complete*.

Actually, in addition, we must require that $f(x)$ satisfies the *homogeneous boundary conditions* in order to avoid boundary terms. Also, we must make sure that the kernel for the Sturm–Liouville eigenvalue problem is *square-integrable*. Since the set $\{\phi_n(x)\}_n$ is complete, we conclude that there must be an infinite number of eigenvalues to the Sturm–Liouville system. Also, it is possible to prove the asymptotic results,

$$\lambda_n = O(n^2) \quad \text{as} \quad n \to \infty.$$

5.6 Generalization of Hilbert–Schmidt Theory

In this section, we consider the generalization of Hilbert–Schmidt theory in five directions.

Direction 1: So far in our discussion of Hilbert–Schmidt theory, we assumed that $K(x, y)$ is real. It is straightforward to extend to the case when $K(x, y)$ is *complex*. We define the norm $\|K\|$ of the kernel $K(x, y)$ by

$$\|K\|^2 = \int_0^h dx \int_0^h dy \, |K(x, y)|^2 . \tag{5.6.1}$$

The iteration series solution to the Fredholm integral equation of the second kind again converges for

$$|\lambda| < \frac{1}{\|K\|}. \tag{5.6.2}$$

Also, the Fredholm theory still remains valid. If $K(x, y)$ is, in addition, *self-adjoint*, i.e.,

$$K(x, y) = K^*(y, x), \tag{5.6.3}$$

then the Hilbert–Schmidt expansion holds in the form,

$$K(x, y) \approx \sum_n \frac{\phi_n(x)\phi_n^*(y)}{\lambda_n}, \tag{5.6.4}$$

where

$$\int_0^h \phi_n^*(x)\phi_m(x)\, dx = \delta_{nm}, \tag{5.6.5}$$

and

$$\lambda_n = \text{real, and } n \text{ is an integer.} \tag{5.6.6}$$

Direction 2: Next we note that, in all the discussion so far, the variable x is restricted to a finite basic interval,

$$x \in [0, h]. \tag{5.6.7}$$

As the second generalization, we extend the basic interval $[0, h]$ to $[0, \infty)$. We want to solve the following integral equation,

$$\phi(x) = f(x) + \lambda \int_0^{+\infty} K(x, y)\phi(y)\, dy, \tag{5.6.8}$$

with

$$\int_0^{+\infty} dx \int_0^{+\infty} dy\, K^2(x, y) < \infty, \qquad \int_0^{+\infty} dx\, f^2(x) < \infty. \tag{5.6.9}$$

By a change of the independent variable x, it is always possible to transform the interval $[0, \infty)$ of x into $[0, h]$ of t, i.e.,

$$x \in [0, \infty) \Rightarrow t \in [0, h]. \tag{5.6.10}$$

For example, the following transformation will do,

$$x = g(t) = \frac{t}{(h - t)}. \tag{5.6.11}$$

Then, writing

$$\tilde{\phi}(t) = \phi(g(t)), \quad \text{etc.,}$$

we have

$$\tilde{\phi}(t) = \tilde{f}(t) + \lambda \int_0^h \tilde{K}(t, t')\tilde{\phi}(t')\, g'(t')\, dt'.$$

By multiplying $\sqrt{g'(t)}$ on both sides of the equation above, we have

$$\sqrt{g'(t)}\tilde{\phi}(t) = \sqrt{g'(t)}\tilde{f}(t) + \lambda \int_0^h \sqrt{g'(t)}\tilde{K}(t, t')\sqrt{g'(t')}\sqrt{g'(t')}\tilde{\phi}(t')\, dt'.$$

Defining $\psi(t)$ by

$$\psi(t) = \sqrt{g'(t)}\tilde{\phi}(t),$$

we obtain

$$\psi(t) = \sqrt{g'(t)}\tilde{f}(t) + \lambda \int_0^h [\sqrt{g'(t)}\tilde{K}(t, t')\sqrt{g'(t')}]\psi(t')\, dt'. \tag{5.6.12}$$

If the original kernel $K(x, y)$ is symmetric, then the transformed kernel is also symmetric. Furthermore, the transformed kernel $\sqrt{g'(t)}\tilde{K}(t,t')\sqrt{g'(t')}$ and the transformed inhomogeneous term $\sqrt{g'(t)}\tilde{f}(t)$ are *square-integrable* if $K(x, y)$ and $f(x)$ are *square-integrable*, since

$$\int_0^h dt \int_0^h dt' g'(t)\tilde{K}^2(t,t')g'(t') = \int_0^{+\infty} dx \int_0^{+\infty} dy\, K^2(x,y) < \infty, \qquad (5.6.13a)$$

and

$$\int_0^h dt g'(t)\tilde{f}^2(t) = \int_0^{+\infty} dx\, f^2(x) < \infty. \qquad (5.6.13b)$$

Thus, under appropriate conditions, the Fredholm theory and the Hilbert–Schmidt theory both apply to Eq. (5.6.8). Similarly we can extend these theories to the case of infinite range.

Direction 3: As the third generalization, we consider the case where we have *multi-dimensional independent variables.*

$$\phi(\vec{x}) = f(\vec{x}) + \lambda \int_0^{+\infty} K(\vec{x},\vec{y})\phi(\vec{y})\, d\vec{y}. \qquad (5.6.14)$$

As long as the kernel $K(\vec{x},\vec{y})$ is *square-integrable*, i.e.,

$$\int_0^{+\infty}\int_0^{+\infty} K^2(\vec{x},\vec{y})\, d\vec{x}d\vec{y} < \infty, \qquad (5.6.15)$$

all the arguments for establishing the Fredholm theory and the Hilbert–Schmidt theory apply.

Direction 4: As the fourth generalization of the theorem, we will *relax the condition on the square-integrability of the kernel*. When a kernel $K(x, y)$ is not *square-integrable*, the integral equation is said to be *singular*. Some singular integral equations can be transformed into one with a *square-integrable* kernel. One method which may work is to try to *symmetrize* them as much as we can. For example, a kernel of the form $H(x, y)$ with $H(x, y)$ bounded can be made *square-integrable* by symmetrizing it into $H(x, y)/(xy)^{\frac{1}{4}}$. Another way is to *iterate the kernel*. Suppose the kernel is of the form,

$$K(x,y) = \frac{H(x,y)}{|x-y|^{\alpha}}, \qquad \frac{1}{2} \le \alpha < 1, \qquad (5.6.16)$$

where $H(x, y)$ is bounded. We have the integral equation of the form,

$$\phi(x) = f(x) + \lambda \int_0^h K(x,y)\phi(y)\, dy. \qquad (5.6.17)$$

Replacing $\phi(y)$ in the integrand with the right-hand side of Eq. (5.6.17) itself, we obtain

$$\phi(x) = \left[f(x) + \lambda \int_0^h K(x,y)f(y)\, dy \right] + \lambda^2 \int_0^h K_2(x,y)\phi(y)\, dy. \qquad (5.6.18)$$

The kernel in Eq. (5.6.18) is $K_2(x, y)$, which may be *square-integrable* since

$$\int_0^h \frac{1}{|x-z|^\alpha} \frac{1}{|z-y|^\alpha} \, dz = O\left(\frac{1}{|x-y|^{2\alpha - 1}} \right). \tag{5.6.19}$$

Indeed, for those α such that

$$\frac{1}{2} \le \alpha < \frac{3}{4},$$

$K_2(x, y)$ is *square-integrable*. If α is such that

$$\frac{3}{4} \le \alpha < 1,$$

then $K_3(x, y)$, $K_4(x, y)$, $\cdots$, etc. may be *square-integrable*. In general, when α lies in the interval

$$1 - \frac{1}{2(n-1)} \le \alpha < 1 - \frac{1}{2n}, \tag{5.6.20}$$

$K_n(x, y)$ will be *square-integrable*. Thus, for those α such that

$$\frac{1}{2} \le \alpha < 1, \tag{5.6.21}$$

we can transform the kernel into a *square-integrable* kernel by the appropriate number of iterations. However, when $\alpha \ge 1$, we have no hope whatsoever of transforming it into a *square-integrable* kernel in this way.

For a kernel which cannot be made *square-integrable*, which properties remain valid? Does the Fredholm theory hold? Is the spectrum of the eigenvalues discrete? Does the Hilbert–Schmidt expansion hold? The following example gives us some insight into these questions.

❑ **Example 5.3.** Suppose we want to solve the homogeneous equation

$$\phi(x) = \lambda \int_0^{+\infty} e^{-|x-y|} \phi(y) \, dy. \tag{5.6.22}$$

The kernel in the above equation is symmetric, but not *square-integrable*; even so, this equation can be solved in the closed form.

Solution. Writing out Eq. (5.6.22) explicitly, we have

$$\phi(x) = \lambda \int_0^x e^{-(x-y)} \phi(y) \, dy + \lambda \int_x^{+\infty} e^{-(y-x)} \phi(y) \, dy.$$

Multiplying both sides by e^{-x}, we have

$$e^{-x} \phi(x) = \lambda e^{-2x} \int_0^x e^y \phi(y) \, dy + \lambda \int_x^{+\infty} e^{-y} \phi(y) \, dy. \tag{5.6.23}$$

Differentiating the above equation with respect to x, we obtain

$$e^{-x}(-\phi(x)+\phi'(x)) = -2\lambda e^{-2x}\int_0^x e^y\phi(y)\,dy.$$

Multiplying both sides by e^{2x}, we have

$$e^{x}(-\phi(x)+\phi'(x)) = -2\lambda \int_0^x e^y\phi(y)\,dy. \tag{5.6.24}$$

Differentiating the above equation with respect to x and cancelling the factor e^x, we obtain

$$\phi''(x) + (2\lambda - 1)\phi(x) = 0.$$

The solution to the above equation is given by

(i) $1 - 2\lambda > 0$,

$$\phi(x) = C_1 e^{\sqrt{1-2\lambda}x} + C_2 e^{-\sqrt{1-2\lambda}x},$$

and,

(ii) $1 - 2\lambda < 0$,

$$\phi\,(x) = C_1' e^{i\sqrt{2\lambda-1}x} + C_2' e^{-i\sqrt{2\lambda-1}x}.$$

These solutions satisfy the equation (5.6.22) only if

$$\phi'(0) = \phi\,(0)\,,$$

which follows from the once-differentiated equation (5.6.24). Thus we have

$$\sqrt{1-2\lambda}(C_1 - C_2) = C_1 + C_2.$$

In order to satisfy Eq. (5.6.23), we must require that the integral

$$\int_x^{+\infty} e^{-y}\phi(y)\,dy \tag{5.6.25}$$

converges. If $\frac{1}{2} > \lambda > 0$, $\phi\,(x)$ grows exponentially but the integral in Eq. (5.6.25) converges. If $\lambda > \frac{1}{2}$, $\phi\,(x)$ oscillates and the integral in Eq. (5.6.25) converges. If $\lambda < 0$, however, the integral in Eq. (5.6.25) diverges and no solution exists.

In summary, a solution exists for $\lambda > 0$, but no solution exists for $\lambda < 0$. We note that:

1. the spectrum is not discrete;
2. the eigenfunctions for $\lambda > \frac{1}{2}$ alone constitute a complete set (very much like a Fourier sine or cosine expansion). Thus not all of the eigenfunctions are necessary to represent an $\mathbb{L}_2$ function.

Direction 5: As the last generalization of the theorem, we shall retain the *square-integrability of the kernel*, but consider the case of the *non-symmetric kernel*. Since this generalization is not always possible, we shall illustrate the point by presenting one example.

❑ **Example 5.4.** We consider the following Fredholm integral equation of the second kind,

$$\phi(x) = f(x) + \lambda \int_0^1 K(x,y)\phi(y)\,dy, \tag{5.6.26}$$

where the kernel is non-symmetric,

$$K(x,y) = \begin{cases} 2, & 0 \le y < x \le 1, \\ 1, & 0 \le x < y \le 1, \end{cases} \tag{5.6.27}$$

but is *square-integrable*.

Solution. We first consider the *homogeneous equation*.

$$\phi(x) = \lambda \int_0^x 2\phi(y)\,dy + \lambda \int_x^1 \phi(y)\,dy = \lambda \int_0^1 \phi(y)\,dy + \lambda \int_0^x \phi(y)\,dy. \tag{5.6.28}$$

Differentiating Eq. (5.6.28) with respect to x, we obtain

$$\phi'(x) = \lambda\phi(x). \tag{5.6.29}$$

From this, we obtain

$$\phi(x) = C\exp[\lambda x], \quad 0 \le x \le 1, \quad C \ne 0. \tag{5.6.30}$$

From Eq. (5.6.28), we get the boundary conditions,

$$\left\{ \begin{array}{l} \phi(0) = \lambda \int_0^1 \phi(y)\,dy, \\ \phi(1) = 2\lambda \int_0^1 \phi(y)\,dy, \end{array} \right. \Rightarrow \phi(1) = 2\phi(0). \tag{5.6.31}$$

Hence, we require that

$$C\exp[\lambda] = 2C \Rightarrow \exp[\lambda] = 2 = \exp[\ln(2) + i2n\pi], \quad n \text{ integer}.$$

Thus we should have the eigenvalues,

$$\lambda = \lambda_n = \ln(2) + i2n\pi, \quad n = 0, \pm 1, \pm 2, \cdots. \tag{5.6.32}$$

The corresponding eigenfunctions are

$$\phi_n(x) = C_n \exp[\lambda_n x], \quad C_n \text{ real}.$$

Finally,

$$\phi_n(x) = C_n \exp[\{\ln(2) + i2n\pi\}x], \quad n \text{ integer}. \tag{5.6.33}$$

Clearly, the kernel is non-symmetric, $K(x,y) \neq K(y,x)$. The transposed kernel $K^{\mathrm{T}}(x,y)$ is given by

$$K^{\mathrm{T}}(x,y) = K(y,x) = \begin{cases} 2, & 0 \le x < y \le 1, \\ 1, & 0 \le y < x \le 1. \end{cases} \tag{5.6.34}$$

We next consider the *homogeneous equation for the transposed kernel* $K^{\mathrm{T}}(x,y)$.

$$\psi(x) = \lambda \int_0^x \psi(y)\,dy + 2\lambda \int_x^1 \psi(y)\,dy = 2\lambda \int_0^1 \psi(y)\,dy - \lambda \int_0^x \psi(y)\,dy. \tag{5.6.35}$$

Differentiating Eq. (5.6.35) with respect to x, we obtain

$$\psi'(x) = -\lambda\psi(x). \tag{5.6.36}$$

From this, we obtain

$$\psi(x) = F\exp[-\lambda x], \quad 0 \le x \le 1, \quad F \neq 0. \tag{5.6.37}$$

From Eq. (5.6.35), we get the boundary conditions,

$$\begin{cases} \psi(0) = 2\lambda \displaystyle\int_0^1 \psi(y)\,dy, \\ \psi(1) = \lambda \displaystyle\int_0^1 \psi(y)\,dy, \end{cases} \Rightarrow \psi(1) = \frac{1}{2}\psi(0). \tag{5.6.38}$$

Hence, we require that

$$F\exp[-\lambda] = \frac{1}{2}F \Rightarrow \exp[\lambda] = 2 = \exp[\ln(2) + i2n\pi], \quad n \text{ integer.}$$

Thus we should have the same eigenvalues,

$$\lambda = \lambda_n = \ln(2) + i2n\pi, \quad n = 0, \pm 1, \pm 2, \cdots. \tag{5.6.39}$$

The corresponding eigenfunctions are

$$\psi_n(x) = F_n \exp[-\lambda_n x], \quad F_n \text{ real.}$$

Finally,

$$\psi_n(x) = F_n \exp[-\{\ln(2) + i2n\pi\}x], \quad n \text{ integer.} \tag{5.6.40}$$

These $\psi_n(x)$ are the solution to the transposed problem. For $n \neq m$, we have

$$\int_0^1 \phi_n(x)\psi_m(x)\,dx = C_n F_m \int_0^1 \exp[i2\pi(n-m)x]dx = 0, \quad n \neq m.$$

The spectral representation of the kernel $K(x,y)$ is given by

$$K(x,y) = \sum_{n=-\infty}^{\infty} \frac{\phi_n(x)\psi_n(y)}{\lambda_n} = \sum_{n=-\infty}^{\infty} \frac{\exp[\{\ln(2)+i2n\pi\}(x-y)]}{\ln(2)+i2n\pi}, \tag{5.6.41}$$

$$C_n = F_n = 1,$$

which we obtain from the following orthogonality,

$$\int_0^1 \phi_n(x)\psi_m(x)\,dx = \delta_{nm}. \tag{5.6.42}$$

In establishing Eq. (5.6.41), we first write

$$R(x,y) \equiv K(x,y) - \sum_{n=1}^{\infty} \frac{\phi_n(x)\psi_n(y)}{\lambda_n}, \tag{5.6.43}$$

and demonstrate the fact that the *remainder* $R(x,y)$ cannot have any eigenfunctions, by exhausting all of the possible eigenfunctions. By explicit solution, we already know that the kernel has at least one eigenvalue.

Crucial to this generalization is the fact that the original integral equation and the transposed integral equation have the same eigenvalues and that the eigenfunctions of the transposed kernel are orthogonal to the eigenfunctions of the original kernel. This last generalization is not always possible for the general nonsymmetric kernel.

5.7 Generalization of Sturm–Liouville System

In Section 5.5, we have shown that, if $p(x) > 0$ and $r(x) > 0$, the eigenvalue equation

$$\frac{d}{dx}\left[p(x)\frac{d}{dx}\phi(x)\right] - q(x)\phi(x) = \lambda r(x)\phi(x) \quad \text{where} \quad x \in [0,h], \tag{5.7.1}$$

with appropriate boundary conditions has the eigenfunctions which form a complete set $\{\phi_n(x)\}_n$ belonging to the discrete eigenvalues λ_n. In this section, we shall relax the conditions on $p(x)$ and $r(x)$. In particular, we shall consider the case in which $p(x)$ has simple or double zeros at the end points, which therefore may be regular singular points of the differential equation (5.7.1).

Let L_x be a second-order differential operator,

$$L_x \equiv a_0(x)\frac{d^2}{dx^2} + a_1(x)\frac{d}{dx} + a_2(x), \quad \text{where} \quad x \in [0,h], \tag{5.7.2}$$

which is, in general, *non self-adjoint*. As a matter of fact, we can always transform a second-order differential operator L_x into a self-adjoint form by multiplying $p(x)/a_0(x)$ on L_x, with

$$p(x) \equiv \exp\left[\int^x \frac{a_1(y)}{a_0(y)}\,dy\right].$$

However, it is instructive to see what happens when L_x is non self-adjoint. So, we shall not transform the differential operator L_x, (5.7.2), into a self-adjoint form. Let us assume that certain boundary conditions are given at $x = 0$ and $x = h$.

Consider the Green's functions $G(x, y)$ and $G(x, y; \lambda)$ defined by

$$\begin{aligned} L_x G(x, y) &= \delta(x - y), \\ (L_x - \lambda) G(x, y; \lambda) &= \delta(x - y), \\ G(x, y; \lambda = 0) &= G(x, y). \end{aligned} \tag{5.7.3}$$

We would like to find a representation of $G(x, y; \lambda)$ in a form similar to $H(x, y; \lambda)$ given by Eq. (5.2.9). Symbolically, we write $G(x, y; \lambda)$ as

$$G(x, y; \lambda) = (L_x - \lambda)^{-1}. \tag{5.7.4}$$

Since the defining equation of $G(x, y; \lambda)$ depends on λ analytically, we expect $G(x, y; \lambda)$ to be an analytic function of λ by the *Poincaré theorem*. There are two possible exceptions: (1) At a regular singular point, the indicial equation yields an exponent which, considered as a function λ, may have branch cuts; (2) For some value of λ, it may be impossible to match the discontinuity at $x = y$.

To elaborate on the second point (2), let ϕ_1 and ϕ_2 be the solution of

$$(L_x - \lambda)\phi_i(x; \lambda) = 0 \quad (i = 1, 2). \tag{5.7.5}$$

We can construct $G(x, y; \lambda)$ to be

$$G(x, y; \lambda) \propto \begin{cases} \phi_1(x; \lambda)\phi_2(y; \lambda), & 0 \le x \le y, \\ \phi_2(x; \lambda)\phi_1(y; \lambda), & y < x \le h, \end{cases} \tag{5.7.6}$$

where $\phi_1(x; \lambda)$ satisfies the boundary condition at $x = 0$, and $\phi_2(x; \lambda)$ satisfies the boundary condition at $x = h$. The constant of proportionality of Eq. (5.7.6) is given by

$$\frac{C}{W(\phi_1(y; \lambda), \phi_2(y; \lambda))}. \tag{5.7.7}$$

When the Wronskian vanishes as a function of λ, $G(x, y; \lambda)$ develops a singularity. It may be a pole or a branch point in λ. However, the vanishing of the Wronskian $W(\phi_1(y; \lambda), \phi_2(y; \lambda))$ implies that $\phi_2(x; \lambda)$ is proportional to $\phi_1(x; \lambda)$ for such λ; namely, we have an eigenfunction of L_x. Thus the singularities of $G(x, y; \lambda)$ as a function of λ are associated with the eigenfunctions of L_x. Hence we shall treat $G(x, y; \lambda)$ as an analytic function of λ, except at poles located at $\lambda = \lambda_i$ $(i = 1, \cdots, n, \cdots)$ and at a branch point located at $\lambda = \lambda_B$, from which a branch cut is extended to $-\infty$. Assuming that $G(x, y; \lambda)$ behaves as

$$G(x, y; \lambda) = O\left(\frac{1}{\lambda}\right) \quad \text{as} \quad |\lambda| \to \infty, \tag{5.7.8a}$$

we obtain

$$\lim_{R \to \infty} \frac{1}{2\pi i} \oint_{C_R} \frac{G(x, y; \lambda')}{\lambda' - \lambda} \, d\lambda' = 0, \tag{5.7.8b}$$

where C_R is the circle of radius R, centered at the origin of the complex λ plane. Invoking the Cauchy Residue Theorem, we have

$$\lim_{R\to\infty}\frac{1}{2\pi i}\oint_{C_R}\frac{G(x,y;\lambda')}{\lambda'-\lambda}\,d\lambda' = G(x,y;\lambda)+\sum_n\frac{R_n(x,y)}{\lambda_n-\lambda}$$
$$-\frac{1}{2\pi i}\int_{-\infty}^{\lambda_B}\frac{G(x,y;\lambda'+i\varepsilon)-G(x,y;\lambda'-i\varepsilon)}{\lambda'-\lambda}\,d\lambda', \quad (5.7.9)$$

where

$$R_n(x,y) = \text{Res}\, G(x,y;\lambda')|_{\lambda'=\lambda_n}\,. \quad (5.7.10)$$

Combining Eqs. (5.7.8b) and (5.7.9), we obtain

$$G(x,y;\lambda) = -\sum_n\frac{R_n(x,y)}{\lambda_n-\lambda}$$
$$+\frac{1}{2\pi i}\int_{-\infty}^{\lambda_B}\frac{G(x,y;\lambda'+i\varepsilon)-G(x,y;\lambda'-i\varepsilon)}{\lambda'-\lambda}\,d\lambda'. \quad (5.7.11)$$

Let us concentrate on the first term in Eq. (5.7.11). Multiplying the second equation in Eq. (5.7.3) by $(\lambda-\lambda_n)$ and letting $\lambda\to\lambda_n$,

$$\lim_{\lambda\to\lambda_n}(\lambda-\lambda_n)\,(L_x-\lambda)\,G(x,y;\lambda) = \lim_{\lambda\to\lambda_n}(\lambda-\lambda_n)\,\delta(x-y),$$

from which we obtain

$$(L_x-\lambda_n)\,R_n(x,y) = 0. \quad (5.7.12)$$

Thus we obtain

$$R_n(x,y)\propto\psi_n(y)\phi_n(x), \quad (5.7.13)$$

where $\phi_n(x)$ is the eigenfunction of L_x belonging to the eigenvalue λ_n (assuming that the eigenvalue is not degenerate),

$$(L_x-\lambda_n)\,\phi_n(x) = 0. \quad (5.7.14)$$

We claim that $\psi_n(x)$ is the eigenfunction of L_x^{T}, belonging to the same eigenvalue λ_n,

$$(L_x^{\mathrm{T}}-\lambda_n)\psi_n(x) = 0, \quad (5.7.15)$$

where L_x^{T} is defined by

$$L_x^{\mathrm{T}} \equiv \frac{d^2}{dx^2}a_0(x)-\frac{d}{dx}a_1(x)+a_2(x). \quad (5.7.16)$$

Suppose we want to solve the following equation,

$$(L_x^{\mathrm{T}}-\lambda)h(x) = f(x). \quad (5.7.17)$$

We construct the following expression,

$$\int_0^h dx\, G(x,y;\lambda)(L_x^{\mathrm{T}}-\lambda)h(x) = \int_0^h dx\, G(x,y;\lambda)f(x), \tag{5.7.18}$$

and perform the integral by parts on the left-hand side of Eq. (5.7.18). We obtain,

$$\int_0^h dx[(L_x-\lambda)G(x,y;\lambda)]h(x) = \int_0^h dx\, G(x,y;\lambda)f(x). \tag{5.7.19}$$

The expression in the square bracket on the left-hand side of Eq. (5.7.19) is $\delta(x-y)$, and we have (after exchanging the variables x and y)

$$h(x) = \int_0^h dy\, G(y,x;\lambda)f(y). \tag{5.7.20}$$

Operating $(L_x^{\mathrm{T}}-\lambda)$ on both sides of Eq. (5.7.20), recalling Eq. (5.7.17), we have

$$f(x) = (L_x^{\mathrm{T}}-\lambda)h(x) = \int_0^h dy(L_x^{\mathrm{T}}-\lambda)G(y,x;\lambda)f(y). \tag{5.7.21}$$

This is true if and only if

$$(L_x^{\mathrm{T}}-\lambda)G(y,x;\lambda) = \delta(x-y). \tag{5.7.22}$$

Multiplying $(\lambda-\lambda_n)$ on both sides of Eq. (5.7.22), and letting $\lambda\to\lambda_n$,

$$\lim_{\lambda\to\lambda_n}(\lambda-\lambda_n)(L_x^{\mathrm{T}}-\lambda)G(y,x;\lambda) = \lim_{\lambda\to\lambda_n}(\lambda-\lambda_n)\delta(x-y), \tag{5.7.23}$$

from which, we obtain

$$(L_x^{\mathrm{T}}-\lambda_n)R_n(y,x) = 0. \tag{5.7.24}$$

Since we know from Eq. (5.7.13),

$$R_n(y,x) \propto \psi_n(x)\phi_n(y), \tag{5.7.13b}$$

equation (5.7.24) indeed demonstrates that $\psi_n(x)$ is the eigenfunction of L_x^{T}, belonging to the eigenvalue λ_n as claimed in Eq. (5.7.15). Thus $R_n(x,y)$ is a product of the eigenfunctions of L_x and L_x^{T}, i.e., a product of $\phi_n(x)$ and $\psi_n(y)$. Incidentally, this also proves that the eigenvalues of L_x^{T} are the same as the eigenvalues of L_x.

Let us now analyze the second term in Eq. (5.7.11),

$$\frac{1}{2\pi i}\int_{-\infty}^{\lambda_B}\frac{G(x,y;\lambda'+i\varepsilon)-G(x,y;\lambda'-i\varepsilon)}{\lambda'-\lambda}\,d\lambda'.$$

In the limit $\varepsilon\to 0$, we have, from Eq. (5.7.3),

$$(L_x-\lambda)G(x,y;\lambda+i\varepsilon) = \delta(x-y), \tag{5.7.25a}$$

$$(L_x-\lambda)G(x,y;\lambda-i\varepsilon) = \delta(x-y). \tag{5.7.25b}$$

Taking the difference of the two expressions above, we have

$$(L_x - \lambda)\left[G(x,y;\lambda+i\varepsilon) - G(x,y;\lambda-i\varepsilon)\right] = 0. \tag{5.7.26}$$

Thus we conclude

$$G(x,y;\lambda+i\varepsilon) - G(x,y;\lambda-i\varepsilon) \propto \psi_\lambda(y)\phi_\lambda(x). \tag{5.7.27}$$

Hence we finally obtain, by choosing proper normalization for ψ_n and ϕ_n,

$$G(x,y;\lambda) = +\sum_n \frac{\psi_n(y)\phi_n(x)}{(\lambda_n - \lambda)} + \int_{-\infty}^{\lambda_B} \frac{d\lambda' \psi_{\lambda'}(y)\phi_{\lambda'}(x)}{(\lambda' - \lambda)}. \tag{5.7.28}$$

The first term in Eq. (5.7.28) represents a *discrete contribution* to $G(x,y;\lambda)$ from the *poles* at $\lambda = \lambda_n$, while the second term represents a *continuum contribution* from the *branch cut* starting at $\lambda = \lambda_B$ and extending to $-\infty$ along the negative real axis. Equation (5.7.28) is the generalization of the formula, Eq. (5.2.9), for the resolvent kernel $H(x,y;\lambda)$. Equation (5.7.28) is consistent with the assumption (5.7.8a). Setting $\lambda = 0$ in Eq. (5.7.28), we obtain

$$G(x,y) = \sum_n \frac{\psi_n(y)\phi_n(x)}{\lambda_n} + \int_{-\infty}^{\lambda_B} \frac{d\lambda' \psi_{\lambda'}(y)\phi_{\lambda'}(x)}{\lambda'}, \tag{5.7.29}$$

which is the generalization of the formula, Eq. (5.2.7), for the kernel $K(x,y)$.

We now anticipate that the completeness of the eigenfunctions will hold with minor modification to take care of the fact that L_x is non self-adjoint. In order to see this, we operate $(L_x - \lambda)$ on $G(x,y;\lambda)$ in Eq. (5.7.28).

$$\begin{aligned}(L_x - \lambda)G(x,y;\lambda) &= (L_x - \lambda)\sum_n \frac{\psi_n(y)\phi_n(x)}{(\lambda_n - \lambda)} \\ &\quad + (L_x - \lambda)\int_{-\infty}^{\lambda_B} \frac{d\lambda' \psi_{\lambda'}(y)\phi_{\lambda'}(x)}{(\lambda' - \lambda)},\end{aligned}$$

from which, we obtain

$$\delta(x-y) = \sum_n \psi_n(y)\phi_n(x) + \int_{-\infty}^{\lambda_B} d\lambda' \psi_{\lambda'}(y)\phi_{\lambda'}(x). \tag{5.7.30}$$

This is a *statement of the completeness of the eigenfunctions*; discrete eigenfunctions $\{\phi_n(x), \psi_n(y)\}$ and continuum eigenfunctions $\{\phi_{\lambda'}(x), \psi_{\lambda'}(y)\}$ together form a *complete set*.

We further anticipate that the *orthogonality* of the eigenfunctions will survive with minor modification. We consider first the following integral,

$$\begin{aligned}\int_0^h \psi_n(x)L_x\phi_m(x)\,dx = \lambda_m \int_0^h \psi_n(x)\phi_m(x)\,dx &= \int_0^h (L_x^{\mathrm{T}}\psi_n(x))\phi_m(x)\,dx \\ &= \lambda_n \int_0^h \psi_n(x)\phi_m(x)\,dx,\end{aligned}$$

from which, we obtain

$$(\lambda_n - \lambda_m)\int_0^h \psi_n(x)\phi_m(x)\,dx = 0. \tag{5.7.31}$$

Hence we have

$$\int_0^h \psi_n(x)\phi_m(x)\,dx = 0 \quad \text{when} \quad \lambda_n \neq \lambda_m. \tag{5.7.32}$$

Thus the eigenfunctions $\{\phi_m(x), \psi_n(x)\}$ belonging to the distinct eigenvalues are orthogonal to each other. Secondly we multiply $\psi_n(x)$ on the completeness relation (5.7.30) and integrate over x. Since $\lambda_n \neq \lambda'$, we have by Eq. (5.7.32),

$$\int_0^h dx\,\psi_n(x)\delta(x-y) = \sum_m \psi_m(y)\int_0^h dx\,\psi_n(x)\phi_m(x),$$

i.e.,

$$\psi_n(y) = \sum_m \psi_m(y)\int_0^h \psi_n(x)\phi_m(x)\,dx.$$

Then we must have

$$\int_0^h \psi_n(x)\phi_m(x)\,dx = \delta_{mn}. \tag{5.7.33}$$

Thirdly, we multiply the completeness relation (5.7.30) by $\psi_{\lambda''}(x)$ and integrate over x. Since $\lambda'' \neq \lambda_n$, we have by Eq. (5.7.32),

$$\psi_{\lambda''}(y) = \int_{-\infty}^{\lambda_B} d\lambda'\psi_{\lambda'}(y)\int_0^h dx\,\psi_{\lambda''}(x)\phi_{\lambda'}(x).$$

Then we must have

$$\int_0^h \psi_{\lambda''}(x)\phi_{\lambda'}(x)\,dx = \delta(\lambda' - \lambda''). \tag{5.7.34}$$

Thus the discrete eigenfunctions $\{\phi_m(x), \psi_n(x)\}$ and the continuum eigenfunctions $\{\phi_{\lambda'}(x), \psi_{\lambda''}(x)\}$ are normalized respectively as Eqs. (5.7.33) and (5.7.34).

The statement of the completeness of the discrete and continuum eigenfunctions (5.7.30) is often assumed at the outset in the standard treatment of Nonrelativistic Quantum Mechanics in the Schrödinger Picture. The discrete eigenfunctions are identified with the bound state wave functions, whereas the continuum eigenfunctions are identified with the scattering state wave functions. For some potential problems, there exist no scattering states. Simple harmonic oscillator potential is such an example.

5.8 Problems for Chapter 5

5.1. (Due to H. C.) Find an upper bound and a lower bound for the first eigenvalue of

$$K(x,y) = \begin{cases} (1-x)y, & 0 \le y \le x \le 1, \\ (1-y)x, & 0 \le x \le y \le 1. \end{cases}$$

5.2. (Due to H. C.) Consider the Bessel equation

$$(xu')' + \lambda x u = 0,$$

with the boundary conditions

$$u'(0) = u(1) = 0.$$

Transform this differential equation into an integral equation and find approximately the lowest eigenvalue.

5.3. Obtain an upper limit for the lowest eigenvalue of

$$\nabla^2 u + \lambda r u = 0,$$

where, in three dimensions,

$$0 < r < a, \quad \text{and} \quad u = 0 \quad \text{on} \quad r = a.$$

5.4. (Due to H. C.) Consider the Gaussian kernel $K(x,y)$ given by

$$K(x,y) = e^{-x^2-y^2}, \quad -\infty < x, y < +\infty.$$

a) Find the eigenvalues and the eigenfunctions of this kernel.

b) Verify the Hilbert–Schmidt expansion of this kernel.

c) By calculating A_2 and A_4, obtain the exact lowest eigenvalue.

d) Solve the integro-differential equation

$$\frac{\partial}{\partial t}\phi(x,t) = \int_{-\infty}^{+\infty} K(x,y)\phi(y,t)\,dy, \quad \text{with} \quad \phi(x,0) = f(x).$$

5.5. Show that the boundary condition (5.5.2) of the Sturm–Liouville system can be replaced by

$$\alpha_1\phi(0) + \alpha_2\phi'(0) = 0 \quad \text{and} \quad \beta_1\phi(h) + \beta_2\phi'(h) = 0$$

where α_1, α_2, β_1 and β_2 are some constants and the corresponding boundary condition (5.5.5) on $G(x,y)$ is replaced accordingly.

5.6. Verify the Hilbert–Schmidt expansion for the case of Direction 1 in Section 5.6, when the kernel $K(x, y)$ is *self-adjoint* and *square-integrable*.

5.7. (Due to H. C.) Reproduce all the results of Section 5.7 with the Green's function $G(x, y; \lambda)$ defined by

$$(L_x - \lambda r(x))G(x, y; \lambda) = \delta(x - y),$$

by the weight function,

$$r(x) > 0 \quad \text{on} \quad x \in [0, h].$$

5.8. (Due to H. C.) Consider the eigenvalue problem of the fourth-order ordinary differential equation of the form,

$$\left(\frac{d^4}{dx^4} + 1\right) \phi(x) = -\lambda x \phi(x), \quad 0 < x < 1,$$

with the boundary conditions,

$$\phi(0) = \phi'(0) = 0,$$

$$\phi(1) = \phi'(1) = 0.$$

Do the eigenfunctions form a complete set?

Hint: Check whether or not the differential operator L_x defined by

$$L_x \equiv -\left(\frac{d^4}{dx^4} + 1\right)$$

is self-adjoint under the specified boundary conditions.

5.9. (Due to D. M.) Show that, if $\tilde{\lambda}$ is an eigenvalue of the symmetric kernel $K(x, y)$, the inhomogeneous Fredholm integral equation of the second kind,

$$\phi(x) = f(x) + \tilde{\lambda} \int_a^b K(x, y)\phi(y)\, dy, \quad a \le x \le b,$$

has no solution, unless the inhomogeneous term $f(x)$ is orthogonal to all of the eigenfunctions $\phi(x)$ corresponding to the eigenvalue $\tilde{\lambda}$.

Hint: You may suppose that $\{\lambda_n\}$ is the sequence of the eigenvalues of the symmetric kernel $K(x, y)$ (ordered by increasing magnitude), with the corresponding eigenfunctions $\{\phi_n(x)\}$. You may assume that the eigenvalue $\tilde{\lambda}$ has multiplicity 1. The extension to the higher multiplicity k, $k \ge 2$, is immediate.

5.10. (Due to D. M.) Consider the integral equation,

$$\phi(x) = f(x) + \lambda \int_0^1 \sin^2[\pi(x-y)]\phi(y)\,dy, \qquad 0 \le x \le 1.$$

a) Solve the homogeneous equation by setting

$$f(x) = 0.$$

Determine all the eigenfunctions and the eigenvalues. What is the spectral representation of the kernel?

Hint: Express the kernel,

$$\sin^2[\pi(x-y)],$$

which is translationally invariant, periodic and symmetric, in terms of the powers of

$$\exp[\pi(x-y)].$$

b) Find the resolvent kernel of this equation.

c) Is there a solution to the given inhomogeneous integral equation when

$$f(x) = \exp[im\pi x], \quad m \text{ integer},$$

and $\lambda = 2$?

5.11. (Due to D. M.) Consider the kernel of the Fredholm integral equation of the second kind, which is given by

$$K(x,y) = \begin{cases} 3, & 0 \le y < x \le 1, \\ 2, & 0 \le x < y \le 1. \end{cases}$$

a) Find the eigenfunctions $\phi_n(x)$ and the corresponding eigenvalues λ_n of the kernel.

b) Is $K(x,y)$ symmetric? Determine the transposed kernel $K^T(x,y)$, and find its eigenfunctions $\psi_n(x)$ and the corresponding eigenvalues λ_n.

c) Show by an explicit calculation that any $\phi_n(x)$ is orthogonal to any $\psi_m(x)$ if $m \neq n$.

d) Derive the spectral representation of $K(x,y)$ in terms of $\phi_n(x)$ and $\psi_n(x)$.

5.12. (Due to D. M.) Consider the Fredholm integral equation of the second kind,

$$\phi(x) = f(x) + \lambda \int_0^1 \left[\frac{1}{2}(x+y) - \frac{1}{2}|x-y|\right] \phi(y)\,dy, \quad 0 \le x \le 1.$$

a) Find all non-trivial solutions $\phi_n(x)$ and corresponding eigenvalues λ_n for $f(x) \equiv 0$.

Hint: Obtain a differential equation for $\phi(x)$ with the suitable conditions for $\phi(x)$ and $\phi(x)'$.

b) For the original inhomogeneous equation ($f(x) \neq 0$), will the iteration series converge?

c) Evaluate the series $\sum_n \lambda_n^{-2}$ by using an appropriate integral.

5.13. If $|h| < 1$, find the non-trivial solutions of the homogeneous integral equation,

$$\phi(x) = \frac{\lambda}{2\pi} \int_{-\pi}^{\pi} \frac{1-h^2}{1-2h\cos(x-y)+h^2} \phi(y)\, dy.$$

Evaluate the corresponding values of the parameter λ.

Hint: The kernel of this equation is translationally invariant. Write $\cos(x-y)$ as a sum of two exponentials, express the kernel in terms of the complex variable $\zeta = \exp[i(y-x)]$, use the partial fractions, and then expand each fraction in powers of ζ.

5.14. If $|h| < 1$, find the solution of the integral equation,

$$f(x) = \frac{\lambda}{2\pi} \int_{-\pi}^{\pi} \frac{1-h^2}{1-2h\cos(x-y)+h^2} \phi(y)\, dy,$$

where $f(x)$ is the periodic and square-integrable known function.

6 Singular Integral Equations of Cauchy Type

6.1 Hilbert Problem

Suppose that, rather than the discontinuity $f^+(x) - f^-(x)$ across a branch cut, the ratio of the values on either side is known. That is, suppose that we wish to find $H(z)$ which has a branch cut on $[a, b]$ with

$$H^+(x) = R(x)H^-(x) \quad \text{on} \quad x \in [a, b]. \tag{6.1.1}$$

Solution. Take the logarithm of both sides of Eq. (6.1.1) to obtain

$$\ln H^+(x) - \ln H^-(x) = \ln R(x) \quad \text{on} \quad x \in [a, b]. \tag{6.1.2}$$

Define

$$h(z) \equiv \ln H(z), \quad \text{and} \quad r(x) \equiv \ln R(x). \tag{6.1.3}$$

Then we have

$$h^+(x) - h^-(x) = r(x) \quad \text{on} \quad x \in [a, b]. \tag{6.1.4}$$

Hence we can apply the discontinuity formula (1.8.5) to determine $h(z)$.

$$h(z) = \frac{1}{2\pi i} \int_a^b \frac{r(x)}{x - z}\, dx + g(z) \tag{6.1.5}$$

where $g(z)$ is an arbitrary function with no cut on $[a, b]$, i.e.,

$$H(z) = G(z) \exp\left[\frac{1}{2\pi i} \int_a^b \frac{\ln R(x)}{x - z}\, dx\right]. \tag{6.1.6}$$

We check this result. Supposing $G(z)$ to be continuous across the cut, we have from this result that

$$H^+(x) = G(x) \exp\left[\frac{1}{2\pi i} \mathrm{P} \int_a^b \frac{\ln R(y)}{y - x}\, dy + \frac{1}{2} \ln R(x)\right], \tag{6.1.7a}$$

$$H^-(x) = G(x) \exp\left[\frac{1}{2\pi i} \mathrm{P} \int_a^b \frac{\ln R(y)}{y - x}\, dy - \frac{1}{2} \ln R(x)\right]. \tag{6.1.7b}$$

Applied Mathematics in Theoretical Physics. Michio Masujima

ISBN: 3-527-40534-8

Here, we used the Plemelj formula (1.8.14a) and (1.8.14b) to take the limit as $z \to x$ from above and below. Dividing $H^+(x)$ by $H^-(x)$, we obtain

$$\frac{H^+(x)}{H^-(x)} = \exp(\ln R(x)) = R(x),$$

as expected.

Thus, in general, $G(z)$ may be of any form as long as it is continuous across the cut. In particular, we can choose $G(z)$ so as to make $H(z)$ have any desired behavior as $z \to a$ or b, by taking it of the form

$$G(z) = (a-z)^n (b-z)^m, \tag{6.1.8}$$

wherein $G(z) = \exp(g(z))$ has no branch points on $[a, b]$. In particular, we can always take $G(z)$ of the form above with n and m integers, and choose n and m to obtain a desired behavior in $H(z)$ as $z \to a$, $z \to b$ and $z \to \infty$.

We now address the inhomogeneous Hilbert problem.

Inhomogeneous Hilbert problem

Obtain $\Phi(z)$ which has a branch cut on $[a, b]$ with

$$\Phi^+(x) = R(x)\Phi^-(x) + f(x) \quad \text{on} \quad x \in [a, b]. \tag{6.1.9}$$

Solution. To solve the inhomogeneous problem, we begin with the solution to the corresponding *homogeneous Hilbert problem*,

$$H^+(x) = R(x)H^-(x) \quad \text{on} \quad x \in [a, b], \tag{6.1.10}$$

whose solution, we know, is given by

$$H(z) = G(z) \exp\left[\frac{1}{2\pi i}\int_a^b \frac{\ln R(x)}{x-z}\,dx\right]. \tag{6.1.11}$$

We then seek a solution of the form

$$\Phi(z) = H(z)K(z), \tag{6.1.12}$$

hence $K(z)$ must satisfy

$$H^+(x)K^+(x) = R(x)H^-(x)K^-(x) + f(x) \tag{6.1.13}$$

or, from Eq. (6.1.10), $H^+(x)K^+(x) = H^+(x)K^-(x) + f(x)$. Therefore, we have

$$K^+(x) - K^-(x) = \frac{f(x)}{H^+(x)}. \tag{6.1.14}$$

The problem reduces to an ordinary application of the discontinuity theorem provided that $H(z)$ is chosen so that $f(x)/H^+(x)$ is less singular than a pole at both endpoints. With this choice for $H(z)$, we then get

$$K(z) = \frac{1}{2\pi i}\int_a^b \frac{f(x)}{H^+(x)(x-z)}\,dx + L(z), \tag{6.1.15}$$

where $L(z)$ has no branch points on $[a, b]$. We thus find the *solution to the original inhomogeneous problem* to be

$$\Phi\,(z) = H\,(z)\,K\,(z)\,, \tag{6.1.16}$$

i.e.,

$$\Phi\,(z) = G\,(z)\exp\left[\frac{1}{2\pi i}\int_a^b \frac{\ln R(x)}{x-z}\,dx\right]\left\{\frac{1}{2\pi i}\int_a^b \frac{f(x)}{H^+(x)(x-z)}\,dx + L(z)\right\}, \tag{6.1.17}$$

with $L(z)$ arbitrary and $G(z)$ chosen such that $f(x)/H^+(x)$ is integrable on $[a, b]$. Neither $L(z)$ nor $G(z)$ can have branch points on $[a, b]$.

❑ **Example 6.1.** Find $\Phi\,(z)$, satisfying

$$\Phi^+(x) = -\Phi^-\,(x) + f(x) \quad \text{on} \quad x \in [a, b]. \tag{6.1.18}$$

Solution. Step 1. Solve the corresponding *homogeneous problem* first.

$$H^+(x) = -H^-(x). \tag{6.1.19}$$

Taking the logarithm on both sides to obtain

$$\ln H^+(x) = \ln H^-(x) + \ln(-1).$$

Setting

$$h(z) = \ln H(z),$$

we have

$$h^+(x) - h^-(x) = i\pi. \tag{6.1.20}$$

Then we can solve for $h(z)$ with the use of the discontinuity formula as

$$h(z) = \frac{1}{2\pi i}\int_a^b \frac{i\pi}{x-z}\,dx + g(z) = \frac{1}{2}\ln\left(\frac{b-z}{a-z}\right) + g(z),$$

whence we obtain the homogeneous solution to be

$$H(z) = G(z)\exp\left[\frac{1}{2}\ln\left(\frac{b-z}{a-z}\right)\right] = G(z)\sqrt{\frac{b-z}{a-z}}. \tag{6.1.21}$$

Suppose we choose

$$G(z) = \frac{1}{(b-z)},$$

so that

$$H(z) = \frac{1}{\sqrt{(b-z)(a-z)}}. \tag{6.1.22}$$

Therefore,

$$\frac{1}{H(z)} = \sqrt{(b-z)(a-z)}$$

does not blow up at $z = a$ or b. So, it is a good choice, since it does not make $f(x)/H^{+}(x)$ blow up any faster than $f(x)$ itself. Hence we take Eq. (6.1.22) for $H(z)$. Take that branch of $H(z)$ for which, on the upper bank of the cut, we have

$$H^{+}(x) = \frac{1}{i}\sqrt{(b-x)(x-a)},$$

where we now have the usual real square-root. To do this, we set

$$z = a + r_1 e^{i\theta_1} \quad \text{and} \quad z = b + r_2 e^{i\theta_2},$$

so that

$$\sqrt{(b-z)(z-a)} = \sqrt{r_1 r_2}\exp\left(i\frac{\theta_1+\theta_2}{2}\right).$$

On $x \in [a,b]$, we have $\sqrt{r_1 r_2} = \sqrt{(b-x)(x-a)}$ which requires, on the upper bank,

$$\exp\left(i\frac{\theta_1+\theta_2}{2}\right) = i.$$

We take, on the upper bank, $\theta_1 = 0$ and $\theta_2 = \pi$. Thus we have

$$H^{+}(x) = \frac{1}{i}\sqrt{(b-x)(x-a)} \quad \text{and} \quad H^{-}(x) = -\frac{1}{i}\sqrt{(b-x)(x-a)}. \tag{6.1.23}$$

Step 2. Now look at the *inhomogeneous problem*.

$$\Phi^{+}(x) = -\Phi^{-}(x) + f(x). \tag{6.1.24}$$

Look for the solution of the form

$$\Phi(z) = H(z)\,K(z). \tag{6.1.25}$$

Then we have

$$H^{+}(x)K^{+}(x) = -H^{-}(x)K^{-}(x) + f(x) = H^{+}(x)K^{-}(x) + f(x),$$

where Eq. (6.1.19) has been used. Dividing both sides of the equation above by $H^+(x)$, we have

$$K^+(x) - K^-(x) = \frac{f(x)}{H^+(x)} = i\sqrt{(b-x)(x-a)}f(x). \tag{6.1.26}$$

By applying the discontinuity formula, we obtain

$$K(z) = \frac{1}{2\pi i}\int_a^b \frac{i\sqrt{(b-x)(x-a)}f(x)}{x-z}\,dx + g(z), \tag{6.1.27}$$

where $g(z)$ is an arbitrary function with no cut on $[a,b]$. Thus the final solution is given by

$$\begin{aligned}\Phi(z) &= H(z)K(z)\\ &= \frac{1}{\sqrt{(b-z)(a-z)}}\left\{\frac{1}{2\pi}\int_a^b \frac{\sqrt{(b-x)(x-a)}f(x)}{x-z}\,dx + g(z)\right\},\end{aligned}$$

or equivalently

$$\Phi(z) = H(z)\left\{\frac{1}{2\pi i}\int_a^b \frac{f(x)}{H^+(x)(x-z)}\,dx + g(z)\right\}. \tag{6.1.28}$$

6.2 Cauchy Integral Equation of the First Kind

Consider now the *inhomogeneous Cauchy integral equation of the first kind*:

$$\frac{1}{\pi i}\mathrm{P}\int_a^b \frac{\phi(y)}{y-x}\,dy = f(x), \qquad a < x < b. \tag{6.2.1}$$

Define

$$\Phi(z) \equiv \frac{1}{2\pi i}\int_a^b \frac{\phi(y)}{y-z}\,dy \quad \text{for} \quad z \text{ not on } [a,b]. \tag{6.2.2}$$

As z approaches the branch cut from above and below, we find

$$\Phi^+(x) = \frac{1}{2\pi i}\mathrm{P}\int_a^b \frac{\phi(y)}{y-x}\,dy + \frac{1}{2}\phi(x), \tag{6.2.3a}$$

$$\Phi^-(x) = \frac{1}{2\pi i}\mathrm{P}\int_a^b \frac{\phi(y)}{y-x}\,dy - \frac{1}{2}\phi(x). \tag{6.2.3b}$$

Adding (6.2.3a) and (6.2.3b) and subtracting (6.2.3b) from (6.2.3a) results in

$$\Phi^+(x) + \Phi^-(x) = \frac{1}{\pi i}\mathrm{P}\int_a^b \frac{\phi(y)}{y-x}\,dy = f(x), \tag{6.2.4}$$

$$\Phi^+(x) - \Phi^-(x) = \phi(x). \tag{6.2.5}$$

Our strategy is as follows. Solve the first equation (6.2.4) (which is the inhomogeneous Hilbert problem) to find $\Phi(z)$. Then use the second equation (6.2.5) to obtain $\phi(x)$. We note that $\Phi(z)$, defined above, is analytic in $\mathbb{C} - [a, b]$, so it can have no other singularities. Also note that it behaves as A/z as $|z| \to \infty$.

Solution.

$$\Phi^+(x) = -\Phi^-(x) + f(x), \qquad a < x < b.$$

This is the inhomogeneous Hilbert problem which we solved in Section 6.2. We know

$$\Phi(z) = H(z)\left\{\frac{1}{2\pi i}\int_a^b \frac{f(x)}{H^+(x)(x-z)}\,dx + g(z)\right\}. \tag{6.2.6}$$

However, we can discover something about the form of $g(z)$ in this case by examining the behavior of $\Phi(z)$ as $|z| \to \infty$. By our original definition of $\Phi(z)$, we have

$$\Phi(z) \sim \frac{A}{z} \quad \text{as} \quad |z| \to \infty, \tag{6.2.7a}$$

with

$$A = -\frac{1}{2\pi i}\int_a^b \phi(y)\,dy. \tag{6.2.8}$$

If we examine the above solution, noting that

$$H(z) \sim \frac{1}{z} \quad \text{as} \quad |z| \to \infty,$$

we find that it has the asymptotic form

$$\Phi(z) \sim O\left(\frac{1}{z^2}\right) + \frac{g(z)}{z} \quad \text{as} \quad |z| \to \infty. \tag{6.2.7b}$$

Now, since $\Phi(z)$ is analytic everywhere away from the branch cut $[a, b]$, we conclude that $g(z)H(z)$ must also be analytic away from the cut $[a, b]$. Other singularities of $g(z)$ on $[a, b]$ can also be excluded. Hence $g(z)$ must be entire. Comparing the asymptotic forms Eqs. (6.2.7a) and (6.2.7b), we conclude $g(z) \to A$ as $|z| \to \infty$. Therefore, by Liouville's theorem, we must have $g(z) = A$ identically. Therefore, we have

$$\Phi(z) = H(z)\left\{\frac{1}{2\pi i}\int_a^b \frac{f(x)}{H^+(x)(x-z)}\,dx + A\right\}. \tag{6.2.9}$$

Thus we have

$$\Phi^+(x) = H^+(x)\left\{\frac{1}{2\pi i}\mathrm{P}\int_a^b \frac{f(y)}{H^+(y)(y-x)}\,dy + \frac{1}{2}\frac{f(x)}{H^+(x)} + A\right\}, \tag{6.2.10a}$$

$$\Phi^-(x) = -H^+(x)\left\{\frac{1}{2\pi i}\mathrm{P}\int_a^b \frac{f(y)}{H^+(y)(y-x)}\,dy - \frac{1}{2}\frac{f(x)}{H^+(x)} + A\right\}. \tag{6.2.10b}$$

Hence $\phi(x)$ is given by

$$\phi(x) = \Phi^+(x) - \Phi^-(x) = H^+(x)\frac{1}{\pi i}\mathrm{P}\int_a^b \frac{f(y)}{H^+(y)(y-x)}\,dy + 2AH^+(x). \quad (6.2.11)$$

Note that

$$H^+(x) = \frac{1}{\sqrt{(b-x)(a-x)}} = \frac{1}{i}\sqrt{(b-x)(x-a)}. \quad (6.2.12)$$

The second term on the right-hand side of $\phi(x)$ turns out to be the solution to the homogeneous problem. So A is arbitrary. In order to understand this point more fully, it makes sense to examine the homogeneous problem separately.

Consider the *homogeneous Cauchy integral equation of the first kind.*

$$\frac{1}{\pi i}\mathrm{P}\int_a^b \frac{\phi(y)}{y-x}\,dy = 0. \quad (6.2.13)$$

Define

$$\Phi(z) \equiv \frac{1}{2\pi i}\int_a^b \frac{\phi(y)}{y-z}\,dy.$$

$\Phi(z)$ is analytic everywhere away from $[a, b]$, and it has the asymptotic behavior

$$\Phi(z) \sim \frac{A}{z} \quad \text{as} \quad |z| \to \infty,$$

with

$$A = -\frac{1}{2\pi i}\int_a^b \phi(y)\,dy. \quad (6.2.14)$$

We have

$$\Phi^+(x) = \frac{1}{2\pi i}\mathrm{P}\int_a^b \frac{\phi(y)}{y-x}\,dy + \frac{1}{2}\phi(x),$$

$$\Phi^-(x) = \frac{1}{2\pi i}\mathrm{P}\int_a^b \frac{\phi(y)}{y-x}\,dy - \frac{1}{2}\phi(x),$$

from which, we obtain

$$\Phi^+(x) + \Phi^-(x) = 0, \quad (6.2.15a)$$

$$\Phi^+(x) - \Phi^-(x) = \phi(x). \quad (6.2.15b)$$

Solve

$$\Phi^+(x) = -\Phi^-(x),$$

which is the homogeneous Hilbert problem whose solution $\Phi(z)$ we already know from Section 6.2. Namely, we have

$$\Phi(z) = G(z)\sqrt{\frac{b-z}{a-z}}, \tag{6.2.16}$$

with $G(z)$ having no branch cuts on $[a, b]$. But in this case, we can determine the form of $G(z)$.

(i) We know that $\Phi(z)$ is analytic away from $[a, b]$, so $G(z)$ can only have singularities on $[a, b]$, but has no branch cuts there.

(ii) We know

$$\Phi(z) \sim \frac{A}{z} \quad \text{as} \quad |z| \to \infty,$$

so $G(z)$ must also behave as

$$G(z) \sim \frac{A}{z} \quad \text{as} \quad |z| \to \infty,$$

taking that branch of $\sqrt{(b-z)/(a-z)}$ which goes to $+1$ as $|z| \to \infty$.

(iii) $\Phi(z)$ can only be singular at the endpoints and even then not as bad as a pole, since it is of the form

$$\frac{1}{2\pi i}\int_a^b \frac{\phi(y)}{y-z}\,dy.$$

So, $G(z)$ may only be singular at the endpoints a or b.

This last condition (iii), together with the asymptotic form of $G(z)$ as $|z| \to \infty$, suggests functions of the form

$$G(z) = \frac{(a-z)^n}{(b-z)^{n+1}} \quad \text{or} \quad G(z) = \frac{(b-z)^n}{(a-z)^{n+1}}.$$

The only choice for which the resulting $\Phi(z)$ does not blow up as much as a pole at the two endpoints is

$$G(z) = \frac{A}{(b-z)} \quad \text{with} \quad A = \text{arbitrary constant}. \tag{6.2.17}$$

Therefore, the form of $\Phi(z)$ is determined to be

$$\Phi(z) = \frac{A}{\sqrt{(b-z)(a-z)}}. \tag{6.2.18}$$

Then the homogeneous solution is given by

$$\phi(x) = \Phi^+(x) - \Phi^-(x) = 2\Phi^+(x) = \frac{2A}{\sqrt{(b-x)(a-x)}} = 2AH^+(x), \tag{6.2.19}$$

where $H^+(x)$ is given by Eq. (6.2.12). To verify that any A works, recall Eq. (6.2.14),

$$A = -\frac{1}{2\pi i}\int_a^b \phi(y)\,dy.$$

Substituting $\phi(y)$ just obtained as Eq. (6.2.19), into the above expression for A, we have

$$\begin{aligned} A &= -\frac{2A}{2\pi i}\int_a^b \frac{dy}{\sqrt{(b-y)(a-y)}} = \frac{A}{\pi}\int_a^b \frac{dy}{\sqrt{(b-y)(y-a)}} \\ &= \frac{A}{\pi}\int_{-1}^{+1} \frac{d\xi}{\sqrt{1-\xi^2}} = A\,, \end{aligned}$$

which is true for any A.

6.3 Cauchy Integral Equation of the Second Kind

We now consider the *inhomogeneous Cauchy integral equation of the second kind*,

$$\phi(x) = f(x) + \frac{\lambda}{\pi}\mathrm{P}\int_0^1 \frac{\phi(y)}{y-x}\,dy, \qquad 0 < x < 1. \tag{6.3.1}$$

Define $\Phi(z)$

$$\Phi(z) \equiv \frac{1}{2\pi i}\int_0^1 \frac{\phi(y)}{y-z}\,dy. \tag{6.3.2}$$

The boundary values of $\Phi(z)$ as z approaches the cut $[0, 1]$ from above and below are given, respectively, by

$$\Phi^+(x) = \frac{1}{2\pi i}\mathrm{P}\int_0^1 \frac{\phi(y)}{y-x}\,dy + \frac{1}{2}\phi(x), \qquad 0 < x < 1,$$

$$\Phi^-(x) = \frac{1}{2\pi i}\mathrm{P}\int_0^1 \frac{\phi(y)}{y-x}\,dy - \frac{1}{2}\phi(x), \qquad 0 < x < 1,$$

so that

$$\Phi^+(x) - \Phi^-(x) = \phi(x), \qquad 0 < x < 1, \tag{6.3.3}$$

$$\Phi^+(x) + \Phi^-(x) = \frac{1}{\pi i}\mathrm{P}\int_0^1 \frac{\phi(y)}{y-x}\,dy, \qquad 0 < x < 1. \tag{6.3.4}$$

Equation (6.3.1) now reads

$$\Phi^+(x) - \Phi^-(x) = f(x) + i\lambda(\Phi^+(x) + \Phi^-(x)),$$

or,

$$\Phi^+(x) - \frac{1+i\lambda}{1-i\lambda}\Phi^-(x) = \frac{1}{1-i\lambda}f(x). \tag{6.3.5}$$

This we recognize as the *inhomogeneous Hilbert problem.*

Consider the case $\lambda > 0$ and set λ equal to

$$\lambda = \tan\pi\gamma, \qquad 0 < \gamma \leq \frac{1}{2}. \tag{6.3.6}$$

Then we have

$$\frac{1+i\lambda}{1-i\lambda} = e^{2\pi i\gamma}.$$

First, we solve an homogeneous problem;

$$H^+(x) = H^-(x)e^{2\pi i\gamma}, \qquad 0 < x < 1. \tag{6.3.7}$$

In terms of $h(z)$ defined by

$$H(z) = e^{h(z)},$$

we have

$$h^+(x) - h^-(x) = 2\pi i\gamma.$$

Hence we obtain

$$h(z) = \frac{1}{2\pi i}\int_0^1 \frac{2\pi i\gamma}{y-z}dz = \gamma\ln\left(\frac{1-z}{-z}\right),$$

and

$$H(z) = \left(\frac{1-z}{-z}\right)^\gamma, \qquad 0 < \gamma \leq \frac{1}{2}. \tag{6.3.8}$$

Since we can add any function with no cut on $[0, 1]$ to $h(z)$, we can multiply any function with no cut on $[0, 1]$ onto $H(z)$. By multiplying $e^{i\pi\gamma}/(1-z)$ on $H(z)$ above, we choose

$$H(z) = \frac{1}{(1-z)^{1-\gamma}z^\gamma}, \qquad 0 < \gamma \leq \frac{1}{2}. \tag{6.3.9}$$

Returning to the inhomogeneous Hilbert problem, Eq. (6.3.5), we write

$$\Phi(z) = H(z)G(z). \tag{6.3.10}$$

Then Eq. (6.3.5) reads

$$H^+(x)G^+(x) - H^+(x)G^-(x) = \frac{1}{1-i\lambda}f(x).$$

Dividing through the above equation by $H^+(x)$, we have

$$G^+(x) - G^-(x) = \frac{1}{1 - i\lambda} \frac{f(x)}{H^+(x)} = \frac{1}{1 - i\lambda} f(x)(1 - x)^{1-\gamma} x^{\gamma}. \tag{6.3.11}$$

Since, for our choice of $H(z)$, Eq. (6.3.9), we have

$$\frac{1}{H^+(0)} = \frac{1}{H^+(1)} = 0,$$

we did not bring in extra singular behavior at $x = 0$ and 1; rather we made $f(x)/H^+(x)$ better behaved than $f(x)$ at $x = 0$ and 1. By the discontinuity formula, we have

$$G(z) = \frac{1}{2\pi i} \frac{1}{1 - i\lambda} \int_0^1 \frac{f(y)(1 - y)^{1-\gamma} y^{\gamma}}{(y - z)} dy + g(z). \tag{6.3.12}$$

The integral on the right-hand side of Eq. (6.3.12) takes care of the discontinuity of $G(z)$ across the cut on $[0, 1]$ so that $g(z)$ does not have any cut on $[0, 1]$. We know, furthermore, that

(1) $G(z)$ is analytic in the cut z plane.

(2) As $z \to 0$, $G(z)$ is bounded by

$$|G(z)| < \left| \frac{1}{z^{\alpha}} \cdot z^{\gamma} \right|,$$

and similarly for $z \to 1$.

(3) As $|z| \to \infty$, $G(z)$ is bounded by constant. Thus we know from Eq. (6.3.12) that $g(z)$ is analytic everywhere and

$$g(z) \sim \text{constant} \quad \text{as} \quad |z| \to \infty.$$

By Liouville's theorem, we obtain

$$g(z) = \text{constant} = k. \tag{6.3.13}$$

Hence we obtain

$$\Phi(z) = \frac{1}{(1 - z)^{1-\gamma} z^{\gamma}} \frac{1}{2\pi i} \frac{1}{1 - i\lambda} \int_0^1 \frac{(1 - y)^{1-\gamma} y^{\gamma}}{y - z} f(y)\, dy + \frac{k}{(1 - z)^{1-\gamma} z^{\gamma}}. \tag{6.3.14}$$

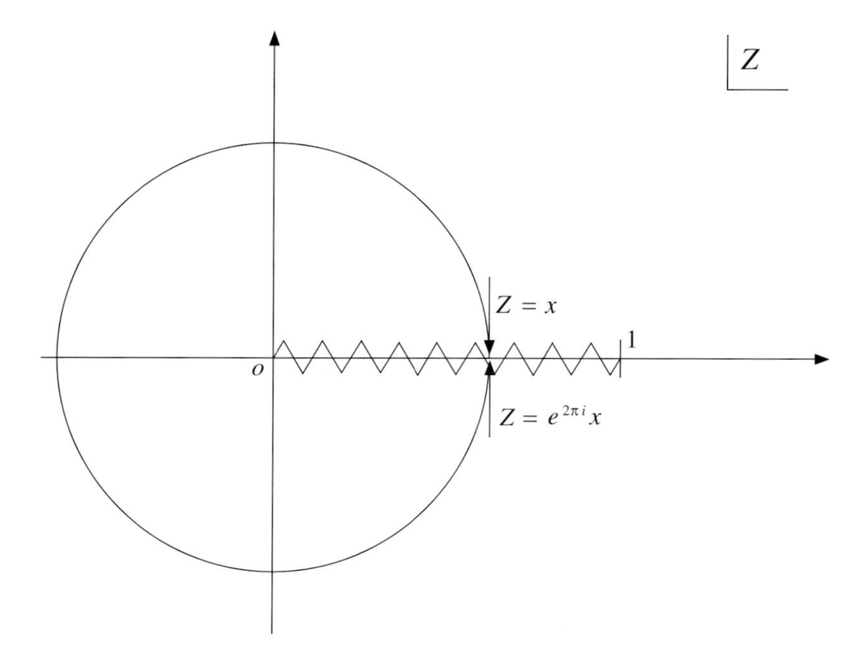

Fig. 6.1: The branch of $H(z)$ chosen for Eqs. (6.3.15a) and (6.3.15b).

Then by choosing the branch of $H(z)$ as indicated in Figure 6.1, we have from the Plemelj formula,

$$\begin{aligned}\Phi^+(x) = & \frac{1}{2\pi i}\frac{1}{1-i\lambda}\frac{1}{(1-x)^{1-\gamma}x^\gamma}\mathrm{P}\int_0^1 \frac{(1-y)^{1-\gamma}y^\gamma}{y-x}f(y)\,dy \\ & +\frac{1}{2}\frac{1}{1-i\lambda}f(x)+\frac{k}{(1-x)^{1-\gamma}x^\gamma},\end{aligned} \tag{6.3.15a}$$

$$\begin{aligned}\Phi^-(x) = & \frac{1}{2\pi i}\frac{1}{1-i\lambda}\frac{e^{-2\pi i\gamma}}{(1-x)^{1-\gamma}x^\gamma}\mathrm{P}\int_0^1 \frac{(1-y)^{1-\gamma}y^\gamma}{y-x}f(y)\,dy \\ & -\frac{1}{2}\frac{e^{-2\pi i\gamma}}{1-i\lambda}f(x)+\frac{ke^{-2\pi i\gamma}}{(1-x)^{1-\gamma}x^\gamma}.\end{aligned} \tag{6.3.15b}$$

By Eq. (6.3.3),

$$\begin{aligned}\phi(x) &= \Phi^+(x)-\Phi^-(x) \\ &= \frac{1}{2\pi i}\frac{1-e^{-2\pi i\gamma}}{1-i\lambda}\frac{1}{(1-x)^{1-\gamma}x^\gamma}\mathrm{P}\int_0^1 \frac{(1-y)^{1-\gamma}y^\gamma}{y-x}f(y)\,dy \\ &\quad +\frac{1}{2}\frac{1+e^{-2\pi i\gamma}}{1-i\lambda}f(x)+\frac{c}{(1-x)^{1-\gamma}x^\gamma}.\end{aligned} \tag{6.3.16}$$

Since we have

$$1-e^{-2\pi i\gamma}=\frac{2i\lambda}{1+i\lambda}, \qquad 1+e^{-2\pi i\gamma}=\frac{2}{1+i\lambda},$$

we finally obtain

$$\begin{aligned}\phi(x) &= \frac{1}{\pi}\frac{\lambda}{1+\lambda^2}\frac{1}{(1-x)^{1-\gamma}x^{\gamma}}\mathrm{P}\int_0^1 \frac{(1-y)^{1-\gamma}y^{\gamma}}{y-x}f(y)\,dy \\ &\quad + \frac{1}{1+\lambda^2}f(x) + \frac{c}{(1-x)^{1-\gamma}x^{\gamma}},\end{aligned} \tag{6.3.17}$$

where c is an arbitrary constant. We note that the solution (6.3.17) involves one arbitrary constant and the spectrum of the eigenvalue λ of Eq. (6.3.1) with $f(x) \equiv 0$ is continuous. We note that the homogeneous solution of Eq. (6.3.1) with $f(x) \equiv 0$ comes from the entire function $g(z)$.

6.4 Carleman Integral Equation

Consider the *inhomogeneous Carleman integral equation*

$$a(x)\phi(x) = f(x) + \lambda\mathrm{P}\int_{-1}^{+1}\frac{\phi(y)}{y-x}\,dy, \qquad -1 < x < 1, \tag{6.4.1}$$

which has a *Cauchy kernel* and is a *generalization of the Cauchy integral equation of the second kind*. Assume that λ is real and that $f(x)$ and $a(x)$ are prescribed real functions. Without loss of generality, we can take $\lambda > 0$ (otherwise, we just change the sign of $a(x)$ and $f(x)$).

As usual, define

$$\Phi(z) \equiv \frac{1}{2\pi i}\int_{-1}^{+1}\frac{\phi(y)}{y-z}\,dy. \tag{6.4.2}$$

Then $\Phi(z)$ is analytic in $\mathbb{C} - [-1, 1]$. The asymptotic behavior of $\Phi(z)$ as $|z| \to \infty$ is given by $\Phi(z) \sim \frac{A}{z}$ as $|z| \to \infty$, with A given by

$$A = -\frac{1}{2\pi i}\int_{-1}^{+1}\phi(y)\,dy.$$

The endpoint behavior of $\Phi(z)$ is that $\Phi(z)$ is less singular than a pole as $z \to \pm 1$.

The boundary values of $\Phi(z)$ as z approaches the cut from above and below are given by Plemelj formulas, namely

$$\Phi^{+}(x) = \frac{1}{2\pi i}\mathrm{P}\int_{-1}^{+1}\frac{\phi(y)}{y-x}\,dy + \frac{1}{2}\phi(x) \quad \text{for} \quad -1 < x < 1, \tag{6.4.3a}$$

$$\Phi^{-}(x) = \frac{1}{2\pi i}\mathrm{P}\int_{-1}^{+1}\frac{\phi(y)}{y-x}\,dy - \frac{1}{2}\phi(x) \quad \text{for} \quad -1 < x < 1. \tag{6.4.3b}$$

Adding (6.4.3a) and (6.4.3b), and subtracting (6.4.3b) from (6.4.3a), we obtain

$$\Phi^{+}(x) + \Phi^{-}(x) = \frac{1}{\pi i}P\int_{-1}^{+1}\frac{\phi(y)}{y-x}\,dy, \qquad \Phi^{+}(x) - \Phi^{-}(x) = \phi(x). \tag{6.4.4}$$

Using the Carleman integral equation (6.4.1), we rewrite Eq. (6.4.4) as

$$[a(x) - \lambda\pi i]\,\Phi^+(x) = [a(x) + \lambda\pi i]\,\Phi^-(x) + f(x),$$

which is an *inhomogeneous Hilbert problem*,

$$\Phi^+(x) = R(x)\Phi^-(x) + \frac{f(x)}{a(x) - \lambda\pi i}, \tag{6.4.5a}$$

with

$$R(x) \equiv \frac{a(x) + \lambda\pi i}{a(x) - \lambda\pi i}. \tag{6.4.5b}$$

Since we need to take $\ln R(x)$, we represent $R(x)$ in the polar form. Recalling that

$$x + iy = \sqrt{x^2 + y^2}e^{i\tan^{-1}(y/x)},$$

we have

$$R(x) = \exp\left[2i\tan^{-1}\left(\frac{\lambda\pi}{a(x)}\right)\right]. \tag{6.4.6}$$

Since the function $a(x) + \lambda\pi i$ is always in the upper half-plane, $\tan^{-1}\left(\frac{\lambda\pi}{a(x)}\right)$ is in the range $(0, \pi)$. Define

$$\theta(x) \equiv \tan^{-1}\left(\frac{\lambda\pi}{a(x)}\right). \tag{6.4.7}$$

Then we have

$$R(x) = e^{2i\theta(x)} \quad \text{with} \quad 0 < \theta(x) < \pi,$$

and we obtain

$$\Phi^+(x) = e^{2i\theta(x)}\Phi^-(x) + \frac{f(x)}{a(x) - \lambda\pi i}. \tag{6.4.8}$$

Homogeneous problem: We first solve the homogeneous problem. Namely, we solve

$$H^+(x) = e^{2i\theta(x)}H^-(x). \tag{6.4.9}$$

Taking the logarithm of both sides, we have

$$\ln H^+(x) - \ln H^-(x) = 2i\theta(x).$$

By the, now familiar, method of solving the homogeneous Hilbert problem, we have

$$\ln H(z) = \frac{1}{\pi}\int_{-1}^{+1}\frac{\theta(x)\,dx}{x - z} + g(z),$$

or,

$$H(z) = G(z)\exp\left[\frac{1}{\pi}\int_{-1}^{+1}\frac{\theta(x)\,dx}{x - z}\right]. \tag{6.4.10}$$

Behavior at the endpoints: Since $\theta(x)$ is bounded at the endpoints, the integral in the square bracket above has logarithmic singularities as $z \to \pm 1$.

$$\frac{1}{\pi}\int_{-1}^{+1}\frac{\theta(x)\,dx}{x-z} \sim \frac{-\theta(-1)}{\pi}\ln(-1-z) \quad \text{as} \quad z \to -1,$$
$$\exp\left[\frac{1}{\pi}\int_{-1}^{+1}\frac{\theta(x)\,dx}{x-z}\right] \sim (-1-z)^{-\theta(-1)/\pi} \quad \text{as} \quad z \to -1, \tag{6.4.11a}$$

and

$$\frac{1}{\pi}\int_{-1}^{+1}\frac{\theta(x)\,dx}{x-z} \sim \frac{\theta(1)}{\pi}\ln(1-z) \quad \text{as} \quad z \to +1,$$
$$\exp\left[\frac{1}{\pi}\int_{-1}^{+1}\frac{\theta(x)\,dx}{x-z}\right] \sim (1-z)^{\theta(1)/\pi} \quad \text{as} \quad z \to +1. \tag{6.4.11b}$$

Since $0 < \theta(x) < \pi$, we have $0 < \frac{\theta(1)}{\pi} < 1$, and $-1 < \frac{-\theta(-1)}{\pi} < 0$. So, disregarding $G(z)$, $H(z)$ could be singular as $z \to -1$ so that $1/H(z) \to 0$ as $z \to -1$, but $H(z)$ could be zero as $z \to 1$ so that $1/H(z)$ becomes singular as $z \to +1$. Since we wish to ensure that $1/H^{+}(x)$ is not singular, in order to facilitate the solution of the inhomogeneous problem, we choose $G(z)$ so that $1/H(z)$ is not singular as $z \to +1$. A good choice for $G(z)$ is $G(z) = 1/(1-z)$.

Then we obtain

$$H(z) = \frac{1}{1-z}\exp\left[\frac{1}{\pi}\int_{-1}^{+1}\frac{\theta(x)\,dx}{x-z}\right]. \tag{6.4.12}$$

With this choice,

$$H(z) \sim \frac{1}{(-1-z)^{\theta(-1)/\pi}} \quad \text{as} \quad z \to -1 \quad \left(0 < \frac{\theta(-1)}{\pi} < 1\right), \tag{6.4.13a}$$

$$H(z) \sim \frac{1}{(1-z)^{1-\theta(1)/\pi}} \quad \text{as} \quad z \to +1 \quad \left(0 < 1 - \frac{\theta(1)}{\pi} < 1\right), \tag{6.4.13b}$$

so that $1/H^{+}(x)$ is not singular as $x \to \pm 1$.

Inhomogeneous problem: We now solve the inhomogeneous problem.

$$\Phi^{+}(x) = e^{2i\theta(x)}\Phi^{-}(x) + \frac{f(x)}{a(x) - \lambda\pi i}. \tag{6.4.14}$$

We seek the solution of the form

$$\Phi(z) = H(z)K(z). \tag{6.4.15}$$

We obtain

$$H^{+}(x)K^{+}(x) = H^{+}(x)K^{-}(x) + \frac{f(x)}{a(x) - \lambda\pi i}.$$

We then have

$$K^+(x) - K^-(x) = \frac{f(x)}{H^+(x)(a(x) - \lambda\pi i)}, \tag{6.4.16}$$

from which we immediately obtain, by the discontinuity theorem,

$$K(z) = \frac{1}{2\pi i}\int_{-1}^{+1} \frac{f(x)}{H^+(x)(a(x) - \lambda\pi i)(x - z)}\,dx + L(z). \tag{6.4.17}$$

Hence we obtain

$$\Phi(z) = H(z)\left[\frac{1}{2\pi i}\int_{-1}^{+1} \frac{f(x)}{H^+(x)(a(x) - \lambda\pi i)(x - z)}\,dx\right] + H(z)L(z). \tag{6.4.18}$$

Determination of $L(z)$ from its behavior at infinity: We know that $\Phi(z)$ is analytic in $\mathbb{C} - [-1, 1]$. Our choice of $H(z)$ is also analytic in $\mathbb{C} - [-1, 1]$. So, $L(z)$ is analytic in $\mathbb{C} - [-1, 1]$. Furthermore, $L(z)$ can have no branch cut in $[-1, 1]$, so it can, at worst, have poles on $[-1, 1]$. But the poles cannot occur inside $(-1, 1)$ since $\Phi(z)$ has no poles there. Hence, at worst, $L(z)$ has poles at the endpoints $z \to \pm 1$. However, we know that $\Phi(z)$ is less singular than a pole as $z \to \pm 1$. This is true of the first term in (6.4.18), which can be shown to be as singular as $f(x)$ at the endpoints (the contributions from $H(z)$ and $1/H^+(x)$ cancelling). Hence $H(z)L(z)$ must be less singular than a pole as $z \to \pm 1$. This implies that $L(z)$ cannot have poles at ± 1. Therefore, $L(z)$ is entire. We know

$$\Phi(z) \sim \frac{A}{z}, \quad H(z) \sim -\frac{1}{z} \quad \text{as} \quad |z| \to \infty.$$

From our solution above, we have

$$\Phi(z) \sim O\left(\frac{1}{z^2}\right) - \frac{L(z)}{z} \quad \text{as} \quad |z| \to \infty.$$

Thus we conclude that $L(z) \sim -A$ as $|z| \to \infty$. By Liouville's theorem, we must have

$$L(z) = -A. \tag{6.4.19}$$

Then our solution for $\Phi(z)$ is given by

$$\Phi(z) = H(z)\left[\frac{1}{2\pi i}\int_{-1}^{+1} \frac{f(x)}{H^+(x)(a(x) - \lambda\pi i)(x - z)}\,dx\right] - AH(z), \tag{6.4.20}$$

with

$$H(z) = \frac{1}{1 - z}\exp\left[\frac{1}{\pi}\int_{-1}^{+1} \frac{\theta(x)\,dx}{x - z}\right]. \tag{6.4.21}$$

With that choice for $H(z)$, $H^+(x)$ can be chosen as

$$H^+(x) = \frac{1}{1 - x}\exp\left[\frac{1}{\pi}\mathrm{P}\int_{-1}^{+1} \frac{\theta(y)}{y - x}\,dy + i\theta(x)\right]. \tag{6.4.22}$$

To obtain $\phi(x)$, we use

$$\Phi^+(x) - \Phi^-(x) = \phi(x).$$

From $\Phi(z)$ obtained above, by the Plemelj formula, we have

$$\Phi^+(x) = H^+(x)\left[\frac{1}{2\pi i}\mathrm{P}\int_{-1}^{+1}\frac{f(y)\,dy}{H^+(y)(a(y)-\lambda\pi i)(y-x)} + \frac{1}{2}\frac{f(x)}{H^+(x)(a(x)-\lambda\pi i)}\right] - AH^+(x), \quad (6.4.23a)$$

$$\Phi^-(x) = H^-(x)\left[\frac{1}{2\pi i}\mathrm{P}\int_{-1}^{+1}\frac{f(y)\,dy}{H^+(y)(a(y)-\lambda\pi i)(y-x)} - \frac{1}{2}\frac{f(x)}{H^+(x)(a(x)-\lambda\pi i)}\right] - AH^-(x). \quad (6.4.23b)$$

Then our solution is given by

$$\phi(x) = H^+(x)\left[1-e^{-2i\theta(x)}\right]\left[\frac{1}{2\pi i}\mathrm{P}\int_{-1}^{+1}\frac{f(y)\,dy}{H^+(y)(a(y)-\lambda\pi i)(y-x)}\right] + \left[1+e^{-2i\theta(x)}\right]\frac{f(x)}{2(a(x)-\lambda\pi i)} - AH^+(x)\left[1-e^{-2i\theta(x)}\right]. \quad (6.4.24)$$

Simplification: We set

$$H^+(x) = \frac{1}{1-x}\exp\left[\frac{1}{\pi}\mathrm{P}\int_{-1}^{+1}\frac{\theta(y)\,dy}{y-x}\right]e^{i\theta(x)} \equiv B(x)e^{i\theta(x)}, \quad (6.4.25)$$

where we define

$$B(x) \equiv \frac{1}{1-x}\exp\left[\frac{1}{\pi}\mathrm{P}\int_{-1}^{+1}\frac{\theta(y)\,dy}{y-x}\right]. \quad (6.4.26)$$

Also, using the definition of $\theta(x)$, Eq. (6.4.7), we note

$$e^{i\theta(x)}(1-e^{-2i\theta(x)}) = 2\pi i\lambda/\sqrt{a^2(x)+\lambda^2\pi^2},$$

$$e^{i\theta(x)}(a(x)-\lambda\pi i) = \sqrt{a^2(x)+\lambda^2\pi^2},$$

$$\frac{1+e^{-2i\theta(x)}}{2(a(x)-\lambda\pi i)} = \frac{a(x)}{a^2(x)+\lambda^2\pi^2}.$$

Thus the final form of the solution is given by

$$\begin{aligned}\phi(x) = &\frac{\lambda B(x)}{\sqrt{a^2(x)+\lambda^2\pi^2}}\mathrm{P}\int_{-1}^{+1}\frac{f(y)\,dy}{B(y)\sqrt{a^2(y)+\lambda^2\pi^2}(y-x)}\\ &+\frac{a(x)f(x)}{a^2(x)+\lambda^2\pi^2}+C\frac{B(x)}{\sqrt{a^2(x)+\lambda^2\pi^2}},\end{aligned} \tag{6.4.27}$$

with $B(x)$ defined by Eq. (6.4.26).

We remark that when $f(x)=0$, i.e., for the *homogeneous Carleman integral equation*,

$$a(x)\phi(x) = \lambda\mathrm{P}\int_{-1}^{+1}\frac{\phi(y)}{y-x}\,dy,$$

then by setting $f(x)=f(y)\equiv 0$ in Eq. (6.4.27), the homogeneous solution is

$$\phi_{\mathrm{H}}(x) = C\frac{B(x)}{\sqrt{a^2(x)+\lambda^2\pi^2}},$$

being well defined for all $\lambda > 0$. Thus there is a *continuous spectrum of eigenvalues* for the homogeneous Carleman integral equation.

Setting $a(x)\equiv 1$, the solution (6.4.27) to the inhomogeneous Carleman integral equation of the second kind reduces to the solution (6.3.17) for the inhomogeneous Cauchy integral equation of the second kind, because the latter is the special case ($a(x)\equiv 1$) of the former.

6.5 Dispersion Relations

Dispersion relations in classical electrodynamics are closely related to the Cauchy integral equations.

Suppose that the complex function $f(z)$ is analytic in the upper half-plane, $\operatorname{Im} z > 0$, and that $f(z)\to 0$ as $|z|\to\infty$. We let its boundary value on the real x-axis in the complex z plane to be given by

$$F(x) = \lim_{\varepsilon\to 0} f(x+i\varepsilon), \quad \text{with} \quad \varepsilon = \text{positive infinitesimal}. \tag{6.5.1}$$

From the Cauchy integral formula,

$$f(z) = \frac{1}{2\pi i}\int_{C_+}\frac{f(\varsigma)}{\varsigma - z}\,d\varsigma,$$

where C_+ is the infinite semicircle in the upper half-plane, we immediately obtain

$$f(z) = \frac{1}{2\pi i}\int_{\text{real axis}}\frac{f(\varsigma)}{\varsigma - z}\,d\varsigma. \tag{6.5.2}$$

From Eqs. (6.5.1) and (6.5.2), we obtain

$$2\pi i F(x) = \lim_{\varepsilon\to 0}\int_{-\infty}^{\infty}\frac{f(x')}{x'-x-i\varepsilon}\,dx' = \mathrm{P}\int_{-\infty}^{\infty}\frac{F(x')}{x'-x}\,dx' + \pi i F(x).$$

Thus we have

$$F(x) = \frac{1}{\pi i} \mathrm{P} \int_{-\infty}^{\infty} \frac{F(x')}{x' - x} \, dx'. \tag{6.5.3}$$

Equating the real part and the imaginary part of Eq. (6.5.3), we obtain the following dispersion relations,

$$\operatorname{Re} F(x) = \frac{1}{\pi} \mathrm{P} \int_{-\infty}^{\infty} \frac{\operatorname{Im} F(x')}{x' - x} \, dx', \quad \operatorname{Im} F(x) = -\frac{1}{\pi} \mathrm{P} \int_{-\infty}^{\infty} \frac{\operatorname{Re} F(x')}{x' - x} \, dx'. \tag{6.5.4}$$

When $f(z)$ has an even symmetry,

$$f(-z) = f^*(z^*), \tag{6.5.5}$$

we write

$$\begin{aligned}
F(x) &= \frac{1}{\pi i} \mathrm{P} \int_{-\infty}^{\infty} \frac{F(x')}{x' - x} \, dx' = \frac{1}{\pi i} \mathrm{P} \int_{\infty}^{0} \frac{F^*(x')}{x' + x} \, dx' + \frac{1}{\pi i} \mathrm{P} \int_{0}^{\infty} \frac{F(x')}{x' - x} \, dx' \\
&= -\frac{1}{\pi i} \mathrm{P} \int_{0}^{\infty} \frac{\operatorname{Re} F(x') - i \operatorname{Im} F(x')}{x' + x} \, dx' + \frac{1}{\pi i} \mathrm{P} \int_{0}^{\infty} \frac{\operatorname{Re} F(x') + i \operatorname{Im} F(x')}{x' - x} \, dx' \\
&= \frac{i}{\pi} \mathrm{P} \int_{0}^{\infty} \left(\frac{\operatorname{Re} F(x')}{x' + x} - \frac{\operatorname{Re} F(x')}{x' - x} \right) dx' + \frac{1}{\pi} \mathrm{P} \int_{0}^{\infty} \left(\frac{\operatorname{Im} F(x')}{x' + x} + \frac{\operatorname{Im} F(x')}{x' - x} \right) dx' \\
&= \frac{-2i}{\pi} \mathrm{P} \int_{0}^{\infty} \frac{x \operatorname{Re} F(x')}{x'^2 - x^2} \, dx' + \frac{2}{\pi} \mathrm{P} \int_{0}^{\infty} \frac{x' \operatorname{Im} F(x')}{x'^2 - x^2} \, dx'.
\end{aligned}$$

Thus, we obtain the following dispersion relations,

$$\operatorname{Re} F(x) = \frac{2}{\pi} \mathrm{P} \int_{0}^{\infty} \frac{x' \operatorname{Im} F(x')}{x'^2 - x^2} \, dx', \quad \operatorname{Im} F(x) = -\frac{2}{\pi} \mathrm{P} \int_{0}^{\infty} \frac{x \operatorname{Re} F(x')}{x'^2 - x^2} \, dx'. \tag{6.5.6}$$

When $f(z)$ has an odd symmetry,

$$f(-z) = -f^*(z^*), \tag{6.5.7}$$

we write

$$\begin{aligned}
F(x) &= \frac{1}{\pi i} \mathrm{P} \int_{-\infty}^{\infty} \frac{F(x')}{x' - x} \, dx' = -\frac{1}{\pi i} \mathrm{P} \int_{\infty}^{0} \frac{F^*(x')}{x' + x} \, dx' + \frac{1}{\pi i} \mathrm{P} \int_{0}^{\infty} \frac{F(x')}{x' - x} \, dx' \\
&= \frac{1}{\pi i} \mathrm{P} \int_{0}^{\infty} \frac{\operatorname{Re} F(x') - i \operatorname{Im} F(x')}{x' + x} \, dx' + \frac{1}{\pi i} \mathrm{P} \int_{0}^{\infty} \frac{\operatorname{Re} F(x') + i \operatorname{Im} F(x')}{x' - x} \, dx' \\
&= \frac{-i}{\pi} \mathrm{P} \int_{0}^{\infty} \left(\frac{\operatorname{Re} F(x')}{x' + x} + \frac{\operatorname{Re} F(x')}{x' - x} \right) dx' - \frac{1}{\pi} \mathrm{P} \int_{0}^{\infty} \left(\frac{\operatorname{Im} F(x')}{x' + x} - \frac{\operatorname{Im} F(x')}{x' - x} \right) dx' \\
&= \frac{-2i}{\pi} \mathrm{P} \int_{0}^{\infty} \frac{x' \operatorname{Re} F(x')}{x'^2 - x^2} \, dx' + \frac{2}{\pi} \mathrm{P} \int_{0}^{\infty} \frac{x \operatorname{Im} F(x')}{x'^2 - x^2} \, dx'.
\end{aligned}$$

Thus, we obtain the following dispersion relations,

$$\operatorname{Re} F(x) = \frac{2}{\pi} \mathrm{P} \int_0^\infty \frac{x \operatorname{Im} F(x')}{x'^2 - x^2} dx', \quad \operatorname{Im} F(x) = -\frac{2}{\pi} \mathrm{P} \int_0^\infty \frac{x' \operatorname{Re} F(x')}{x'^2 - x^2} dx'. \quad (6.5.8)$$

These dispersion relations were derived independently by H.A. Kramers in 1927 and R. de L. Kronig in 1926 for the X-ray dispersion and the optical dispersion. Kramers-Kronig dispersion relations are of very general validity and only depend on the assumption of the causality. The analyticity of $f(z)$ assumed at the outset is identical to the requirement of the causality.

In the mid 1950s, these dispersion relations were derived from quantum field theory and applied to strong interaction physics, where the requirement for the causality and the unitarity of the S matrix are mandatory.

The application of the covariant perturbation theory to strong interaction physics was impossible due to the large coupling constant, despite the fact that the pseudo-scalar meson theory is renormalizable by power counting.

For some time, the dispersion-theoretic approach to strong interaction physics was the only realistic approach which provided many sum rules. To cite a few, we have the Goldberger–Treiman relation, the Goldberger–Miyazawa–Oehme formula and the Adler–Weisberger sum rule.

In the dispersion-theoretic approach to strong interaction physics, the experimentally observed data were directly used in the sum rules.

The situation changed dramatically in the early 1970s when the quantum field theory of strong interaction physics (quantum chromodynamics, QCD for short) was invented by the use of the asymptotically free non-Abelian gauge field theory.

We now present two examples of the integral equation in the dispersion theory in quantum mechanics.

❑ **Example 6.2.** Solve

$$F(x) = \frac{1}{\pi} \int \frac{|F(x')|^2 h(x')}{x' - x - i\varepsilon} dx', \quad (6.5.9)$$

where $h(x)$ is a given real function.

Solution. We set

$$f(z) = \frac{1}{\pi} \int \frac{g(x')}{x' - z} dx' \quad \text{with} \quad g(x) = |F(x)|^2 h(x). \quad (6.5.10)$$

Since $g(x)$ is real, it is the imaginary part of $F(x)$,

$$\operatorname{Im} F(x) = |F(x)|^2 h(x). \quad (6.5.11)$$

Assuming that $f(z)$ never vanishes anywhere, we set

$$\phi(z) = \frac{1}{f(z)}. \quad (6.5.12)$$

We compute the discontinuity,

$$\phi(x+i\varepsilon)-\phi(x-i\varepsilon)=2i\,\mathrm{Im}\,\phi(x+i\varepsilon)=-2i\frac{\mathrm{Im}\,f(x+i\varepsilon)}{|f(x+i\varepsilon)|^2}=-2ih(x). \quad (6.5.13)$$

Thus we have

$$\phi(z)=-\frac{1}{\pi}\int\frac{h(x')}{x'-z}\,dx', \quad (6.5.14)$$

and

$$F(x)=f(x+i\varepsilon)=\frac{1}{\phi(x+i\varepsilon)}=-\left[\frac{1}{\pi}\int\frac{h(x')}{x'-x-i\varepsilon}\,dx'\right]^{-1}, \quad (6.5.15)$$

provided that $\phi(z)$ never vanishes anywhere.
This example originates from the unitarity of the elastic scattering amplitude in the potential scattering problem in quantum mechanics. The unitarity of the elastic scattering amplitude often requires $F(x)$ to be of the form,

$$F(x)=\frac{\exp[i\delta(x)]\sin\delta(x)}{h(x)} \quad \text{with} \quad \delta(x)\text{ and }h(x) \quad \text{real}, \quad (6.5.16)$$

where

$$\mathrm{Im}\,F(x)=|F(x)|^2\,h(x). \quad (6.5.17)$$

In the three-dimensional scattering problem in quantum mechanics, we can obtain the dispersion relations for the scattering amplitude in the forward direction. We shall now address this problem for the spherically symmetric potential in three dimensions.

❑ **Example 6.3.** Potential scattering from the spherically symmetric potential in three dimensions in quantum mechanics.

The Schrödinger equation for the spherically symmetric potential in three dimensions is given by

$$\vec{\nabla}^2\psi(\vec{r})+k^2\psi(\vec{r})-U(\vec{r})\psi(\vec{r})=0, \quad (6.5.18)$$

where $U(\vec{r})$ is the potential $V(\vec{r})$ multiplied by $2m/\hbar^2$. Using the Green's function $\exp[ikR]/4\pi R$ for the Helmholtz equation, we obtain

$$\psi(\vec{r})=\exp[i\vec{k}_0\vec{r}]-\frac{1}{4\pi}\int\frac{\exp[ik\,|\vec{r}-\vec{r}_1|]}{|\vec{r}-\vec{r}_1|}U(\vec{r}_1)\psi(\vec{r}_1)\,d^3\vec{r}_1, \quad (6.5.19)$$

where $\exp[i\vec{k}_0\vec{r}]$ is the incident wave with $\vec{k}_0$ along the z-axis in the spherical polar coordinate. In the region with $|\vec{r}|\gg 1$, the scattering amplitude $f(\theta,k)$ is obtained as

$$f(\theta,k)=-\frac{1}{4\pi}\int\exp[-i\vec{k}_s\vec{r}_1]U(\vec{r}_1)\psi(\vec{r}_1)d^3\vec{r}_1, \quad (6.5.20)$$

where θ is the polar angle between $\vec{r}$ and the z-axis, and $\vec{k}_s$ is in the direction of $\vec{r}$ with magnitude $\left|\vec{k}_s\right| = k$.

Solving Eq. (6.5.18) for $\psi(\vec{r})$ iteratively for the case of the weak potential, we have

$$f(\theta, k) = \sum f_n(\theta, k),$$

where

$$f_n(\theta, k) = \left(\frac{-1}{4\pi}\right)^n \int \cdots \int d^3\vec{r}_1 \cdots d^3\vec{r}_n \frac{U(\vec{r}_1)U(\vec{r}_2)\cdots U(\vec{r}_n)}{|\vec{r} - \vec{r}_1|\,|\vec{r}_1 - \vec{r}_2| \cdots |\vec{r}_{n-1} - \vec{r}_n|} \times \exp[-i\vec{k}_s\vec{r}_1 + ik\,|\vec{r}_1 - \vec{r}_2| + \cdots + ik\,|\vec{r}_{n-1} - \vec{r}_n| + i\vec{k}_0\vec{r}_n], \tag{6.5.21}$$

which has the k-dependence only in the exponent. In the case of forward scattering ($\theta = 0$), we have $\vec{k}_s = \vec{k}_0$. The exponent of Eq. (6.5.21) becomes

$$i\vec{k}_0(\vec{r}_n - \vec{r}_1) + ik\,|\vec{r}_1 - \vec{r}_2| + ik\,|\vec{r}_2 - \vec{r}_3| + \cdots + ik\,|\vec{r}_{n-1} - \vec{r}_n|\,.$$

Since the sum $|\vec{r}_1 - \vec{r}_2| + |\vec{r}_2 - \vec{r}_3| + \cdots + |\vec{r}_{n-1} - \vec{r}_n|$ is greater than $|\vec{r}_n - \vec{r}_1|$, we have the exponent of Eq. (6.5.21) as

$$ik \times (\text{positive number}).$$

Thus $f_n(0, k)$ is analytic in the upper half-plane, $\operatorname{Im} k > 0$. In order to apply the dispersion relation to $f(0, k)$, we must define $f_n(0, k)$ for the real negative k. From Eq. (6.5.21), we have

$$f_n(0, -k) = f_n^*(0, k) \quad \text{with} \quad k \quad \text{real and positive.} \tag{6.5.22}$$

We note that

$$f_1(0, 0) = -\frac{1}{4\pi} \int U(\vec{r}_1) d^3\vec{r}_1 = f_1(0, k)$$

is a real constant so that we can invoke the dispersion relation,

$$\begin{aligned} \operatorname{Re} F(x) &= \frac{2}{\pi} \mathrm{P} \int_0^\infty \frac{x' \operatorname{Im} F(x')}{x'^2 - x^2}\, dx', \\ \operatorname{Im} F(x) &= -\frac{2}{\pi} \mathrm{P} \int_0^\infty \frac{x \operatorname{Re} F(x')}{x'^2 - x^2}\, dx', \end{aligned} \tag{6.5.23}$$

to the present case with the minor modification,

$$F(k) = f(0, k) - f_1(0, k) \quad \text{with} \quad k \quad \text{real and positive.} \tag{6.5.24}$$

Thus we have

$$\operatorname{Re}[f(0, k) - f_1(0, 0)] = \frac{2}{\pi} \mathrm{P} \int_0^\infty \frac{k' \operatorname{Im} f(0, k')}{k'^2 - k^2}\, dk'. \tag{6.5.25}$$

We have tacitly assumed the square-integrability of $[f(0,k) - f_1(0,0)]$ in the present discussion. If this assumption fails, we may replace

$$[f(0,k) - f_1(0,0)] \quad \text{with} \quad [f(0,k) - f_1(0,0)]/k^2$$

where the latter is assumed to be bounded at $k = 0$ and *square-integrable*. We obtain the dispersion relation,

$$\mathrm{Re}[f(0,k) - f_1(0,0)] = \frac{2k^2}{\pi}\mathrm{P}\int_0^\infty \frac{\mathrm{Im}\, f(0,k')}{k'(k'^2 - k^2)}\,dk'. \tag{6.5.26}$$

We shall now replace the incident plane wave in Eq. (6.5.19) with the incident spherical wave originating from the point on the negative z axis, $\vec{r} = \vec{r}_0$,

$$\psi_0(\vec{r}) = \frac{r_0}{|\vec{r} - \vec{r}_0|}\exp[ik\,|\vec{r} - \vec{r}_0| - ikr_0],$$

and consider the scattered wave. The exponent in Eq. (6.5.21) is replaced with

$$-i\vec{k}_s\vec{r}_1 + ik\,|\vec{r}_1 - \vec{r}_2| + \cdots + ik\,|\vec{r}_{n-1} - \vec{r}_n| + ik\,|\vec{r}_n - \vec{r}_0| - ikr_0.$$

In the case of forward scattering, we have $\vec{k}_s = \vec{k}_0$ and $\vec{k}_s$ points in the positive z direction. Since we have $kr_0 = -\vec{k}_0\vec{r}_0$, the exponent written out above becomes

$$i\vec{k}_0(\vec{r}_0 - \vec{r}_1) + ik\{|\vec{r}_1 - \vec{r}_2| + \cdots + |\vec{r}_{n-1} - \vec{r}_n| + |\vec{r}_n - \vec{r}_0|\},$$

so that we have the exponent as

$$ik \times (\text{positive number}).$$

Since we have $\vec{r}_0$ on the z axis, we have

$$\frac{\exp[ik\,|\vec{r} - \vec{r}_0|]}{|\vec{r} - \vec{r}_0|} = ik\sum_{l=0}^{\infty}(2l+1)P_l(\cos\theta)j_l(kr)h_l^{(1)}(kr_0), \qquad r_0 > r. \tag{6.5.27}$$

The total wave must assume the form,

$$ik\sum_{l=0}^{\infty}(2l+1)P_l(\cos\theta)h_l^{(1)}(kr_0)r_0\exp[-ikr_0](j_l(kr) + i\exp[i\delta_l]\sin\delta_l h_l^{(1)}(kr)),$$

and the corresponding forward scattering amplitude $f'(0,k)$ becomes

$$f'(0,k) = \sum_{l=0}^{\infty}(2l+1)h_l^{(1)}(kr_0)r_0\exp[-ikr_0]\cdot i\exp[i\delta_l]\sin\delta_l\cdot i^{-l}$$

$$= \frac{1}{2ik}\sum_{l=0}^{\infty}(2l+1)(\exp[i\delta_l] - 1)q_l(kr_0), \tag{6.5.28}$$

where

$$q_l(kr_0) = kr_0 \cdot h_l^{(1)}(kr_0) \cdot i^{-l+1} \exp[-ikr_0] = \sum_{m=0}^{l} \frac{i^m(l+m)!}{m!(l-m)!} \frac{1}{(2kr_0)^m}. \quad (6.5.29)$$

We note that $f'(0,k)$ is a function of r_0. Since r_0 is arbitrary, we can obtain many relations from the dispersion relation for $f'(0,k)$ by equating the same power of r_0. Picking up the $1/r_0^m$ term, we obtain

$$f_m(k) = \frac{1}{2ik^{2m+1}} \sum_{l=m}^{\infty} (2l+1) \frac{(l+m)!}{(l-m)!} (\exp[i\delta_l] - 1). \quad (6.5.30)$$

Using the identity

$$\sum_{m=l}^{l'} \frac{(-1)^{m-l}(l'+m)!}{(m-l)!(m+l+1)!(l'-m)!} = \begin{cases} \dfrac{1}{2l+1}, & \text{for } l'=l, \\ 0, & \text{for } l'>l, \end{cases}$$

we invert Eq. (6.5.30) for $\exp[i\delta_l] - 1$, obtaining

$$\frac{\exp[i\delta_l] - 1}{2i} = \sum_{m=l}^{\infty} \frac{(-1)^{m-l} k^{2m+1}}{(m-l)!(m+l+1)!} f_m(k).$$

The dispersion relation for the phase shift δ_l becomes

$$\operatorname{Im} \exp[2i\delta_l] = \sum_{m=l}^{\infty} \frac{(-1)^{m-l} k^{2m+1}}{(m-l)!(m+l+1)!}$$

$$\times \left[\frac{2}{\pi} \mathrm{P} \int \frac{\sum_{l'=m}^{\infty} \left[\dfrac{(2l'+1)(l'+m)!}{(l'-m)!} \right] \operatorname{Re}(1 - \exp[i\delta_{l'}(k')])}{k'^2 - k^2} dk' + C_m \right], \quad (6.5.31)$$

where

$$C_m = \frac{1}{2} \frac{d^m}{dk^m} \operatorname{Im} \exp[2i\delta_m(k)] \bigg|_{k=0}.$$

We refer the reader to the following book for the mathematical details of the dispersion relations in classical electrodynamics, quantum mechanics and relativistic quantum field theory. Goldberger, M.L., and Watson, K.M.: "*Collision Theory*", John Wiley & Sons, New York, (1964). Chapter 10 and Appendix G.2.

6.6 Problems for Chapter 6

6.1. Solve

$$\frac{1}{\pi i}\mathrm{P}\int_{-1}^{+1}\frac{1}{y-x}\phi(y)\,dy=0,\qquad -1\leq x\leq 1.$$

6.2. (Due to H. C.) Solve

$$\phi(x)=\frac{\tan x}{\pi}\mathrm{P}\int_{0}^{1}\frac{1}{y-x}\phi(y)\,dy+f(x),\qquad 0\leq x\leq 1.$$

6.3. (Due to H. C.) Solve

$$\phi(x)=\frac{\tan(x^3)}{\pi}\mathrm{P}\int_{0}^{1}\frac{1}{y-x}\phi(y)\,dy+f(x),\qquad 0\leq x\leq 1.$$

6.4. (Due to H. C.) Solve

$$\int_{0}^{x}\frac{1}{\sqrt{x-y}}\phi(y)\,dy+A\int_{x}^{1}\frac{1}{\sqrt{y-x}}\phi(y)\,dy=1,$$

where

$$0\leq x\leq 1.$$

6.5. (Due to H. C.) Solve

$$\lambda\mathrm{P}\int_{0}^{+\infty}\frac{1}{y-x}\phi(y)\,dy=\phi(x),\qquad 0\leq x<\infty.$$

Find the eigenvalues and the corresponding eigenfunctions.

6.6. (Due to H. C.) Solve

$$\mathrm{P}\int_{-1}^{+1}\frac{1+\alpha(y-x)}{y-x}\phi(y)\,dy=0,\qquad -1\leq x\leq 1.$$

6.7. (Due to H. C.) Obtain the eigenvalues and the eigenfunctions of

$$\phi(x)=\frac{\lambda}{\pi}\mathrm{P}\int_{-1}^{+1}\frac{1}{y-x}\phi(y)\,dy,\qquad -1\leq x\leq 1.$$

6.8. (Due to H. C.) Solve

$$\sqrt{x}\psi(x)=\frac{1}{\pi}\mathrm{P}\int_{0}^{+\infty}\frac{1}{y-x}\psi(y)\,dy,\qquad 0\leq x<\infty.$$

6.9. (Due to H. C.) Solve

$$\frac{1}{\pi}\mathrm{P}\int_{-\infty}^{+\infty}\frac{1}{y-x}\phi(y)\,dy = f(x), \quad -\infty < x < \infty.$$

6.10. (Due to H. C.) Solve

$$\frac{1}{\pi}\mathrm{P}\int_{0}^{+\infty}\frac{1}{y-x}\phi(y)\,dy = f(x), \qquad 0 \le x < \infty.$$

6.11. (Due to H. C.) Solve

$$\int_{0}^{1}\ln|x-y|\,\phi(y)\,dy = 1, \qquad 0 \le x \le 1.$$

6.12. Solve

$$\frac{\Gamma(x)}{B(x)} + \frac{1}{\pi}\mathrm{P}\int_{-a}^{+a}\frac{\Gamma'(y)}{x-y}\,dy = f(x)$$

with $\Gamma(x)$, $B(x)$ and $f(x)$ even and

$$\Gamma(-a) = \Gamma(a) = 0.$$

Hint: The unknown $\Gamma(x)$ is the circulation of the thin wing in Prandtl's thin-wing theory.

Kondo, J.: "*Integral Equations* ", Kodansha Ltd., Tokyo, (1991). p412.

6.13. (Due to H. C.) Prove that

$$\mathrm{P}\int_{0}^{\infty}\frac{dy}{\sqrt{y}(y-x)} = 0 \quad \text{for} \quad x > 0.$$

6.14. (Due to H. C.) Prove that

$$\mathrm{P}\int_{0}^{1}\frac{dy}{\sqrt{y(1-y)}(y-x)} = 0, \qquad 0 < x < 1.$$

6.15. (Due to H. C.) Prove that

$$\frac{1}{\pi}\mathrm{P}\int_{0}^{1}\frac{\sqrt{y(1-y)}}{y-x}\,dy = \frac{1}{2} - x, \qquad 0 < x < 1.$$

6.16. (Due to H. C.) Prove that

$$\frac{1}{\pi}\mathrm{P}\int_{0}^{1}\frac{\ln y}{\sqrt{y(1-y)}(y-x)}\,dy = \frac{\pi}{\sqrt{x(1-x)}} - \int_{1}^{\infty}\frac{1}{\sqrt{y(y-1)}(y-x)}\,dy$$

$$0 \le x \le 1.$$

6.17. Solve the integral equation of the Cauchy type,

$$F(x) = \frac{1}{\pi} \int \frac{F(x')h^*(x')}{x' - x - i\varepsilon} \, dx',$$

where $h(x)$ is given by

$$h(x) = \exp[i\delta(x)] \sin \delta(x),$$

and $\delta(x)$ is a given real function.

Hint: Consider $f(z)$ defined by

$$f(z) = \frac{1}{\pi} \int \frac{F(x')h^*(x')}{x' - z} \, dx',$$

and compute the discontinuity across the real x axis. Also use the identity,

$$1 - 2ih^*(x) = \exp[-2i\delta(x)].$$

The function $h(x)$ originates from the phase shift analysis of the potential scattering problem in quantum mechanics.

7 Wiener–Hopf Method and Wiener–Hopf Integral Equation

7.1 The Wiener–Hopf Method for Partial Differential Equations

In Sections 6.3, 6.4 and 6.5, we reduced the singular integral equations of Cauchy type and their variants to the inhomogeneous Hilbert problem by the introduction of the function $\Phi(z)$ appropriately defined. We found that the boundary values $\Phi^{\pm}(x)$ across the cut $[a, b]$ satisfy a linear equation,

$$\Phi^{+}(x) = R(x)\Phi^{-}(x) + f(x), \qquad a \leq x \leq b,$$

which we are able to solve using the argument in Sections 6.1 and 6.2.

Suppose now we are given one linear equation involving two unknown functions, $\phi_{-}(k)$ and $\psi_{+}(k)$, in the complex k plane,

$$\phi_{-}(k) = \psi_{+}(k) + F(k), \tag{7.1.1}$$

where $\phi_{-}(k)$ is analytic in the lower half-plane ($\operatorname{Im} k < \tau_{-}$) and $\psi_{+}(k)$ is analytic in the upper half-plane ($\operatorname{Im} k \geq \tau_{+}$). Can we solve Eq. (7.1.1) for $\phi_{-}(k)$ and $\psi_{+}(k)$? As long as $\phi_{-}(k)$ and $\psi_{+}(k)$ have a common region of analyticity as in Figure 7.1, namely

$$\tau_{+} \leq \tau_{-}, \tag{7.1.2}$$

we can solve for $\phi_{-}(k)$ and $\psi_{+}(k)$. In the most stringent case, the common region of analyticity can be an arc below which (excluding the arc) $\phi_{-}(k)$ is analytic and above which (including the arc) $\psi_{+}(k)$ is analytic.

We proceed to split $F(k)$ into a sum of two functions, one analytic in the upper half-plane and the other analytic in the lower half-plane,

$$F(k) = F_{+}(k) + F_{-}(k). \tag{7.1.3}$$

This *sum-splitting* can be carried out either by inspection or by the general method utilizing the Cauchy integral formula, to be discussed in Section 7.3. Once the sum-splitting is accomplished, we write Eq. (7.1.1) in the following form,

$$\phi_{-}(k) - F_{-}(k) = \psi_{+}(k) + F_{+}(k) \equiv G(k). \tag{7.1.4}$$

Applied Mathematics in Theoretical Physics. Michio Masujima

ISBN: 3-527-40534-8

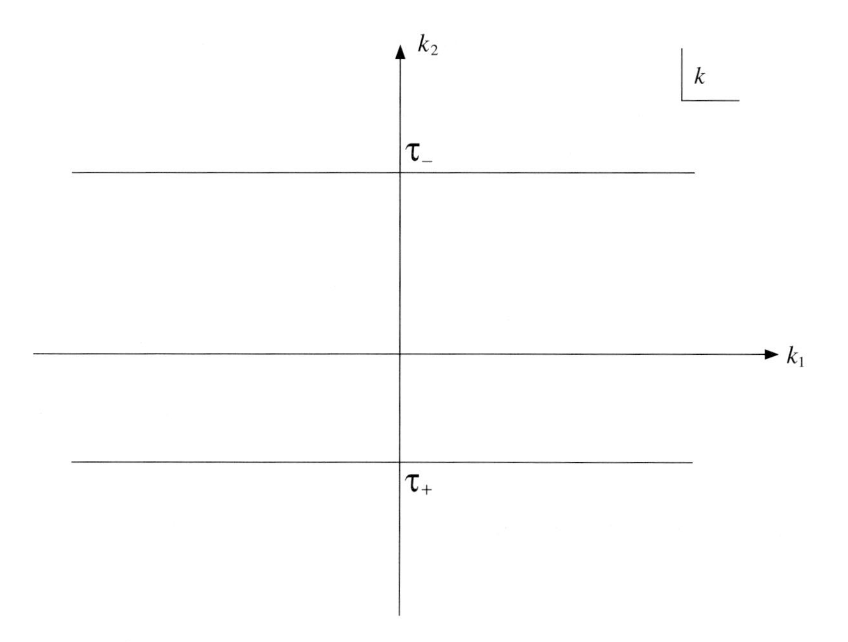

Fig. 7.1: The common region of analyticity for $\phi_-(k)$ and $\psi_+(k)$ in the complex k plane. $\psi_+(k)$ and $F_+(k)$ are analytic in the upper half-plane, $\mathrm{Im}\, k \geq \tau_+$. $\phi_-(k)$ and $F_-(k)$ are analytic in the lower half-plane, $\mathrm{Im}\, k < \tau_-$. The common region of analyticity for $\phi_-(k)$ and $\psi_+(k)$ is inside the strip, $\tau_+ \leq \mathrm{Im}\, k < \tau_-$ in the complex k plane.

We immediately note that $G(k)$ is *entire* in k. If the asymptotic behavior of $F_\pm(k)$ as $|k| \to \infty$ is such that

$$F_\pm(k) \to 0 \quad \text{as} \quad |k| \to \infty, \tag{7.1.5}$$

and on physical grounds,

$$\phi_-(k), \psi_+(k) \to 0 \quad \text{as} \quad |k| \to \infty, \tag{7.1.6}$$

then the entire function $G(k)$ must vanish by Liouville's theorem. Thus we obtain

$$\phi_-(k) = F_-(k), \tag{7.1.7a}$$

$$\psi_+(k) = -F_+(k). \tag{7.1.7b}$$

We call this method the *Wiener–Hopf method.*

In the following two examples we apply this method to the *mixed boundary value problem* of a partial differential equation.

❑ **Example 7.1.** Find the solution to the *Laplace Equation in the half-plane*.

$$\left(\frac{\partial^2}{\partial x^2} + \frac{\partial^2}{\partial y^2}\right)\phi(x,y) = 0, \qquad y \geq 0, \quad -\infty < x < \infty, \tag{7.1.8}$$

subject to the boundary condition on $y = 0$, as displayed in Figure 7.2,

$$\phi(x,0) = e^{-x}, \qquad x > 0, \tag{7.1.9a}$$

$$\phi_y(x,0) = ce^{bx}, \qquad b > 0, \quad x < 0. \tag{7.1.9b}$$

We further assume

$$\phi(x,y) \to 0 \quad \text{as} \quad x^2 + y^2 \to \infty. \tag{7.1.9c}$$

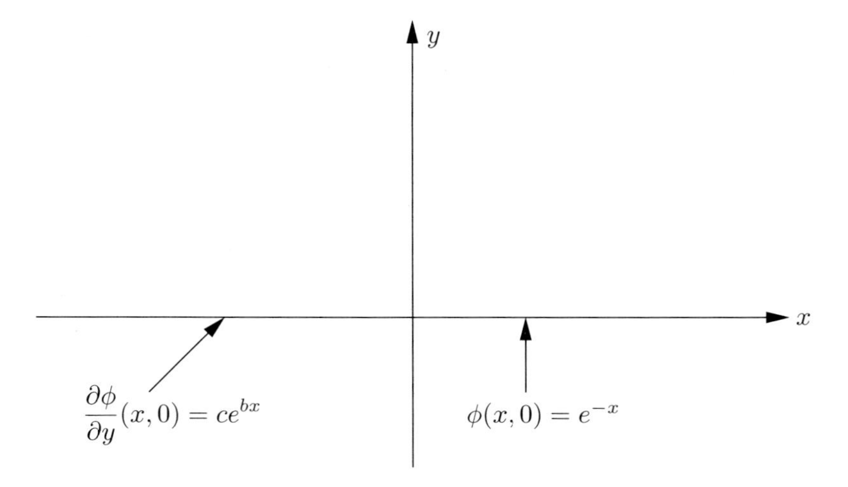

Fig. 7.2: Boundary conditions of $\phi(x,y)$ for the Laplace Equation (7.1.8) in the half-plane, $y \geq 0$ and $-\infty < x < \infty$.

Solution. Since $-\infty < x < \infty$, we may take the *Fourier transform* with respect to x, i.e.,

$$\hat{\phi}(k,y) \equiv \int_{-\infty}^{+\infty} dx\, e^{-ikx}\phi(x,y), \tag{7.1.10a}$$

$$\phi(x,y) = \frac{1}{2\pi}\int_{-\infty}^{+\infty} dk\, e^{ikx}\hat{\phi}(k,y). \tag{7.1.10b}$$

So, if we can obtain $\hat{\phi}(k,y)$, we will have the solution. Taking the Fourier transform of the partial differential equation (7.1.8) with respect to x, we have

$$\left(-k^2 + \frac{\partial^2}{\partial y^2}\right)\hat{\phi}(k,y) = 0,$$

i.e.,

$$\frac{\partial^2}{\partial y2}\hat{\phi}(k,y) = k^2\hat{\phi}(k,y), \tag{7.1.11}$$

from which we obtain

$$\hat{\phi}(k,y) = C_1(k)e^{ky} + C_2(k)e^{-ky}.$$

The boundary condition at infinity,

$$\hat{\phi}(k,y) \to 0 \quad \text{as} \quad y \to +\infty,$$

implies

$$\begin{cases} C_1(k) = 0 & \text{for} \quad k > 0, \\ C_2(k) = 0 & \text{for} \quad k < 0. \end{cases}$$

We can write more generally

$$\hat{\phi}(k,y) = A(k)e^{-|k|y}. \tag{7.1.12}$$

To obtain $A(k)$, we need to apply the boundary conditions at $y = 0$, but this is not trivial.

Take the Fourier transform of $\phi(x,0)$ to find

$$\begin{aligned} \hat{\phi}(k,0) &\equiv \int_{-\infty}^{+\infty} dx\, e^{-ikx}\phi(x,0) \\ &= \int_{-\infty}^{0} dx\, e^{-ikx}\phi(x,0) + \int_{0}^{+\infty} dx\, e^{-ikx}e^{-x}, \end{aligned} \tag{7.1.13}$$

where the first term is unknown for $x < 0$, while the second term is equal to $1/(1+ik)$.

Recall that

$$\left|e^{-ikx}\right| = \left|e^{-i(k_1+ik_2)x}\right| = e^{k_2 x} \quad \text{with} \quad k = k_1 + ik_2,$$

which vanishes in the upper half-plane ($k_2 > 0$) as $x \to -\infty$. So, define

$$\phi_+(k) \equiv \int_{-\infty}^{0} dx\, e^{-ikx}\phi(x,0), \tag{7.1.14}$$

which is a $+$ function. (Assuming that $\phi(x,0) \sim O(e^{bx})$ as $x \to -\infty$, $\phi_+(k)$ is analytic for $k_2 > -b$. So, we have from Eqs. (7.1.12) and (7.1.13),

$$\hat{\phi}(k,0) = \phi_+(k) + \frac{1}{1+ik} \quad \text{or} \quad A(k) = \phi_+(k) + \frac{1}{1+ik}. \tag{7.1.15}$$

Also, take the Fourier transform of $\partial\phi(x,0)/\partial y$ to find

$$\begin{aligned}\frac{\partial\hat{\phi}}{\partial y}(k,0) &\equiv \int_{-\infty}^{+\infty} dx\, e^{-ikx}\frac{\partial\phi(x,0)}{\partial y} \\ &= \int_{-\infty}^{0} dx\, e^{-ikx} c e^{bx} + \int_{0}^{+\infty} dx\, e^{-ikx}\frac{\partial\phi(x,0)}{\partial y},\end{aligned} \tag{7.1.16}$$

where the first term is equal to $c/(b-ik)$, and the second term is unknown for $x>0$. Then define

$$\psi_-(k) \equiv \int_0^{+\infty} dx\, e^{-ikx}\frac{\partial\phi(x,0)}{\partial y}, \tag{7.1.17}$$

which is a $-$ function. Assuming that $\partial\phi(x,0)/\partial y \sim O(e^{-x})$ as $x\to\infty$, $\psi_-(k)$ is analytic for $k_2<1$. Thus we obtain

$$-|k|\,A(k) = \frac{c}{b-ik} + \psi_-(k),$$

or we have

$$A(k) = \frac{-c}{|k|\,(b-ik)} - \frac{\psi_-(k)}{|k|}. \tag{7.1.18}$$

Equating the two expressions (7.1.15) and (7.1.18) for $A(k)$, (assuming that they are both valid in some common region, say $k_2=0$), we get

$$\phi_+(k) + \frac{1}{1+ik} = \frac{-c}{|k|\,(b-ik)} - \frac{\psi_-(k)}{|k|}. \tag{7.1.19}$$

Now, the function $|k|$ is not analytic and so we cannot proceed with the Wiener–Hopf method unless we express $|k|$ in a suitable form. One such form, often suitable for application, is

$$|k| = \lim_{\varepsilon\to 0+}(k^2+\varepsilon^2)^{1/2} = \lim_{\varepsilon\to 0+}(k+i\varepsilon)^{1/2}(k-i\varepsilon)^{1/2}, \tag{7.1.20}$$

where, in the last expression, $(k+i\varepsilon)^{1/2}$ is a $+$ function and $(k-i\varepsilon)^{1/2}$ is a $-$ function, as displayed in Figure 7.3.

We can verify that on the real axis,

$$\sqrt{k^2+\varepsilon^2} = |k| \quad \text{as} \quad \varepsilon\to 0^+. \tag{7.1.21}$$

We thus have

$$\phi_+(k) - \frac{i}{k-i} = \frac{-ci}{(k+ib)(k+i\varepsilon)^{1/2}(k-i\varepsilon)^{1/2}} - \frac{\psi_-(k)}{(k+i\varepsilon)^{1/2}(k-i\varepsilon)^{1/2}}. \tag{7.1.22}$$

Since $(k+i\varepsilon)^{1/2}$ is a $+$ function (i.e., is analytic in the upper half-plane), we multiply the whole equation (7.1.22) by $(k+i\varepsilon)^{1/2}$ to get

$$(k+i\varepsilon)^{1/2}\phi_+(k) - \frac{i(k+i\varepsilon)^{1/2}}{k-i} = \frac{-ci}{(k+ib)(k-i\varepsilon)^{1/2}} - \frac{\psi_-(k)}{(k-i\varepsilon)^{1/2}}, \tag{7.1.23}$$

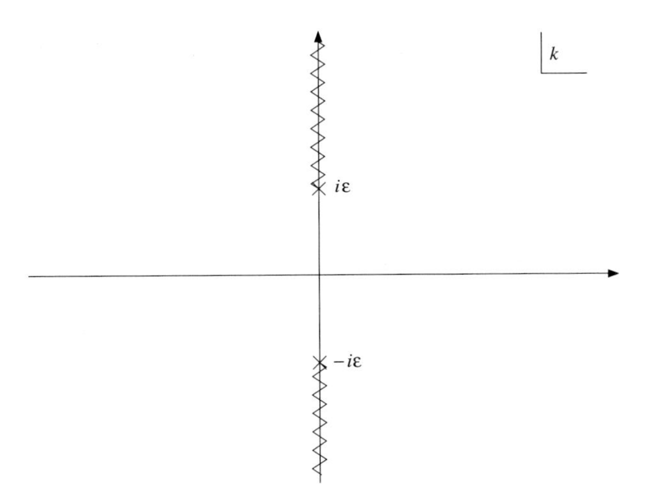

Fig. 7.3: Factorization of $|k|$ into a + function and a − function. The points, $k = \pm i\varepsilon$, are the branch points of $\sqrt{k^2+\varepsilon^2}$, from which the branch cuts are extended to $\pm i\infty$ along the imaginary k axis in the complex k plane.

where the first term on the left-hand side is a + function and the second term on the right-hand side is a − function. We shall decompose the remaining terms into a sum of the + function and the − function. Consider the second term of the left-hand side of Eq. (7.1.23). The numerator is a + function but the denominator is a − function. We rewrite

$$-\frac{i(k+i\varepsilon)^{1/2}}{k-i} = \frac{-i(k+i\varepsilon)^{1/2} + i(i+i\varepsilon)^{1/2}}{k-i} - \frac{i(i+i\varepsilon)^{1/2}}{k-i},$$

where, in the first term on the right-hand side of the above, we removed the singularity at $k = i$ by making the numerator vanish at $k = i$ so that it is a + function. The second term on the right-hand side has a pole at $k = i$ so that it is a − function. Similarly, the first term on the right-hand side of Eq. (7.1.23) is rewritten in the following form,

$$\frac{-ci}{(k+ib)(k-i\varepsilon)^{1/2}} = \frac{-ci}{k+ib}\left[\frac{1}{(k-i\varepsilon)^{1/2}} - \frac{1}{(-ib-i\varepsilon)^{1/2}}\right] - \frac{ci}{(k+ib)(-ib-i\varepsilon)^{1/2}},$$

where the first term on the right-hand side of the above is no longer singular at $k = -ib$, but has a branch point at $k = i\varepsilon$ so that it is a − function. The second term on the right-hand side has a pole at $k = -ib$ so that it is a + function.

Collating all this, we have the following equation,

$$(k+i\varepsilon)^{1/2}\phi_+(k)+\frac{-i(k+i\varepsilon)^{1/2}+i(i+i\varepsilon)^{1/2}}{k-i}+\frac{ci}{(k+ib)(-ib-i\varepsilon)^{1/2}}$$
$$=-\frac{\psi_-(k)}{(k-i\varepsilon)^{1/2}}-\frac{ci}{(k+ib)}\left[\frac{1}{(k-i\varepsilon)^{1/2}}-\frac{1}{(-ib-i\varepsilon)^{1/2}}\right]$$
$$+\frac{i(i+i\varepsilon)^{1/2}}{k-i}, \quad (7.1.24)$$

where each term on the left-hand side is a $+$ function while each term on the right-hand side is a $-$ function. Applying the Wiener–Hopf method, and noting that the left-hand side and the right-hand side both go to 0 as $|k|\to\infty$, we have

$$\phi_+(k)=\frac{i}{k-i}\left[1-\frac{(i+i\varepsilon)^{1/2}}{(k+i\varepsilon)^{1/2}}\right]-\frac{ci}{(k+ib)(-ib-i\varepsilon)^{1/2}(k+i\varepsilon)^{1/2}}.$$

We can simplify somewhat,

$$(i+i\varepsilon)^{1/2}=(1+\varepsilon)^{1/2}e^{i\pi/4}=e^{i\pi/4} \quad \text{as} \quad \varepsilon\to 0^+.$$

Similarly

$$(-ib-i\varepsilon)^{1/2}=\sqrt{b+\varepsilon}e^{-i\pi/4}=\sqrt{b}e^{-i\pi/4} \quad \text{as} \quad \varepsilon\to 0^+.$$

Hence we have

$$\phi_+(k)=\frac{i}{k-i}\left[1-\frac{e^{i\pi/4}}{(k+i\varepsilon)^{1/2}}\right]-\frac{(c/\sqrt{b})ie^{i\pi/4}}{(k+ib)(k+i\varepsilon)^{1/2}}. \quad (7.1.25)$$

So finally, we obtain $A(k)$ from Eqs. (7.1.15) and (7.1.25) as,

$$A(k)=\frac{-ie^{i\pi/4}}{(k+i\varepsilon)^{1/2}}\left[\frac{c}{\sqrt{b}(k+ib)}+\frac{1}{k-i}\right], \quad (7.1.26)$$

where we note

$$(k+i\varepsilon)^{1/2}=\begin{cases}\sqrt{k} & \text{for} \quad k>0,\\ \sqrt{|k|}e^{i\pi/2}=i\sqrt{|k|} & \text{for} \quad k<0.\end{cases}$$

Therefore,

$$\hat{\phi}(k,y)=A(k)e^{-|k|y}$$

can be inverted with respect to k to obtain $\phi(x,y)$.

We now discuss another example.

❑ **Example 7.2. Sommerfeld diffraction problem.**
Solve the wave equation in two space dimensions:

$$\frac{\partial^2}{\partial t^2}u(x,y,t)=c^2\nabla^2 u(x,y,t) \tag{7.1.27}$$

with the boundary condition,

$$\frac{\partial}{\partial y}u(x,y,t)=0 \quad \text{at} \quad y=0 \quad \text{for} \quad x<0. \tag{7.1.28}$$

The incident, reflected, and diffracted waves are shown in Figure 7.4.

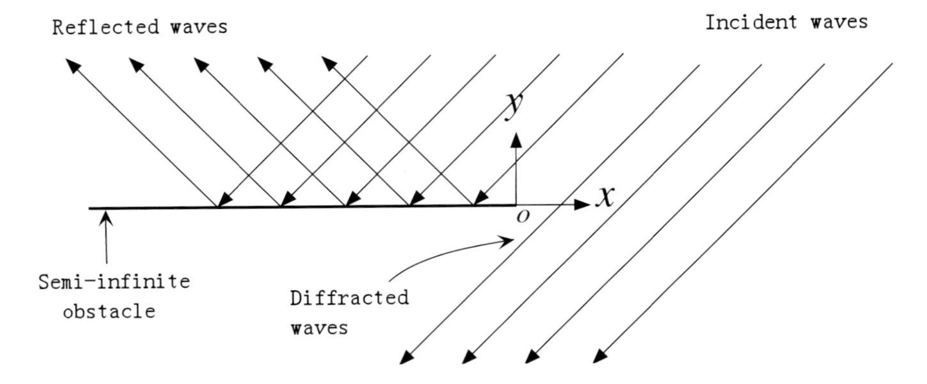

Fig. 7.4: Incident wave, reflected wave and diffracted wave in the *Sommerfeld diffraction problem.*

Solution. We look for the solution of the form,

$$u(x,y,t)=\phi(x,y)e^{-i\omega t}. \tag{7.1.29}$$

Setting

$$p\equiv\omega/c, \tag{7.1.30}$$

the wave equation assumes the following form,

$$\nabla^2\phi+p^2\phi=0. \tag{7.1.31}$$

Letting the forcing increase exponentially in time (so that it is absent as $t\to-\infty$), we must have

$$\mathrm{Im}\,\omega>0,$$

which requires

$$p=p_1+i\varepsilon, \qquad \varepsilon\to 0^+. \tag{7.1.32}$$

So, $\phi(x, y)$ satisfies

$$\begin{cases} \nabla^2 \phi + p^2 \phi = 0, \\ \dfrac{\partial \phi}{\partial y} = 0 \qquad \text{at} \quad y = 0 \quad \text{for} \quad x < 0. \end{cases} \tag{7.1.33}$$

Consider the incident waves,

$$u_{\text{inc}} = e^{i(\vec{p} \cdot \vec{x} - \omega t)} = \phi_{\text{inc}} e^{-i\omega t}, \quad \text{with} \quad |\vec{p}| = p,$$

so that u_{inc} also satisfies the wave equation. The u_{inc} is a plane wave moving in the direction of $\vec{p}$. We take

$$\vec{p} = -p \cos\theta \cdot \vec{e}_x - p \sin\theta \cdot \vec{e}_y,$$

and assume

$$0 < \theta < \pi/2.$$

Then $\phi_{\text{inc}}(x, y)$ is given by

$$\phi_{\text{inc}}(x, y) = e^{-ip(x \cos\theta + y \sin\theta)}. \tag{7.1.34}$$

We seek a *disturbance solution*

$$\psi(x, y) \equiv \phi(x, y) - \phi_{\text{inc}}(x, y). \tag{7.1.35}$$

Thus the governing equation and the boundary condition for $\psi(x, y)$ become

$$\nabla^2 \psi + p^2 \psi = 0, \tag{7.1.36}$$

subject to

$$\frac{\partial \psi(x, 0)}{\partial y} = -\frac{\partial \phi_{\text{inc}}(x, 0)}{\partial y} = ip \sin\theta \cdot e^{-ip(x \cos\theta)} \quad \text{for} \quad x < 0 \quad \text{at} \quad y = 0. \tag{7.1.37}$$

Asymptotic behavior: We replace p by $p_1 + i\varepsilon$,

$$p = p_1 + i\varepsilon. \tag{7.1.38}$$

In the *reflection region*, we have

$$\psi(x, y) \sim e^{-ip(x \cos\theta - y \sin\theta)} \quad \text{for} \quad y > 0, \quad x \to -\infty.$$

We note the change of sign in front of y. Near $y = 0^+$, we have

$$\psi(x, 0^+) \sim e^{-ipx \cos\theta} \quad \text{as} \quad x \to -\infty,$$

or,

$$\left|\psi(x, 0^+)\right| \sim e^{\varepsilon x \cos\theta} \quad \text{as} \quad x \to -\infty.$$

In the *shadow region* ($y < 0$, $x \to -\infty$),

$$\phi(-\infty, y) = 0 \Rightarrow \psi = -\phi_{\text{inc}},$$

so that

$$\psi(x, 0^-) \sim -e^{-ipx\cos\theta},$$

or,

$$\left|\psi(x, 0^-)\right| \sim e^{\varepsilon x\cos\theta} \quad \text{as} \quad x \to -\infty.$$

In summary,

$$|\psi(x, y)| \sim e^{\varepsilon x\cos\theta} \quad \text{as} \quad x \to -\infty, \quad y = 0^{\pm}. \tag{7.1.39}$$

Although $\psi(x, y)$ might be discontinuous across the obstacle, its normal derivative is continuous as given by the boundary condition. On the other side, as $x \to +\infty$, both $\psi(x, 0)$ and $\partial\psi(x, 0)/\partial x$ are continuous, but the asymptotic behavior at infinity is obtained by approximating the effect of the leading edge of the obstacle as a delta function,

$$\nabla^2\psi + p^2\psi = -4\pi\delta(x)\delta(y). \tag{7.1.40}$$

From Eq. (7.1.40), we obtain

$$\psi(x, y) = \pi i H_0^{(1)}(pr), \qquad r = \sqrt{x^2 + y^2}, \tag{7.1.41}$$

where $H_0^{(1)}(pr)$ is the zeroth-order Hankel function of the first kind. It behaves as

$$H_0^{(1)}(pr) \sim \frac{1}{\sqrt{r}}\exp(ipr) \quad \text{as} \quad r \to \infty,$$

which implies that

$$|\psi(x, 0)| \sim \frac{1}{\sqrt{x}}e^{-\varepsilon x} \quad \text{as} \quad x \to \infty \quad \text{near} \quad y = 0. \tag{7.1.42}$$

Now, we try to solve the equation for $\psi(x, y)$ using the Fourier transforms,

$$\hat{\psi}(k, y) \equiv \int_{-\infty}^{+\infty} dx\, e^{-ikx}\psi(x, y), \tag{7.1.43a}$$

$$\psi(x, y) = \frac{1}{2\pi}\int_{-\infty}^{+\infty} dk\, e^{ikx}\hat{\psi}(k, y). \tag{7.1.43b}$$

We take the Fourier transform of the partial differential equation,

$$\nabla^2\psi + p^2\psi = 0,$$

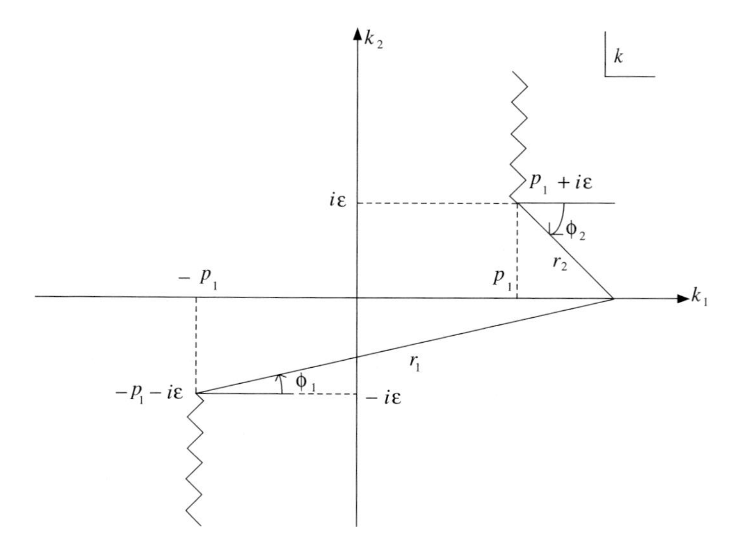

Fig. 7.5: Branch cuts of $(k^2 - p^2)^{1/2}$ in the complex k plane. The points, $k = \pm(p_1 + i\varepsilon)$, are the branch points of $\sqrt{k^2 - p^2}$, from which the branch cuts are extended to $\pm(p_1 + i\infty)$.

resulting in the form,

$$\frac{\partial^2}{\partial y^2}\hat{\psi}(k, y) = (k^2 - p^2)\hat{\psi}(k, y). \tag{7.1.44}$$

Consider the following branch of

$$(k^2 - p^2)^{1/2} = \lim_{\varepsilon \to 0+} [(k - p_1 - i\varepsilon)(k + p_1 + i\varepsilon)]^{1/2}, \tag{7.1.45}$$

where the branch cuts are drawn in Figure 7.5.

Then we have

$$[k^2 - (p_1 + i\varepsilon)^2]^{1/2} = \sqrt{r_1 r_2} e^{i(\phi_1 + \phi_2)/2},$$

so that

$$\mathrm{Re}[k^2 - (p_1 + i\varepsilon)^2]^{1/2} = \sqrt{r_1 r_2} \cos\left(\frac{\phi_1 + \phi_2}{2}\right).$$

On the real axis above,

$$-\pi < \phi_1 + \phi_2 < 0,$$

so that

$$0 < \cos\left(\frac{\phi_1 + \phi_2}{2}\right) < 1.$$

Thus the branch chosen has

$$\mathrm{Re}[k^2 - (p_1 + i\varepsilon)^2]^{1/2} > 0 \tag{7.1.46}$$

on the whole real axis. We can then write the solution to

$$\frac{\partial^2}{\partial y^2}\hat{\psi}(k, y) = [k^2 - (p_1 + i\varepsilon)^2]\hat{\psi}(k, y) \tag{7.1.47}$$

as

$$\hat{\psi}(k, y) = \begin{cases} A(k)\exp(-\sqrt{k^2 - (p_1 + i\varepsilon)^2}y), & y > 0, \\ B(k)\exp(+\sqrt{k^2 - (p_1 + i\varepsilon)^2}y), & y < 0. \end{cases}$$

But since $\psi(x, y)$ is not continuous across $y = 0$ for $x < 0$, the amplitudes $A(k)$ and $B(k)$ need not be identical. However, we know $\partial\psi(x, y)/\partial y$ is continuous across $y = 0$ for all x. We must therefore have

$$\frac{\partial}{\partial y}\hat{\psi}(k, 0^+) = \frac{\partial}{\partial y}\hat{\psi}(k, 0^-),$$

from which we obtain

$$B(k) = -A(k).$$

Thus we have

$$\hat{\psi}(k, y) = \begin{cases} A(k)\exp(-\sqrt{k^2 - (p + i\varepsilon)^2}y), & y > 0, \\ -A(k)\exp(+\sqrt{k^2 - (p + i\varepsilon)^2}y), & y < 0. \end{cases} \tag{7.1.48}$$

If we can determine $A(k)$, we will have arrived at the solution.

Let us recall everything we know. We know that $\psi(x, y)$ is continuous for $x > 0$ when $y = 0$, but is discontinuous for $x < 0$. So, consider the function,

$$\psi(x, 0^+) - \psi(x, 0^-) = \begin{cases} 0 & \text{for} \quad x > 0, \\ \text{unknown} & \text{for} \quad x < 0. \end{cases} \tag{7.1.49}$$

But from one earlier discussion, we know the asymptotic form of the latter unknown function to be like $e^{\varepsilon x \cos\theta}$ as $x \to -\infty$. Take the Fourier transform of the above discontinuity (7.1.49) to obtain

$$\hat{\psi}(k, 0^+) - \hat{\psi}(k, 0^-) = \int_{-\infty}^{0} dx\, e^{-ikx}(\psi(x, 0^+) - \psi(x, 0^-)),$$

the right-hand side of which is a + function, analytic for $k_2 > -\varepsilon\cos\theta$. We define

$$U_+(k) \equiv \hat{\psi}(k, 0^+) - \hat{\psi}(k, 0^-), \quad \text{analytic} \quad \text{for} \quad k_2 > -\varepsilon\cos\theta. \tag{7.1.50}$$

Now consider the derivative $\partial\psi(x,y)/\partial y$. This function is continuous at $y = 0$ for all x ($-\infty < x < \infty$). Furthermore, we know what it is for $x < 0$. Namely,

$$\frac{\partial}{\partial y}\psi(x,0) = \begin{cases} ip\sin\theta\cdot e^{-ipx\cos\theta} & \text{for} \quad x < 0, \\ \text{unknown} & \text{for} \quad x > 0. \end{cases} \tag{7.1.51}$$

But we know the asymptotic form of the latter unknown function to be like $e^{-\varepsilon x}$ as $x \to \infty$. Taking the Fourier transform of Eq. (7.1.51), we obtain

$$\frac{\partial}{\partial y}\hat{\psi}(k,0) = \int_{-\infty}^{0} dx\, e^{-ikx}\cdot ip\sin\theta\cdot^{-ipx\cos\theta} + \int_{0}^{+\infty} dx\, e^{-ikx}\frac{\partial}{\partial y}\psi(x,0).$$

Thus we have

$$\frac{\partial}{\partial y}\hat{\psi}(k,0) = \frac{-p\sin\theta}{(k+p\cos\theta)} + L_-(k), \tag{7.1.52}$$

where in the first term on the right-hand side of the above,

$$p = p_1 + i\varepsilon,$$

and the second term represented as $L_-(k)$ is defined by

$$L_-(k) \equiv \int_{0}^{+\infty} dx\, e^{-ikx}\frac{\partial}{\partial y}\psi(x,0).$$

$L_-(k)$ is a $-$ function, analytic for $k_2 < \varepsilon$.

We shall now use Eqs. (7.1.50) and (7.1.52), where $U_+(k)$ and $L_-(k)$ are defined, together with the Wiener–Hopf method to solve for $A(k)$. We know

$$\hat{\psi}(k,y) = \begin{cases} A(k)\exp(-\sqrt{k^2-p^2}y) & \text{for} \quad y > 0, \\ -A(k)\exp(\sqrt{k^2-p^2}y) & \text{for} \quad y < 0. \end{cases}$$

Inserting these equations into Eq. (7.1.50), we have

$$2A(k) = U_+(k).$$

Inserting these equations into Eq. (7.1.52), we have

$$-\sqrt{k^2-p^2}\,A(k) = \frac{-p\sin\theta}{(k+p\cos\theta)} + L_-(k).$$

Eliminating $A(k)$ from the last two equations, we obtain the Wiener–Hopf problem,

$$-\sqrt{k^2-p^2}U_+(k)/2 = \frac{-p\sin\theta}{(k+p\cos\theta)} + L_-(k), \tag{7.1.53}$$

$$\begin{cases} L_-(k) & \text{analytic for} \quad k_2 < \varepsilon, \\ U_+(k) & \text{analytic for} \quad k_2 > -\varepsilon\cos\theta. \end{cases} \tag{7.1.54}$$

We divide through both sides of Eq. (7.1.53) by $\sqrt{k-p}$, obtaining

$$\frac{-\sqrt{k+p}U_+(k)}{2} = \frac{L_-(k)}{\sqrt{k-p}} - \frac{p\sin\theta}{(k+p\cos\theta)\sqrt{k-p}}.$$

The term involving $U_+(k)$ is a + function, while the term involving $L_-(k)$ is a − function. We decompose the last term on the right-hand side.

$$-\frac{p\sin\theta}{(k+p\cos\theta)\sqrt{k-p}} = -\frac{p\sin\theta}{k+p\cos\theta}\left[\frac{1}{\sqrt{k-p}} - \frac{1}{\sqrt{-p\cos\theta-p}}\right] - \frac{p\sin\theta}{(k+p\cos\theta)\sqrt{-p\cos\theta-p}},$$

where the first term on the right-hand side is a − function and the second term a + function. Collecting the + functions and the − functions respectively, we obtain by the Wiener–Hopf method,

$$\frac{U_+(k)}{2} = \frac{p\sin\theta}{\sqrt{k+p}(k+p\cos\theta)\sqrt{-p\cos\theta-p}}, \tag{7.1.55}$$

and hence we finally obtain

$$A(k) = \frac{p\sin\theta}{(k+p\cos\theta)\sqrt{k+p}\sqrt{-p\cos\theta-p}}. \tag{7.1.56}$$

Singularities of $A(k)$ in the complex k plane are drawn in Figure 7.6.

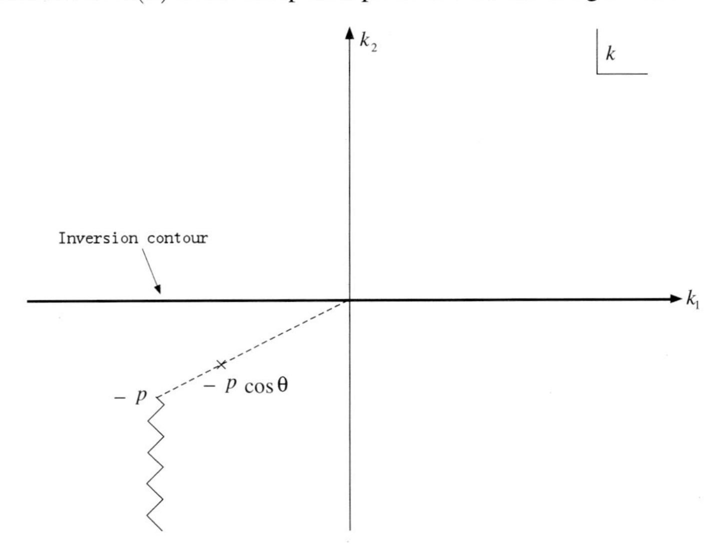

Fig. 7.6: Singularities of $A(k)$ in the complex k plane. $A(k)$ has a simple pole at $k=-p\cos\theta$ and a branch point at $k=-p$, from which the branch cut is extended to $-p-i\infty$.

The final solution for the disturbance function $\psi(x, y)$ in the limit as $\varepsilon \to 0^+$, is given by

$$\psi(x,y) = \frac{\text{sgn}(y)}{2\pi} \int_C dk\, A(k) \exp(-\sqrt{k^2 - p^2}\, |y| + ikx), \tag{7.1.57}$$

where C is the contour specified in Figure 7.7.

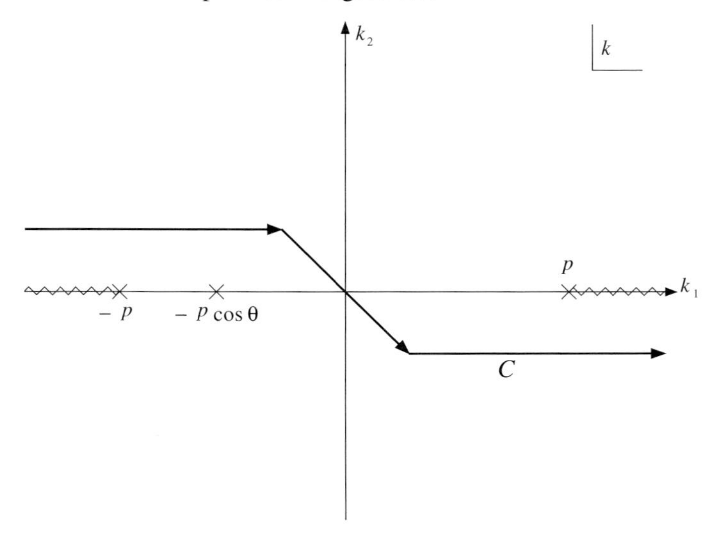

Fig. 7.7: The contour of the complex k integration in Eq. (7.1.57) for $\psi(x, y)$. The integrand has a simple pole at $k = -p\cos\theta$ and the branch points at $k = \pm p$. The branch cuts are extended from $k = \pm p$ to $\pm\infty$ along the real k axis.

We note that the choice of

$$p = p_1 + i\varepsilon, \quad \varepsilon > 0$$

is based on the requirement that

$$u_{\text{inc}}(\vec{x}, t) \to 0 \quad \text{as} \quad t \to -\infty$$

but it grows exponentially in time, i.e.,

$$u_{\text{inc}} = e^{i(\vec{p}\cdot\vec{x} - \omega t)} \quad \text{Im}\,\omega > 0,$$

which is known as *'turning on the perturbation adiabatically'*.

7.2 Homogeneous Wiener–Hopf Integral Equation of the Second Kind

The Wiener–Hopf integral equations are characterized by *translation kernels*, $K(x, y) = K(x - y)$, and the integral is on the *semi-infinite range*, $0 < x, y < \infty$. We list the Wiener–Hopf integral equations of several types.

Wiener–Hopf integral equation of the first kind:

$$F(x) = \int_0^{+\infty} K(x-y)\phi(y)\,dy, \qquad 0 \le x < \infty.$$

Homogeneous Wiener–Hopf integral equation of the second kind:

$$\phi(x) = \lambda \int_0^{+\infty} K(x-y)\phi(y)\,dy, \qquad 0 \le x < \infty.$$

Inhomogeneous Wiener–Hopf integral equation of the second kind:

$$\phi(x) = f(x) + \lambda \int_0^{+\infty} K(x-y)\phi(y)\,dy, \qquad 0 \le x < \infty.$$

Let us begin with the *homogeneous Wiener–Hopf integral equation of the second kind.*

$$\phi(x) = \lambda \int_0^{+\infty} K(x-y)\phi(y)\,dy, \qquad 0 \le x < \infty. \tag{7.2.1}$$

Here, the translation kernel $K(x-y)$ is defined for its argument both positive and negative. Suppose that

$$K(x) \to \begin{cases} e^{-bx} & \text{as} \quad x \to +\infty, \\ e^{ax} & \text{as} \quad x \to -\infty, \end{cases} \qquad a, b > 0, \tag{7.2.2}$$

so that the Fourier transform of $K(x)$, defined by

$$\hat{K}(k) = \int_{-\infty}^{+\infty} dx\, e^{-ikx} K(x), \tag{7.2.3}$$

is analytic for

$$-a < \operatorname{Im} k < b. \tag{7.2.4}$$

The region of analyticity of $\hat{K}(k)$ in the complex k plane is displayed in Figure 7.8.

Now, define

$$\phi(x) = \begin{cases} \phi(x) & \text{given for} \quad x > 0, \\ 0 & \text{for} \qquad\quad x < 0. \end{cases} \tag{7.2.5}$$

But then, although $\phi(x)$ is only known for positive x, since $K(x-y)$ is defined even for negative x, we can certainly define $\psi(x)$ for negative x,

$$\psi(x) \equiv \lambda \int_0^{+\infty} K(x-y)\phi(y)\,dy \quad \text{for} \quad x < 0. \tag{7.2.6}$$

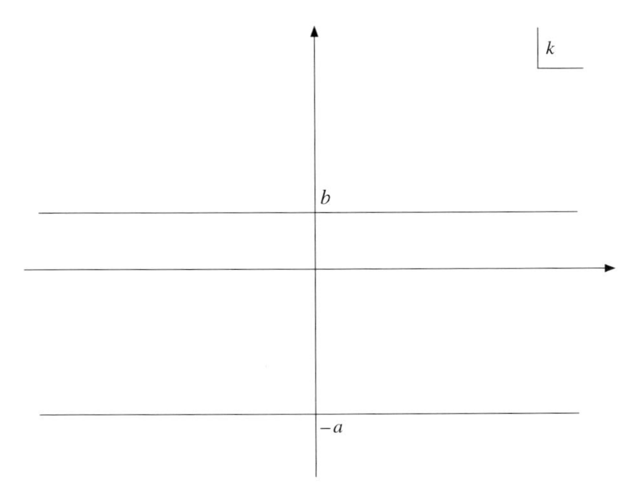

Fig. 7.8: Region of analyticity of $\hat{K}(k)$ in the complex k plane. $\hat{K}(k)$ is defined and analytic inside the strip, $-a < \operatorname{Im} k < b$.

Take the *Fourier transforms* of Eqs. (7.2.1) and (7.2.6). Adding up the results, and using the *convolution property*, we have

$$\int_0^{+\infty} dx\, e^{-ikx}\phi(x) + \int_{-\infty}^{0} dx\, e^{-ikx}\psi(x) = \lambda\hat{K}(k)\hat{\phi}_-(k),$$

i.e., we have

$$\hat{\phi}_-(k) + \hat{\psi}_+(k) = \lambda\hat{K}(k)\hat{\phi}_-(k),$$

or, we have

$$\left[1 - \lambda\hat{K}(k)\right]\hat{\phi}_-(k) = -\hat{\psi}_+(k), \tag{7.2.7}$$

where we have defined

$$\hat{\phi}_-(k) \equiv \int_0^{+\infty} dx\, e^{-ikx}\phi(x), \tag{7.2.8}$$

$$\hat{\psi}_+(k) \equiv \int_{-\infty}^{0} dx\, e^{-ikx}\psi(x). \tag{7.2.9}$$

Since we have

$$\left|e^{-ikx}\right| = e^{k_2 x}, \quad k = k_1 + ik_2,$$

we know that $\hat{\phi}_-(k)$ is analytic in the lower half-plane and $\hat{\psi}_+(k)$ is analytic in the upper half-plane. Thus, once again, we have one equation involving two unknown functions, $\hat{\phi}_-(k)$

and $\hat{\psi}_+(k)$, one analytic in the lower half-plane and the other analytic in the upper half-plane. The precise regions of analyticity for $\hat{\phi}_-(k)$ and $\hat{\psi}_+(k)$ are each determined by the asymptotic behavior of the kernel $K(x)$ as $x \to -\infty$.

In the original equation, Eq. (7.2.1), at the upper limit of the integral, we have $y \to \infty$ so that $x - y \to -\infty$ as $y \to \infty$. By Eq. (7.2.2), we have

$$K(x-y) \sim e^{a(x-y)} \sim e^{-ay} \quad \text{as} \quad y \to \infty.$$

To ensure that the integral in Eq. (7.2.1) converges, we conclude that $\phi(x)$ can grow as fast as

$$\phi(x) \sim e^{(a-\varepsilon)x} \quad \text{with} \quad \varepsilon > 0 \quad \text{as} \quad x \to \infty.$$

By definition of $\hat{\phi}_-(k)$, the region of the analyticity of $\hat{\phi}_-(k)$ is determined by the requirement

$$\left| e^{-ikx}\phi(x) \right| \sim e^{(k_2+a-\varepsilon)x} \to 0 \quad \text{as} \quad x \to \infty.$$

Thus $\hat{\phi}_-(k)$ is analytic in the lower half-plane,

$$\operatorname{Im} k = k_2 < -a + \varepsilon, \qquad \varepsilon > 0,$$

which includes

$$\operatorname{Im} k \leq -a.$$

As for the behavior of $\psi(x)$ as $x \to -\infty$, we observe that $x - y \to -\infty$ as $x \to -\infty$, and

$$K(x-y) \sim e^{a(x-y)} \quad \text{as} \quad x \to -\infty.$$

By definition of $\psi(x)$, we have

$$\psi(x) = \lambda \int_0^{+\infty} K(x-y)\phi(y)\, dy \sim \lambda e^{ax} \int_0^{+\infty} e^{-ay}\phi(y)\, dy \quad \text{as} \quad x \to -\infty,$$

where the integral above is convergent due to the asymptotic behavior of $\phi(x)$ as $x \to \infty$. The region of analyticity of $\hat{\psi}_+(k)$ is determined by the requirement,

$$\left| e^{-ikx}\psi(x) \right| \sim e^{(k_2+a)x} \to 0 \quad \text{as} \quad x \to -\infty.$$

Thus $\hat{\psi}_+(k)$ is analytic in the upper half-plane,

$$\operatorname{Im} k = k_2 > -a.$$

To summarize, we know

$$\begin{cases} \phi(x) \to e^{(a-\varepsilon)x} & \text{as} \quad x \to \infty, \\ \psi(x) \to e^{ax} & \text{as} \quad x \to -\infty. \end{cases} \tag{7.2.10}$$

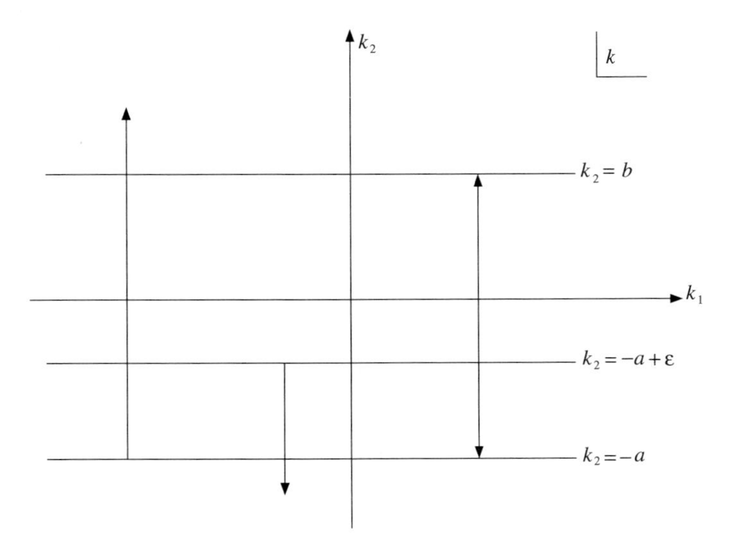

Fig. 7.9: Region of analyticity of $\hat{\phi}_-(k)$, $\hat{\psi}_+(k)$ and $\hat{K}(k)$. $\hat{\phi}_-(k)$ is analytic in the lower half-plane, $\mathrm{Im}\, k < -a + \varepsilon$. $\hat{\psi}_+(k)$ is analytic in the upper half-plane, $\mathrm{Im}\, k > -a$. $\hat{K}(k)$ is analytic inside the strip, $-a < \mathrm{Im}\, k < b$.

Hence we have

$$\hat{\phi}_-(k) = \int_0^{+\infty} dx\, e^{-ikx}\phi(x) \quad \text{analytic} \quad \text{for} \quad \mathrm{Im}\, k = k_2 < -a + \varepsilon, \tag{7.2.11a}$$

$$\hat{\psi}_+(k) = \int_{-\infty}^{0} dx\, e^{-ikx}\psi(x) \quad \text{analytic} \quad \text{for} \quad \mathrm{Im}\, k = k_2 > -a. \tag{7.2.11b}$$

Various regions of the analyticity are drawn in Figure 7.9.

Recalling Eq. (7.2.7), we write $1-\lambda\hat{K}(k)$ as the ratio of the $-$ function and the $+$ function,

$$1 - \lambda\hat{K}(k) = \frac{Y_-(k)}{Y_+(k)}. \tag{7.2.12}$$

From Eqs. (7.2.7) and (7.2.12), we have

$$Y_-(k)\hat{\phi}_-(k) = -Y_+(k)\hat{\psi}_+(k) \equiv F(k), \tag{7.2.13}$$

where $F(k)$ is an entire function. The asymptotic behavior of $F(k)$ as $|k| \to \infty$ will determine $F(k)$ completely.

We know that

$$\begin{aligned} &\hat{\phi}_-(k) \to 0 \quad \text{as} \quad |k| \to \infty, \\ &\hat{\psi}_+(k) \to 0 \quad \text{as} \quad |k| \to \infty, \\ &\hat{K}(k) \to 0 \quad \text{as} \quad |k| \to \infty. \end{aligned} \tag{7.2.14}$$

By Eq. (7.2.12), we know then

$$\frac{Y_-(k)}{Y_+(k)} \to 1, \quad \text{as} \quad |k| \to \infty. \tag{7.2.15}$$

We only need to know the asymptotic behavior of $Y_-(k)$ or $Y_+(k)$ as $|k| \to \infty$ in order to determine the entire function $F(k)$. Once $F(k)$ is determined we have, from Eq. (7.2.13),

$$\hat{\phi}_-(k) = \frac{F(k)}{Y_-(k)} \tag{7.2.16}$$

and we have finished.

We note that it is convenient to choose a function $Y_-(k)$ which is not only analytic in the lower half-plane, but also has no zeros in the lower half-plane, so that $F(k)/Y_-(k)$ is itself a $-$ function for all entire $F(k)$. Otherwise, we need to choose $F(k)$ so as to have zeros exactly at zeros of $Y_-(k)$ to cancel the possible poles in $F(k)/Y_-(k)$ and to yield the $-$ function $\hat{\phi}_-(k)$.

The *factorization of* $1 - \lambda\hat{K}(k)$ is essential in solving the Wiener–Hopf integral equation of the second kind. As noted earlier, it can be done either by inspection or by the general method based on the Cauchy integral formula.

As a general rule, we assign

Any pole in the lower half-plane $(k = p_l)$	to	$Y_+(k)$,	
Any zero in the lower half-plane $(k = z_l)$	to	$Y_+(k)$,	(7.2.17)
Any pole in the upper half-plane $(k = p_u)$	to	$Y_-(k)$,	
Any zero in the upper half-plane $(k = z_u)$	to	$Y_-(k)$.	

We first solve the following simple example where the factorization is carried out by inspection and illustrate the rationale of this general rule for the assignment.

❑ **Example 7.3.** Solve

$$\phi(x) = \lambda \int_0^{+\infty} e^{-|x-y|}\phi(y)\,dy, \qquad x \geq 0. \tag{7.2.18}$$

Solution. Define

$$\psi(x) = \lambda \int_0^{+\infty} e^{-|x-y|}\phi(y)\,dy, \qquad x < 0. \tag{7.2.19}$$

Also, define

$$\phi(x) = \begin{cases} \phi(x) & \text{for} \quad x \geq 0, \\ 0 & \text{for} \quad x < 0. \end{cases}$$

Take the Fourier transform of Eqs. (7.2.18) and (7.2.19) and add the results together to obtain

$$\hat{\phi}_-(k) + \hat{\psi}_+(k) = \lambda \cdot \frac{2}{k^2+1} \cdot \hat{\phi}_-(k). \tag{7.2.20}$$

Now,

$$K(x) = e^{-|x|} \to \begin{cases} e^{-x} & \text{as} \quad x \to \infty, \\ e^{x} & \text{as} \quad x \to -\infty. \end{cases} \tag{7.2.21}$$

Therefore, $\phi(y)$ can be allowed to grow as fast as $e^{(1-\varepsilon)y}$ as $y \to \infty$. Thus $\hat{\phi}_-(k)$ is analytic for $k_2 < -1 + \varepsilon$. We also find that $\psi(x) \to e^x$ as $x \to -\infty$. Thus $\hat{\psi}_+(k)$ is analytic for $k_2 > -1$.

To solve Eq. (7.2.20), we first write

$$\frac{k^2 + 1 - 2\lambda}{k^2 + 1} \hat{\phi}_-(k) = -\hat{\psi}_+(k), \tag{7.2.22}$$

and then decompose

$$\frac{k^2 + 1 - 2\lambda}{k^2 + 1} \tag{7.2.23}$$

into a ratio of a $-$ function to a $+$ function. The common region of analyticity of $\hat{\phi}_-(k)$, $\hat{\psi}_+(k)$ and $\hat{K}(k)$ of this example are drawn in Figure 7.10.

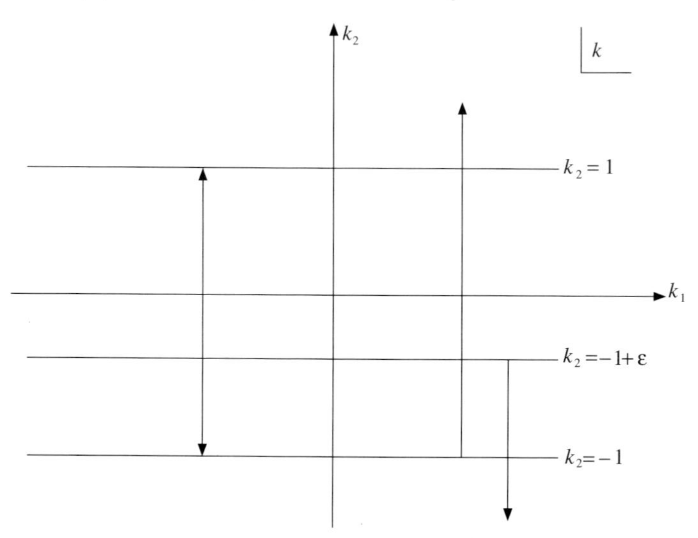

Fig. 7.10: Common region of analyticity of $\hat{\phi}_-(k)$, $\hat{\psi}_+(k)$ and $\hat{K}(k)$ of Example 7.3. $\hat{\phi}_-(k)$ is analytic in the lower half-plane, $\operatorname{Im} k < -1 + \varepsilon$. $\hat{\psi}_+(k)$ is analytic in the upper half-plane, $\operatorname{Im} k > -1$. $\hat{K}(k)$ is analytic inside the strip, $-1 < \operatorname{Im} k < 1$.

The designations, the lower half-plane, and the upper half-plane, must be made relative to a line with

$$-1 < \operatorname{Im} k = k_2 < -1 + \varepsilon. \tag{7.2.24}$$

Referring to Eq. (7.2.23),

$$k^2 + 1 = (k + i)(k - i)$$

so that $k = i$ is a pole of Eq. (7.2.23) in the upper half-plane and $k = -i$ is a pole of Eq. (7.2.23) in the lower half-plane. Now look at the numerator of Eq. (7.2.23),

$$k^2 + 1 - 2\lambda.$$

Case 1. $\lambda < 0. \Rightarrow 1 - 2\lambda > 1.$

$$k^2 + 1 - 2\lambda = (k + i\sqrt{1 - 2\lambda})(k - i\sqrt{1 - 2\lambda}) \tag{7.2.25}$$

The first factor corresponds to a zero in the lower half-plane, while the second factor corresponds to a zero in the upper half-plane.

Case 2. $0 < \lambda < 1/2. \Rightarrow 0 < 1 - 2\lambda < 1.$

$$k^2 + 1 - 2\lambda = (k + i\sqrt{1 - 2\lambda})(k - i\sqrt{1 - 2\lambda}) \tag{7.2.26}$$

Both factors correspond to a zero in the upper half-plane.

Case 3. $\lambda > 1/2. \Rightarrow 1 - 2\lambda < 0.$

$$k^2 + 1 - 2\lambda = (k + \sqrt{2\lambda - 1})(k - \sqrt{2\lambda - 1}) \tag{7.2.27}$$

Both factors correspond to a zero in the upper half-plane.

Now, in general, when we write

$$1 - \lambda \hat{K}(k) = \frac{Y_-(k)}{Y_+(k)},$$

since we will end up with

$$Y_-(k)\hat{\phi}_-(k) = -Y_+(k)\hat{\psi}_+(k) \equiv G(k)$$

which is entire, we wish to have

$$\hat{\phi}_-(k) = \frac{G(k)}{Y_-(k)}$$

analytic in the lower half-plane. So, $Y_-(k)$ must not have any zeros in the lower half-plane. Hence we assign any zeros or poles in the lower half-plane to $Y_+(k)$ so that $Y_-(k)$ has neither poles nor zeros in the lower half-plane.

Now we have

$$\begin{aligned} 1 - \lambda \hat{K}(k) &= \frac{(k^2 + 1 - 2\lambda)}{(k^2 + 1)} \\ &= \frac{(k + i\sqrt{1 - 2\lambda})(k - i\sqrt{1 - 2\lambda})}{(k + i)(k - i)}. \end{aligned}$$

Case 1. $\lambda < 0$.

$$\begin{array}{llll} k + i\sqrt{1-2\lambda} & \Rightarrow \text{zero in the lower half-plane} & \Rightarrow & Y_+(k) \\ k - i\sqrt{1-2\lambda} & \Rightarrow \text{zero in the upper half-plane} & \Rightarrow & Y_-(k) \\ k + i & \Rightarrow \text{pole in the lower half-plane} & \Rightarrow & Y_+(k) \\ k - i & \Rightarrow \text{pole in the upper half-plane} & \Rightarrow & Y_-(k) \end{array}$$

Thus we obtain

$$\begin{cases} Y_-(k) = \dfrac{(k - i\sqrt{1-2\lambda})}{(k-i)}, \\ \\ Y_+(k) = \dfrac{(k+i)}{(k + i\sqrt{1-2\lambda})}. \end{cases} \tag{7.2.28}$$

Hence, in the following equation,

$$Y_-(k)\hat{\phi}_-(k) = -Y_+(k)\hat{\psi}_+(k) = G(k),$$

we know

$$Y_-(k) \to 1, \quad \hat{\phi}_-(k) \to 0, \quad \text{as} \quad k \to \infty,$$

so that

$$G(k) \to 0 \quad \text{as} \quad k \to \infty.$$

By Liouville's theorem, we conclude

$$G(k) = 0,$$

from which it follows that

$$\hat{\phi}_-(k) = 0, \quad \hat{\psi}_+(k) = 0. \tag{7.2.29}$$

So, there exists no nontrivial solution, i.e.,

$$\phi(x) = 0, \quad \text{for} \quad \lambda < 0.$$

Case 2. $0 < \lambda < 1/2$.

With a similar analysis as in Case 1, we obtain

$$\begin{cases} Y_-(k) = \dfrac{(k + i\sqrt{1-2\lambda})(k - i\sqrt{1-2\lambda})}{(k-i)}, \\ Y_+(k) = (k+i). \end{cases} \tag{7.2.30}$$

Noting that

$$Y_-(k) \to k \quad \text{as} \quad k \to \infty,$$

and

$$Y_{-}(k)\hat{\phi}_{-}(k) = G(k),$$

we find that $G(k)$ grows less fast than k as $k \to \infty$. By Liouville's theorem, we find

$$G(k) = A, \quad \text{constant},$$

and thus conclude that

$$\hat{\phi}_{-}(k) = \frac{A(k-i)}{(k+i\sqrt{1-2\lambda})(k-i\sqrt{1-2\lambda})}. \tag{7.2.31}$$

Case 3. $\lambda > 1/2$.

With a similar analysis as in Case 1, we obtain

$$\begin{cases} Y_{-}(k) = \dfrac{(k+\sqrt{2\lambda-1})(k-\sqrt{2\lambda-1})}{(k-i)} & \to k \quad \text{as} \quad k \to \infty, \\ Y_{+}(k) = (k+i) & \to k \quad \text{as} \quad k \to \infty. \end{cases} \tag{7.2.32}$$

Again, we find that

$$G(k) = A, \quad \text{constant},$$

and thus conclude that

$$\hat{\phi}_{-}(k) = \frac{A(k-i)}{(k+\sqrt{2\lambda-1})(k-\sqrt{2\lambda-1})}. \tag{7.2.33}$$

To summarize, we find the following.

$$\text{For } \lambda \leq 0 \Rightarrow \phi(x) = 0. \tag{7.2.34}$$

$$\text{For } \lambda > 0 \Rightarrow \phi(x) = \frac{1}{2\pi}\int_C dk\, e^{ikx} A(k-i)/(k^2+1-2\lambda), \tag{7.2.35}$$

where the inversion contour C is indicated in Figure 7.11.

For $x > 0$, we close the contour in the upper half-plane and get the contribution from both poles in the upper half-plane in either **Case 2** or **Case 3**. The result of inversions are the following.

Case 2. $0 < \lambda < 1/2$.

$$\phi(x) = C\left(\cosh\sqrt{1-2\lambda}x + \frac{\sinh\sqrt{1-2\lambda}x}{\sqrt{1-2\lambda}}\right), \qquad x > 0. \tag{7.2.36}$$

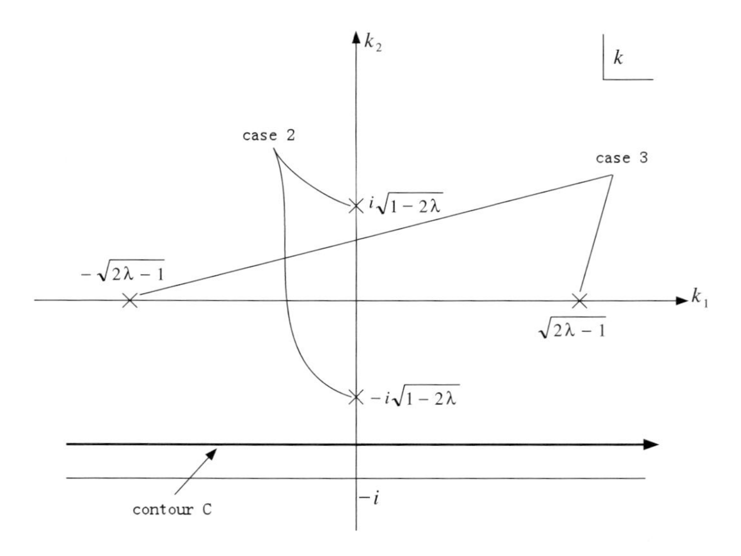

Fig. 7.11: The inversion contour for $\phi(x)$ of Eq. (7.2.35). Simple poles are located at $k = \pm i\sqrt{1-2\lambda}$ for Case 2 and at $k = \pm\sqrt{2\lambda - 1}$ for Case 3.

Case 3. $\lambda > 1/2$.

$$\phi(x) = C\left(\cos\sqrt{2\lambda-1}x + \frac{\sin\sqrt{2\lambda-1}x}{\sqrt{2\lambda-1}}\right), \qquad x > 0. \tag{7.2.37}$$

We shall now consider another example where the factorization also is carried out by inspection after some juggling of the gamma functions.

❑ **Example 7.4.** Solve

$$\phi(x) = \lambda\int_0^{+\infty} \frac{1}{\cosh[\frac{1}{2}(x-y)]}\phi(y)\,dy, \qquad x \geq 0. \tag{7.2.38}$$

Solution. We begin with the Fourier transform of the kernel $K(x)$.

$$K(x) = \frac{1}{\cosh\left(\frac{1}{2}x\right)} \to 2e^{-\frac{1}{2}|x|}, \quad \text{as} \quad |x| \to \infty.$$

Then $\hat{K}(k)$ is analytic inside the strip,

$$-\frac{1}{2} < \operatorname{Im} k = k_2 < \frac{1}{2}. \tag{7.2.39}$$

We calculate $\hat{K}(k)$ as follows.

$$\hat{K}(k) = \int_{-\infty}^{+\infty} dx\, e^{-ikx} \frac{1}{\cosh\left(\frac{1}{2}x\right)} = 2\int_{-\infty}^{+\infty} dx\, \frac{e^{-ikx}e^{\frac{1}{2}x}}{(e^x+1)}.$$

Setting

$$e^x = t, \qquad x = \ln t, \qquad dx = \frac{dt}{t},$$

we have

$$\hat{K}(k) = 2\int_0^{+\infty} dt \, \frac{t^{-ik-\frac{1}{2}}}{(t+1)}.$$

A further change of variable

$$\rho = \frac{1}{(t+1)}, \quad t = \frac{(1-\rho)}{\rho}, \quad dt = \frac{-d\rho}{\rho^2}$$

results in

$$\hat{K}(k) = 2\int_0^1 d\rho \, \rho^{ik-\frac{1}{2}}(1-\rho)^{-ik-\frac{1}{2}}.$$

Recalling the definition of the *Beta function* $B(n,m)$,

$$B(n,m) = \int_0^1 d\rho \, \rho^{n-1}(1-\rho)^{m-1} = \frac{\Gamma(n)\Gamma(m)}{\Gamma(n+m)},$$

we have

$$\hat{K}(k) = \frac{2\Gamma\left(ik+\frac{1}{2}\right)\Gamma\left(-ik+\frac{1}{2}\right)}{\Gamma\left(ik+\frac{1}{2}-ik+\frac{1}{2}\right)} = 2\Gamma\left(ik+\frac{1}{2}\right)\Gamma\left(-ik+\frac{1}{2}\right).$$

Recalling the property of the gamma function,

$$\Gamma(z)\Gamma(1-z) = \frac{\pi}{\sin \pi z}, \tag{7.2.40}$$

we thus obtain the Fourier transform of $K(x)$ as

$$\hat{K}(k) = \frac{2\pi}{\cosh \pi k}. \tag{7.2.41}$$

Defining $\psi(x)$ by

$$\psi(x) = \lambda \int_0^{+\infty} \frac{1}{\cosh\left[\frac{1}{2}(x-y)\right]}\phi(y)\, dy, \quad x < 0, \tag{7.2.42}$$

we obtain the following equation as usual,

$$(1 - \lambda\hat{K}(k))\hat{\phi}_-(k) = -\hat{\psi}_+(k), \tag{7.2.43}$$

where

$$1 - \lambda \hat{K}(k) = 1 - \frac{2\pi\lambda}{\cosh \pi k} = \frac{Y_-(k)}{Y_+(k)}, \tag{7.2.44}$$

and the regions of the analyticity of $\hat{\phi}_-(k)$ and $\hat{\psi}_+(k)$ are such that

$$\begin{aligned} &\text{(i)} \quad \hat{\phi}_-(k) \text{ is analytic in the lower half-plane } (\operatorname{Im} k \leq -1/2), \\ &\text{(ii)} \quad \hat{\psi}_+(k) \text{ is analytic in the upper half-plane } (\operatorname{Im} k > -1/2). \end{aligned} \tag{7.2.45}$$

Rewriting Eq. (7.2.43) in terms of $Y_\pm(k)$, we have

$$Y_-(k)\hat{\phi}_-(k) = -Y_+(k)\hat{\psi}_+(k) \equiv G(k), \tag{7.2.46}$$

where $G(k)$ is entire in k.

Factorizing $1 - \lambda \hat{K}(k)$:

$$\frac{Y_-(k)}{Y_+(k)} = 1 - \frac{2\pi\lambda}{\cosh \pi k} = \frac{\cosh \pi k - 2\pi\lambda}{\cosh \pi k}. \tag{7.2.47}$$

Case 1. $\quad 0 < 2\pi\lambda \leq 1.$

Setting

$$\cos \pi\alpha \equiv 2\pi\lambda, \qquad 0 \leq \alpha < \frac{1}{2}, \tag{7.2.48}$$

we have

$$\begin{aligned} \frac{Y_-(k)}{Y_+(k)} &= \frac{\cos(i\pi k) - \cos \pi\alpha}{\sin \pi \left(ik + \frac{1}{2}\right)} \\ &= \frac{2 \sin\left[\frac{\pi}{2}(\alpha + ik)\right] \sin\left[\frac{\pi}{2}(\alpha - ik)\right]}{\sin \pi \left(ik + \frac{1}{2}\right)}. \end{aligned} \tag{7.2.49}$$

Replacing all sine functions in Eq. (7.2.49) with the appropriate product of the gamma functions by the use of the formula

$$\sin \pi z = \frac{\pi}{\Gamma(z)\Gamma(1-z)}, \tag{7.2.50}$$

we obtain

$$\begin{aligned} &\frac{Y_-(k)}{Y_+(k)} \\ &= \frac{2\pi \Gamma\left(\frac{1}{2} + ik\right) \Gamma\left(\frac{1}{2} - ik\right)}{\Gamma\left(\frac{\alpha+ik}{2}\right) \Gamma\left(\frac{\alpha-ik}{2}\right) \Gamma\left(1 - \frac{\alpha+ik}{2}\right) \Gamma\left(1 - \frac{\alpha-ik}{2}\right)}. \end{aligned} \tag{7.2.51}$$

We note that

$$\begin{cases} \Gamma(z) \text{ has a simple pole} & \text{at} \quad z = 0, -1, -2, \cdots, \\ \Gamma(1-z) \text{ has a simple pole} & \text{at} \quad z = 1, 2, 3, \cdots. \end{cases} \tag{7.2.52}$$

(1) $\Gamma\left(\frac{1}{2}+ik\right)$ has simple poles at $k = i\frac{1}{2}, i\frac{3}{2}, i\frac{5}{2}, i\frac{7}{2}, \cdots$, all of which are assigned to $Y_-(k)$.

(2) $\Gamma\left(\frac{1}{2}-ik\right)$ has simple poles at $k = -i\frac{1}{2}, -i\frac{3}{2}, -i\frac{5}{2}, -i\frac{7}{2}, \cdots$, all of which are assigned to $Y_+(k)$.

(3) $\Gamma\left(\frac{\alpha+ik}{2}\right)$ has simple poles at $k = i\alpha, i(2+\alpha), i(4+\alpha), \cdots$, all of which are assigned to $Y_-(k)$.

(4) $\Gamma\left(\frac{\alpha-ik}{2}\right)$ has simple poles at $k = -i\alpha, -i(2+\alpha), -i(4+\alpha), \cdots$. Since $0 < \alpha < 1/2$, the first pole at $k = -i\alpha$ is assigned to $Y_-(k)$, while the remaining poles are assigned to $Y_+(k)$. Using the property of the gamma function, $\Gamma(z) = \frac{\Gamma(z+1)}{z}$, we rewrite

$$\Gamma\left(\frac{\alpha-ik}{2}\right) = \frac{2}{\alpha-ik}\Gamma\left(1+\frac{\alpha-ik}{2}\right),$$

where $(\alpha-ik)/2$ is assigned to $Y_-(k)$ while $\Gamma\left(1+\frac{\alpha-ik}{2}\right)$ is assigned to $Y_+(k)$.

(5) $\Gamma(1-\frac{\alpha+ik}{2})$ has simple poles at $k = -i(2-\alpha), -i(4-\alpha), -i(6-\alpha), \cdots$, all of which are assigned to $Y_+(k)$.

(6) $\Gamma(1-\frac{\alpha-ik}{2})$ has simple poles at $k = i(2-\alpha), i(4-\alpha), i(6-\alpha), \cdots$, all of which are assigned to $Y_-(k)$.

Then we obtain $Y_\pm(k)$ as follows.

$$Y_-(k) = \frac{-2\pi\Gamma\left(\frac{1}{2}+ik\right)}{\Gamma\left(\frac{\alpha+ik}{2}\right)\Gamma\left(-\frac{\alpha-ik}{2}\right)}, \tag{7.2.53}$$

$$Y_+(k) = \frac{\Gamma\left(1+\frac{\alpha-ik}{2}\right)\Gamma\left(1-\frac{\alpha+ik}{2}\right)}{\Gamma\left(\frac{1}{2}-ik\right)}. \tag{7.2.54}$$

Now follows the determination of $G(k)$, which is determined by the asymptotic behavior of $Y_\pm(k)$ as $k \to \infty$. Making use of the *Duplication formula* and the *Stirling formula*,

$$\Gamma(2z) = \frac{2^{2z-1}\Gamma(z)\Gamma\left(z+\frac{1}{2}\right)}{\sqrt{\pi}}, \tag{7.2.55}$$

$$\lim_{|z|\to\infty} \frac{\Gamma(z+\beta)}{\Gamma(z)} \sim z^\beta, \tag{7.2.56}$$

we find the asymptotic behavior of $Y_-(k)$ to be given by

$$Y_-(k) \sim -i\sqrt{\frac{\pi}{2}}2^{ik} \cdot k. \tag{7.2.57}$$

Defining

$$Z_\pm(k) \equiv 2^{-ik} \cdot Y_\pm(k), \tag{7.2.58}$$

we find

$$Z_{\pm}(k) \sim -i\sqrt{\frac{\pi}{2}}k, \tag{7.2.59}$$

since

$$\frac{Z_{-}(k)}{Z_{+}(k)} = \frac{Y_{-}(k)}{Y_{+}(k)} = 1 - \lambda\hat{K}(k) \to 1 \quad \text{as} \quad k \to \infty.$$

Then Eq. (7.2.46) becomes

$$Z_{-}(k)\hat{\phi}_{-}(k) = -Z_{+}(k)\hat{\psi}_{+}(k) = 2^{-ik}G(k) \equiv g(k), \tag{7.2.60}$$

where $g(k)$ is now entire. Since

$$\hat{\phi}_{-}(k), \hat{\psi}_{+}(k) \to 0 \quad \text{as} \quad k \to \infty,$$

and Eq. (7.2.59) for $Z_{\pm}(k)$, $g(k)$ cannot grow as fast as k. By Liouville's theorem, we then have

$$g(k) = C', \quad \text{constant.}$$

Thus we obtain

$$\hat{\phi}_{-}(k) = \frac{C'}{Z_{-}(k)} = C''2^{ik}\frac{\Gamma\left(\frac{\alpha+ik}{2}\right)\Gamma\left(-\frac{\alpha-ik}{2}\right)}{\Gamma\left(\frac{1}{2}+ik\right)}. \tag{7.2.61}$$

We now invert $\hat{\phi}_{-}(k)$ to obtain $\phi(x)$,

$$\phi(x) = C''\int_{-\infty}^{+\infty}\frac{dk}{2\pi}e^{ikx}2^{ik}\frac{\Gamma\left(\frac{\alpha+ik}{2}\right)\Gamma(-\frac{\alpha-ik}{2})}{\Gamma\left(\frac{1}{2}+ik\right)}, \qquad x \geq 0, \tag{7.2.62}$$

$$\phi(x) = 0, \quad x < 0.$$

For $x > 0$, we close the contour in the upper half-plane, picking up the pole contributions from $\Gamma\left(\frac{\alpha+ik}{2}\right)$ and $\Gamma\left(-\frac{\alpha-ik}{2}\right)$. Poles of $\Gamma\left(\frac{\alpha+ik}{2}\right)$ are located at $k = i(2n+\alpha), n = 0, 1, 2, \cdots$. Poles of $\Gamma\left(-\frac{\alpha-ik}{2}\right)$ are located at $k = i(2n-\alpha), \quad n = 0, 1, 2, \cdots$. Since

$$\text{Res. } \Gamma(z)\,|_{z=-n} = (-1)^{n}\frac{1}{n!}, \tag{7.2.63}$$

we have

$$\phi(x) = C''\sum_{n=0}^{\infty}\left\{e^{-(2n+\alpha)x}2^{-(2n+\alpha)}\frac{(-1)^{n}}{n!}\frac{\Gamma(-n-\alpha)}{\Gamma\left(\frac{1}{2}-2n-\alpha\right)} + (\alpha \to -\alpha)\right\}. \tag{7.2.64}$$

Since

$$\Gamma(z) = \frac{\pi}{\sin\pi z}\frac{1}{\Gamma(1-z)},$$

we have

$$\Gamma(-n-\alpha) = \frac{(-1)^{n+1}\pi}{\sin\alpha\pi}\frac{1}{\Gamma(n+1+\alpha)},$$

and with the use of the Duplication formula,

$$\Gamma\left(\frac{1}{2}-2n-\alpha\right) = \frac{\pi}{\cos\alpha\pi}\cdot\sqrt{2\pi}2^{-(2n+\alpha)}\cdot\frac{1}{\Gamma\left(\frac{1}{4}+\frac{\alpha}{2}+n\right)\Gamma\left(\frac{3}{4}+\frac{\alpha}{2}+n\right)},$$

we have

$$\frac{\Gamma(-n-\alpha)}{\Gamma\left(\frac{1}{2}-2n-\alpha\right)} = \frac{(-1)^{n+1}}{\sqrt{2\pi}}\cdot 2^{(2n+\alpha)}\cdot\frac{\cos\alpha\pi}{\sin\alpha\pi}\cdot\frac{\Gamma\left(\frac{1}{4}+\frac{\alpha}{2}+n\right)\Gamma\left(\frac{3}{4}+\frac{\alpha}{2}+n\right)}{\Gamma(1+\alpha+n)}.$$

Our solution $\phi(x)$ is given by

$$\begin{aligned}\phi(x) &= C'''\left(\frac{\cos\alpha\pi}{\sin\alpha\pi}\right)\\ &\cdot\sum_{n=0}^{\infty}\left\{e^{-\alpha x}\frac{\left(e^{-2x}\right)^n}{n!}\frac{\Gamma(\frac{1}{4}+\frac{\alpha}{2}+n)\Gamma\left(\frac{3}{4}+\frac{\alpha}{2}+n\right)}{\Gamma\left(1+\alpha+n\right)}-(\alpha\to-\alpha)\right\}.\end{aligned} \tag{7.2.65}$$

We recall that the *hypergeometric function* $F(a,b,c;z)$ is given by

$$F(a,b,c;z) = \sum_{n=0}^{\infty}\frac{\Gamma(\alpha+n)}{\Gamma(a)}\cdot\frac{\Gamma(b+n)}{\Gamma(b)}\cdot\frac{\Gamma(c)}{\Gamma(c+n)}\cdot\frac{z^n}{n!}. \tag{7.2.66}$$

Setting

$$a = \frac{1}{4}+\frac{\alpha}{2},\qquad b = \frac{3}{4}+\frac{\alpha}{2},\qquad c = 1+\alpha,$$

we have

$$\begin{aligned}\phi(x) = C'''\left(\frac{\cos\alpha\pi}{\sin\alpha\pi}\right)&\left[e^{-\alpha x}\frac{\Gamma\left(\frac{1}{4}+\frac{\alpha}{2}\right)\Gamma\left(\frac{3}{4}+\frac{\alpha}{2}\right)}{\Gamma\left(1+\alpha\right)}\right.\\ &\left.\cdot F\left(\frac{1}{4}+\frac{\alpha}{2},\frac{3}{4}+\frac{\alpha}{2},1+\alpha;e^{-2x}\right)-(\alpha\to-\alpha)\right].\end{aligned} \tag{7.2.67}$$

Recall that the *Legendre function of the second kind* $Q_{\alpha-\frac{1}{2}}(z)$ is given by

$$\begin{aligned}Q_{\alpha-\frac{1}{2}}(z) &= \frac{\sqrt{\pi}}{2^{\alpha+\frac{1}{2}}}\cdot\frac{\Gamma\left(\frac{1}{2}+\alpha\right)}{\Gamma(1+\alpha)}\cdot z^{-\left(\frac{1}{2}+\alpha\right)}\cdot F\left(\frac{1}{4}+\frac{\alpha}{2},\frac{3}{4}+\alpha,1+\alpha;z^{-2}\right)\\ &= \frac{1}{2}\cdot\frac{\Gamma\left(\frac{1}{4}+\frac{\alpha}{2}\right)\Gamma(\frac{3}{4}+\frac{\alpha}{2})}{\Gamma(1+\alpha)}\cdot z^{-\left(\frac{1}{2}+\alpha\right)}\cdot F\left(\frac{1}{4}+\frac{\alpha}{2},\frac{3}{4}+\frac{\alpha}{2},1+\alpha;z^{-2}\right),\end{aligned} \tag{7.2.68}$$

so that our solution given above can be simplified as

$$\phi(x) = C''' \left(\frac{\cos\alpha\pi}{\sin\alpha\pi}\right) e^{\frac{x}{2}} \left\{Q_{\alpha-\frac{1}{2}}(e^x) - Q_{-\alpha-\frac{1}{2}}(e^x)\right\}. \tag{7.2.69}$$

Recall that the *Legendre function of the first kind* $P_\beta(z)$ is given by

$$P_\beta(z) = \frac{1}{\pi}\left(\frac{\sin\beta\pi}{\cos\beta\pi}\right)\{Q_\beta(z) - Q_{-\beta-1}(z)\}, \tag{7.2.70}$$

so that

$$P_{\alpha-\frac{1}{2}}(z) = -\frac{1}{\pi}\left(\frac{\cos\alpha\pi}{\sin\alpha\pi}\right)\left\{Q_{\alpha-\frac{1}{2}}(z) - Q_{-\alpha-\frac{1}{2}}(z)\right\}. \tag{7.2.71}$$

So, the final expression for $\phi(x)$ is given by

$$\phi(x) = C\cdot\left(\exp\frac{x}{2}\right)\cdot P_{\alpha-\frac{1}{2}}(e^x), \quad x \geq 0, \qquad 0 \leq \alpha < 1/2, \quad 2\pi\lambda = \cos\alpha\pi. \tag{7.2.72}$$

A similar analysis can be carried out for the cases, $2\pi\lambda > 1$, and $\lambda < 0$. We only list the final answers for all cases.

Summary of Example 7.4

Case 1. $0 < 2\pi\lambda \leq 1, \quad 2\pi\lambda = \cos\alpha\pi, \quad 0 \leq \alpha < 1/2.$

$$\phi(x) = C_1\cdot\left(\exp\frac{x}{2}\right)\cdot P_{\alpha-\frac{1}{2}}(e^x), \qquad x \geq 0.$$

Case 2. $2\pi\lambda > 1, \quad 2\pi\lambda = \cosh\alpha\pi, \quad \alpha > 0.$

$$\phi(x) = C_2\cdot\left(\exp\frac{x}{2}\right)\cdot P_{i\alpha-\frac{1}{2}}(e^x), \qquad x \geq 0.$$

Case 3. $2\pi\lambda \leq 0.$

$$\phi(x) = 0, \qquad x \geq 0.$$

7.3 General Decomposition Problem

In the original Wiener–Hopf problem we examined, in Section 7.1,

$$\phi_-(k) = \psi_+(k) + F(k), \tag{7.3.1}$$

we need to make the decomposition,

$$F(k) = F_+(k) + F_-(k). \tag{7.3.2}$$

In the problems we just examined in Section 7.1, i.e., the homogeneous Wiener–Hopf integral equation of the second kind, we need to make the decomposition,

$$1 - \lambda \hat{K}(k) = \frac{Y_-(k)}{Y_+(k)}. \tag{7.3.3}$$

Here we discuss how this can be done in general, rather than by inspection.

Consider the first problem (*sum-splitting*) first. (Figure 7.12.)

$$\phi_-(k) = \psi_+(k) + F(k).$$

Assume that

$$\phi_-(k),\ \psi_+(k),\ F(k) \to 0 \quad \text{as} \quad k \to \infty. \tag{7.3.4}$$

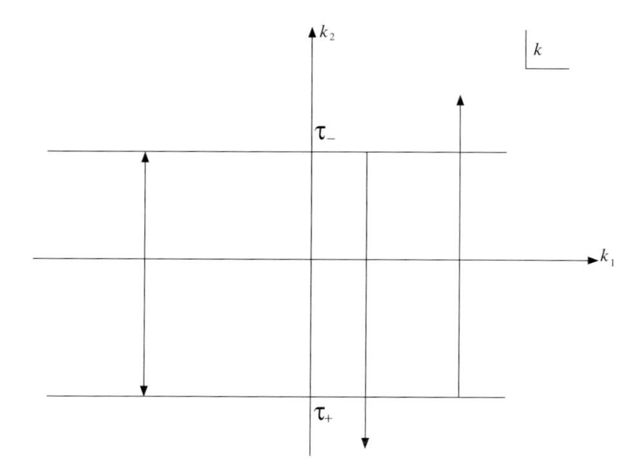

Fig. 7.12: Sum-splitting of $F(k)$. $\phi_-(k)$ is analytic in the lower half-plane, $\mathrm{Im}\, k < \tau_-$. $\psi_+(k)$ is analytic in the upper half-plane, $\mathrm{Im}\, k > \tau_+$. $F(k)$ is analytic inside the strip, $\tau_+ < \mathrm{Im}\, k < \tau_-$.

Examine the decomposition of $F(k)$. Since $F(k)$ is analytic inside the strip,

$$\tau_+ < \mathrm{Im}\, k = k_2 < \tau_-, \tag{7.3.5}$$

by the Cauchy integral formula, we have

$$F(k) = \frac{1}{2\pi i} \int_C \frac{F(\zeta)}{\zeta - k}\, d\zeta \tag{7.3.6}$$

where the complex integration contour C consists of the following path as in Figure 7.13,

$$C = C_1 + C_2 + C_\uparrow + C_\downarrow. \tag{7.3.7}$$

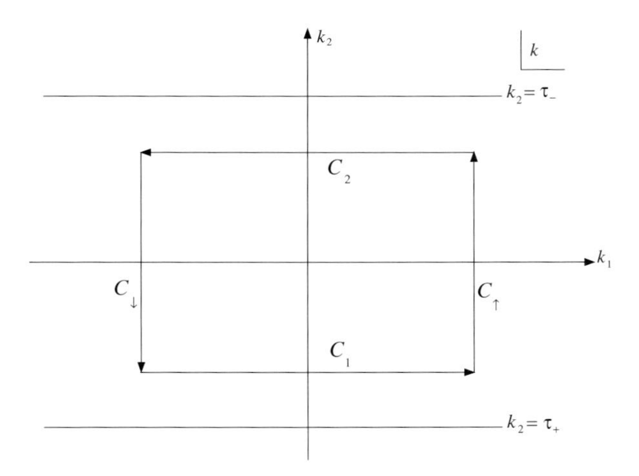

Fig. 7.13: Sum-splitting contour C of Eq. (7.3.6) for $F(k)$ inside the strip, $\tau_+ < \operatorname{Im} k < \tau_-$.

The contributions from $C_\uparrow$ and $C_\downarrow$ vanish as these contours tend to infinity, since

$|F(\zeta)|$ is bounded (actually $\to 0$),

$$\left|\frac{1}{(\zeta - k)}\right| \to 0 \quad \text{as} \quad \zeta \to \infty.$$

Thus we have

$$F(k) = \frac{1}{2\pi i}\int_{C_1}\frac{F(\zeta)}{\zeta - k}\,d\zeta + \frac{1}{2\pi i}\int_{C_2}\frac{F(\zeta)}{\zeta - k}\,d\zeta, \tag{7.3.8}$$

where the contribution from C_1 is a $+$ function, analytic for $\operatorname{Im} k = k_2 > \tau_+$, while the contribution from C_2 is a $-$ function, analytic for $\operatorname{Im} k = k_2 < \tau_-$, i.e.,

$$F_+(k) = \frac{1}{2\pi i}\int_{-\infty + i\tau_+}^{+\infty + i\tau_+}\frac{F(\zeta)}{\zeta - k}\,d\zeta, \tag{7.3.9a}$$

$$F_-(k) = \frac{1}{2\pi i}\int_{-\infty + i\tau_-}^{+\infty + i\tau_-}\frac{F(\zeta)}{\zeta - k}\,d\zeta. \tag{7.3.9b}$$

Consider now the *factorization of* $1 - \lambda\hat{K}(k)$ *into a ratio of the* $-$ *function to the* $+$ *function*. The function $1 - \lambda\hat{K}(k)$ is analytic inside the strip,

$$-a < \operatorname{Im} k = k_2 < b, \tag{7.3.10}$$

and the inversion contour is somewhere inside the strip,

$$-a < \operatorname{Im} k = k_2 < -a + \varepsilon. \tag{7.3.11}$$

The analytic function $1 - \lambda\hat{K}(k)$ may have some zeros inside the strip,

$$-a < \operatorname{Im} k = k_2 < b.$$

Choose a rectangular contour, as indicated in Figure 7.14, below all zeros of $1 - \lambda\hat{K}(k)$ inside the strip,

$$-a < \operatorname{Im} k = k_2 < b.$$

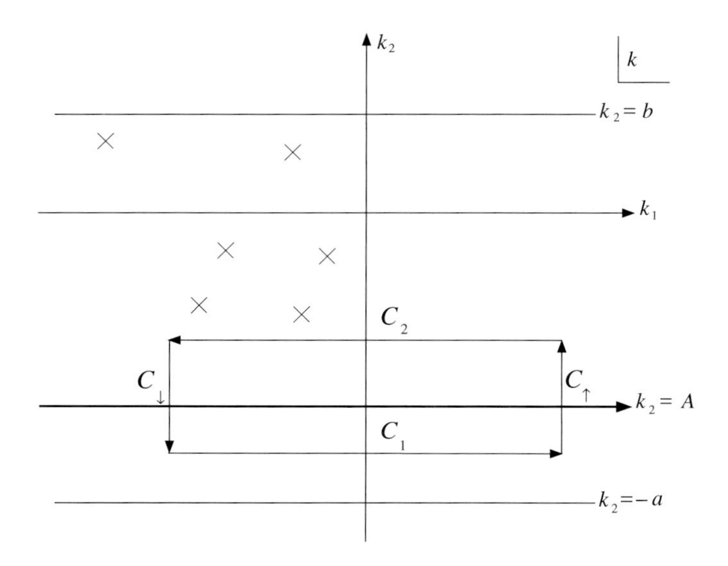

Fig. 7.14: Rectangular contour for the factorization of $1-\lambda\hat{K}(k)$. This contour is chosen below all the zeros of $1 - \lambda\hat{K}(k)$ inside the strip, $-a < \operatorname{Im} k < b$.

Note if $1 - \lambda\hat{K}(k)$ has a zero on $k_2 = -a$, it is all right, since it just remains in the lower half-plane. The inversion contour $k_2 = A$ will be chosen inside this rectangle. Now, $1 - \lambda\hat{K}(k)$ is analytic inside the rectangle

$$C_1 + C_\uparrow + C_2 + C_\downarrow$$

and has no zeros inside this rectangle. Also since

$$\hat{K}(k) \to 0 \quad \text{as} \quad k \to \infty,$$

we know

$$1 - \lambda\hat{K}(k) \to 1 \quad \text{as} \quad k \to \infty.$$

In order to express $1 - \lambda\hat{K}(k)$ as the ratio $Y_-(k)/Y_+(k)$, we take the logarithm of (7.3.3) to find

$$\ln\left[1 - \lambda\hat{K}(k)\right] = \ln\left[\frac{Y_-(k)}{Y_+(k)}\right] = \ln Y_-(k) - \ln Y_+(k). \tag{7.3.12}$$

Now, $\ln[1-\lambda\hat{K}(k)]$ is itself analytic in the rectangle (because it has no branch points since $1-\lambda\hat{K}(k)$ has no zeros there), so we can apply the Cauchy integral formula

$$\ln\left[1-\lambda\hat{K}(k)\right]=\frac{1}{2\pi i}\int_C \frac{\ln\left[1-\lambda\hat{K}(\zeta)\right]}{\zeta-k}\,d\zeta,$$

with k inside the rectangle and C consisting of $C_1+C_2+C_\uparrow+C_\downarrow$. Thus we write

$$\begin{aligned}\ln\left[1-\lambda\hat{K}(k)\right]&=\ln Y_-(k)-\ln Y_+(k)\\&=\frac{1}{2\pi i}\int_{C_1}\frac{\ln\left[1-\lambda\hat{K}(\zeta)\right]}{\zeta-k}\,d\zeta-\frac{1}{2\pi i}\int_{-C_2}\frac{\ln\left[1-\lambda\hat{K}(\zeta)\right]}{\zeta-k}\,d\zeta\\&\quad+\frac{1}{2\pi i}\int_{C\uparrow+C\downarrow}\frac{\ln\left[1-\lambda\hat{K}(\zeta)\right]}{\zeta-k}\,d\zeta.\end{aligned} \tag{7.3.13}$$

In Eq. (7.3.13), it is tempting to drop the contributions from $C_\uparrow$ and $C_\downarrow$ altogether. It is, however, not always possible to do so. Because of the multivaluedness of the logarithm, we may have, in the limit $|\zeta|\to\infty$,

$$\ln\left[1-\lambda\hat{K}(\zeta)\right]\to\ln e^{2\pi in}=2\pi in,\qquad(n=0,\pm1,\pm2,\cdots) \tag{7.3.14}$$

and we have no guarantee that the contributions from $C_\uparrow$ and $C_\downarrow$ cancel each other. In other words, $1-\lambda\hat{K}(\zeta)$ may develop a phase angle as ζ ranges from $-\infty+iA$ to $+\infty+iA$.

Definition of Wiener–Hopf index. Let us define the *index* ν of $1-\lambda\hat{K}(\zeta)$ by

$$\nu\equiv\frac{1}{2\pi i}\ln\left[1-\lambda\hat{K}(\zeta)\right]\Big|_{\zeta=-\infty+iA}^{\zeta=+\infty+iA}. \tag{7.3.15}$$

Graphically we do the following: Plot $z=[1-\lambda\hat{K}(\zeta)]$ as ζ ranges from $-\infty+iA$ to $+\infty+iA$ in the complex z plane, and count the number of counter-clockwise revolutions z makes about the origin. The index ν is equal to the number of these revolutions.

We shall now examine the properties of the index ν; in particular, a *relationship between the index ν and the zeros and the poles of $1-\lambda\hat{K}(k)$ in the complex k plane.* Suppose $1-\lambda\hat{K}(k)$ has a zero in the upper half-plane, say,

$$1-\lambda\hat{K}(k)=k-z_u,\qquad \operatorname{Im}z_u>-a.$$

Then the contribution from this z_u to the index ν is

$$\nu(z_u)=\frac{1}{2\pi i}\left[0-(-i\pi)\right]=\frac{1}{2}.$$

Similar analysis yields the following results:

$$\begin{array}{lcl} \textbf{zero in the upper half-plane} & \Rightarrow & \nu(z_u) = +\frac{1}{2}, \\ \textbf{pole in the upper half-plane} & \Rightarrow & \nu(p_u) = -\frac{1}{2}, \\ \textbf{zero in the lower half-plane} & \Rightarrow & \nu(z_l) = -\frac{1}{2}, \\ \textbf{pole in the lower half-plane} & \Rightarrow & \nu(p_l) = +\frac{1}{2}. \end{array} \tag{7.3.16}$$

In many cases, the translation kernel $K(x-y)$ is of the form

$$K(x-y) = K(|x-y|). \tag{7.3.17}$$

Then $\hat{K}(k)$ is even in k,

$$\hat{K}(k) = \hat{K}(-k), \tag{7.3.18}$$

so that $1-\lambda\hat{K}(k)$ (which is even) has an equal number of zeros (poles) in the upper half-plane and in the lower half-plane,

$$\textbf{number of } z_u = \textbf{number of } z_l, \quad \textbf{number of } p_u = \textbf{number of } p_l.$$

Thus the *index of* $1-\lambda\hat{K}(k)$ *on the real line is equal to zero* ($\nu \equiv 0$) *for* $\hat{K}(k)$ *even.*

Suppose we now lift the path above the real line ($\operatorname{Im} k = 0$) into the upper half-plane. As the path C ($\operatorname{Im} k = A$) passes by a zero of $1-\lambda\hat{K}(k)$ in $\operatorname{Im} k > 0$, the index ν of $1-\lambda\hat{K}(k)$ with respect to the path C ($\operatorname{Im} k = A$) decreases by 1. This is because the point $k = z_0$ is the zero in the upper half-plane with respect to the path $C_<$ ($\operatorname{Im} k = A^-$) while it is the zero in the lower half-plane with respect to the path $C_>$ ($\operatorname{Im} k = A^+$), and hence

$$\triangle\nu = \nu(z_l) - \nu(z_u) = -\frac{1}{2} - \frac{1}{2} = -1. \tag{7.3.19a}$$

Likewise, for a pole of $1-\lambda\hat{K}(k)$ in $\operatorname{Im} k > 0$, we find

$$\Delta\nu = \nu(p_l) - \nu(p_u) = +\frac{1}{2} + \frac{1}{2} = +1. \tag{7.3.19b}$$

Consider first the case when the index ν is equal to *zero*,

$$\nu = 0. \tag{7.3.20}$$

Then we choose a branch so that $\ln[1-\lambda\hat{K}(\zeta)]$ vanishes on $C_\uparrow$ and $C_\downarrow$. We then have

$$\ln Y_-(k) - \ln Y_+(k) = \frac{1}{2\pi i}\int_{C_1} \frac{\ln[1-\lambda\hat{K}(\zeta)]}{\zeta - k} d\zeta - \frac{1}{2\pi i}\int_{-C_2} \frac{\ln[1-\lambda\hat{K}(\zeta)]}{\zeta - k} d\zeta. \tag{7.3.21}$$

In the first integral on the second line of Eq. (7.3.21), we may let k be anywhere above C_1 where C_1 is arbitrarily close to $\operatorname{Im} k = k_2 = -a$ from above. Then we conclude that the integral

$$\frac{1}{2\pi i}\int_{C_1}\frac{\ln[1-\lambda\hat{K}(\zeta)]}{\zeta-k}\,d\zeta, \qquad \operatorname{Im} k = k_2 > -a,$$

is analytic in the upper half-plane, and hence is identified to be a + function,

$$\ln Y_+(k) = -\frac{1}{2\pi i}\int_{C_1}\frac{\ln[1-\lambda\hat{K}(\zeta)]}{\zeta-k}\,d\zeta, \qquad \operatorname{Im} k = k_2 > -a. \tag{7.3.22}$$

It also vanishes as $|k| \to \infty$ in the upper half-plane ($\operatorname{Im} k > -a$). In the second integral on the second line of Eq. (7.3.21), we may let k be anywhere below $-C_2$ where $-C_2$ is arbitrarily close to $\operatorname{Im} k = k_2 = -a$ from above. Then we conclude that the integral

$$\frac{1}{2\pi i}\int_{-C_2}\frac{\ln[1-\lambda\hat{K}(\zeta)]}{\zeta-k}\,d\zeta, \qquad \operatorname{Im} k = k_2 \leq -a,$$

is analytic in the lower half-plane, and hence is identified to be a − function,

$$\ln Y_-(k) = -\frac{1}{2\pi i}\int_{-C_2}\frac{\ln[1-\lambda\hat{K}(\zeta)]}{\zeta-k}\,d\zeta, \qquad \operatorname{Im} k = k_2 \leq -a. \tag{7.3.23}$$

It also vanishes as $|k| \to \infty$ in the lower half-plane ($\operatorname{Im} k \leq -a$). Thus

$$Y_+(k) = \exp\left[-\frac{1}{2\pi i}\int_{C_1}\frac{\ln[1-\lambda\hat{K}(\zeta)]}{\zeta-k}\,d\zeta\right], \tag{7.3.24}$$

$$Y_-(k) = \exp\left[-\frac{1}{2\pi i}\int_{-C_2}\frac{\ln[1-\lambda\hat{K}(\zeta)]}{\zeta-k}\,d\zeta\right]. \tag{7.3.25}$$

We also note that

$$Y_\pm(k) \to 1 \quad \text{as} \quad |k| \to \infty \quad \text{in} \quad \begin{cases} \operatorname{Im} k > -a, \\ \operatorname{Im} k \leq -a. \end{cases} \tag{7.3.26}$$

Then the entire function $G(k)$ in the following equation,

$$Y_-(k)\hat{\phi}_-(k) = -Y_+(k)\hat{\psi}_+(k) = G(k),$$

must vanish identically, by Liouville's theorem. Hence

$$\hat{\phi}_-(k) = 0 \quad \text{or} \quad \phi(x) = 0 \quad \text{when} \quad \nu = 0. \tag{7.3.27}$$

Consider next the case when index ν is *positive*,

$$\nu > 0. \tag{7.3.28}$$

Instead of dealing with $C_\uparrow$ and $C_\downarrow$ of the integral (7.3.13), we *construct the object whose index is equal to zero*,

$$\prod_{i=1}^{\nu}\left(\frac{k-z_l(i)}{k-p_u(i)}\right)\left[1-\lambda\hat{K}(k)\right] = \frac{Z_-(k)}{Z_+(k)}, \tag{7.3.29}$$

where $z_l(i)$ is a point in the lower half-plane ($\operatorname{Im} k \leq A$) which contributes $-\frac{\nu}{2}$ in its totality ($i = 1, \cdots, \nu$) to the index and $p_u(i)$ is a point in the upper half-plane ($\operatorname{Im} k > A$) which contributes $-\frac{\nu}{2}$ in its totality ($i = 1, \cdots, \nu$) to the index. Then the expression (7.3.29) has the index equal to zero with respect to $\operatorname{Im} k = A$,

$$-\frac{\nu}{2}(\text{from } z_l(i)'\text{s}) - \frac{\nu}{2}(\text{from } p_u(i)'\text{s}) + \nu(\text{from } 1-\lambda\hat{K}(k)) = 0. \tag{7.3.30}$$

By factoring of Eq. (7.3.29), using Eqs. (7.3.24) and (7.3.25), we obtain

$$Z_-(k) = \exp\left(-\frac{1}{2\pi i}\int_{-C_2}\frac{\ln\left[(1-\lambda\hat{K}(\zeta))\prod_{i=1}^{\nu}\left(\frac{\zeta-z_l(i)}{\zeta-p_u(i)}\right)\right]}{\zeta-k}\,d\zeta\right), \tag{7.3.31}$$

$$Z_+(k) = \exp\left(-\frac{1}{2\pi i}\int_{C_1}\frac{\ln\left[(1-\lambda\hat{K}(\zeta))\prod_{i=1}^{\nu}\left(\frac{\zeta-z_l(i)}{\zeta-p_u(i)}\right)\right]}{\zeta-k}\,d\zeta\right), \tag{7.3.32}$$

with the properties,

(1) $Z_\pm(k) \to 1$ as $|k| \to \infty$,

(2) $Z_-(k)$ $(Z_+(k))$ is analytic in the lower (upper) half-plane,

(3) $Z_-(k)$ $(Z_+(k))$ has no zero in the lower (upper) half-plane. We write Eq. (7.3.29) as

$$1-\lambda\hat{K}(k) = \frac{Y_-(k)}{Y_+(k)} = \frac{Z_-(k)}{Z_+(k)}\cdot\frac{\prod_{i=1}^{\nu}(k-p_u(i))}{\prod_{i=1}^{\nu}(k-z_l(i))}. \tag{7.3.33}$$

By the formula stated in Eq. (7.2.17), we obtain

$$Y_-(k) = Z_-(k)\cdot\prod_{i=1}^{\nu}(k-p_u(i)), \tag{7.3.34}$$

$$Y_+(k) = Z_+(k)\cdot\prod_{i=1}^{\nu}(k-z_l(i)). \tag{7.3.35}$$

We observe that

$$Y_{\pm}(k) \to k^{\nu} \quad \text{as} \quad |k| \to \infty. \tag{7.3.36}$$

Thus the entire function $G(k)$ in the following equation,

$$Y_{-}(k)\hat{\phi}_{-}(k) = -Y_{+}(k)\hat{\psi}_{+}(k) = G(k),$$

cannot grow as fast as k^{ν} as $k \to \infty$. By Liouville's theorem, we have

$$G(k) = \sum_{j=0}^{\nu-1} C_j k^j, \qquad 0 \le j \le \nu - 1, \tag{7.3.37}$$

where the C_j are arbitrary ν constants. Then we obtain

$$\hat{\phi}_{-}(k) = \frac{G(k)}{Y_{-}(k)} = \sum_{j=0}^{\nu-1} \frac{C_j k^j}{Y_{-}(k)}, \qquad \operatorname{Im} k \le A. \tag{7.3.38}$$

Inverting this expression along $\operatorname{Im} k = A$, we obtain

$$\phi(x) = \frac{1}{2\pi i} \int_{-\infty+iA}^{+\infty+iA} dk\, e^{ikx} \hat{\phi}_{-}(k) = \sum_{j=0}^{\nu-1} C_j \phi_{(j)}(x), \quad \text{when} \quad \nu > 0, \tag{7.3.39}$$

where

$$\phi_{(j)}(x) = \frac{1}{2\pi i} \int_{-\infty+iA}^{+\infty+iA} dk \frac{e^{ikx} k^j}{Y_{-}(k)}, \qquad j = 0, \cdots, \nu - 1. \tag{7.3.40}$$

We have ν *independent homogeneous solutions*, $\phi_{(j)}(x)$, $j = 0, \cdots, \nu - 1$, which are related to each other by differentiation,

$$\left(-i\frac{d}{dx}\right) \phi_{(j)}(x) = \phi_{(j+1)}(x), \qquad 0 \le j \le \nu - 2.$$

Thus it is sufficient to compute $\phi_{(0)}(x)$,

$$\phi_{(j)}(x) = \left(-i\frac{d}{dx}\right)^j \phi_{(0)}(x), \qquad j = 0, 1, \cdots, \nu - 1. \tag{7.3.41}$$

Note that the differentiation under the integral, Eq. (7.3.40), is justified by

$$\frac{k^{j+1}}{Y_{-}(k)} \to \frac{k^{j+1}}{k^{\nu}} \to 0 \quad \text{as} \quad k \to \infty, \quad j \le \nu - 2,$$

so that the integral converges.

Consider thirdly the case when the index ν is *negative*,

$$\nu < 0. \tag{7.3.42}$$

As before, we *construct the object whose index is equal to zero.*

$$\prod_{i=1}^{|\nu|} \frac{(k - z_u(i))}{(k - p_l(i))} \cdot [1 - \lambda \hat{K}(k)] = \frac{Z_-(k)}{Z_+(k)}, \tag{7.3.43}$$

which does indeed have an index of zero as shown below.

$$+\frac{|\nu|}{2}(\text{from } z_u(i)'\text{s}) + \frac{|\nu|}{2}(\text{from } p_l(i)'\text{s}) + \nu(\text{from } 1 - \lambda \hat{K}(k)) = 0. \tag{7.3.44}$$

We apply the factorization to the left-hand side of Eq. (7.3.43). Then we write

$$1 - \lambda \hat{K}(k) = \frac{Y_-(k)}{Y_+(k)} = \frac{Z_-(k)}{Z_+(k)} \cdot \frac{\prod_{i=1}^{|\nu|}(k - p_l(i))}{\prod_{i=1}^{|\nu|}(k - z_u(i))}. \tag{7.3.45}$$

By the formula stated in Eq. (7.2.17), we obtain

$$Y_-(k) = \frac{Z_-(k)}{\prod_{i=1}^{|\nu|}(k - z_u(i))}, \tag{7.3.46}$$

$$Y_+(k) = \frac{Z_+(k)}{\prod_{i=1}^{|\nu|}(k - p_l(i))}. \tag{7.3.47}$$

Then we have

$$Z_\pm(k) \to 1 \quad \text{and} \quad Y_\pm(k) \to \frac{1}{k^{|\nu|}}, \quad \text{as} \quad k \to \infty. \tag{7.3.48}$$

Thus the entire function $G(k)$ in the following equation,

$$Y_-(k)\hat{\phi}_-(k) = -Y_+(k)\hat{\psi}_+(k) = G(k),$$

must vanish identically by Liouville's theorem. Hence we obtain

$$\phi(x) = 0, \quad \text{when} \quad \nu < 0. \tag{7.3.49}$$

7.4 Inhomogeneous Wiener–Hopf Integral Equation of the Second Kind

Let us consider the *inhomogeneous Wiener–Hopf integral equation of the second kind,*

$$\phi(x) = f(x) + \lambda \int_0^{+\infty} K(x - y)\phi(y)\, dy, \quad x \geq 0, \tag{7.4.1}$$

where we assume, as in Section 7.3, that the asymptotic behavior of the kernel $K(x)$ is given by

$$K(x) \sim \begin{cases} O(e^{ax}) & \text{as} \quad x \to -\infty, \\ O(e^{-bx}) & \text{as} \quad x \to +\infty, \end{cases} \quad a, b > 0, \tag{7.4.2}$$

and the asymptotic behavior of the inhomogeneous term $f(x)$ is given by

$$f(x) \to O(e^{cx}) \quad \text{as} \quad x \to +\infty. \tag{7.4.3}$$

We define $\psi(x)$ for $x < 0$ as before

$$\psi(x) = \lambda \int_0^{+\infty} K(x-y)\phi(y)\,dy, \quad x < 0. \tag{7.4.4}$$

We take the Fourier transform of $\phi(x)$ and $\psi(x)$ for $x \geq 0$ and $x < 0$ and add the results together,

$$\hat{\phi}_-(k) + \hat{\psi}_+(k) = \hat{f}_-(k) + \lambda \hat{K}(k)\hat{\phi}_-(k), \tag{7.4.5}$$

where $\hat{K}(k)$ is analytic inside the strip,

$$-a < \operatorname{Im} k = k_2 < b. \tag{7.4.6}$$

Trouble may arise when the inhomogeneous term $f(x)$ grows too fast as $x \to \infty$ so that there may not exist a *common region of analyticity* for Eq. (7.4.5) to hold. The Fourier transform $\hat{f}_-(k)$ is defined by

$$\hat{f}_-(k) = \int_0^{+\infty} dx\, e^{-ikx} f(x), \tag{7.4.7}$$

where

$$\left|e^{-ikx} f(x)\right| \sim e^{(k_2+c)x} \quad \text{as} \quad x \to \infty \quad \text{with} \quad k = k_1 + ik_2.$$

That is, $\hat{f}_-(k)$ is analytic in the lower half-plane,

$$\operatorname{Im} k = k_2 < -c. \tag{7.4.8}$$

We require that a and c satisfy

$$a > c. \tag{7.4.9}$$

In other words, $f(x)$ grows at most as fast as

$$f(x) \sim e^{(a-\varepsilon)x}, \quad \varepsilon > 0, \quad \text{as} \quad x \to \infty.$$

We try to solve Eq. (7.4.5) in the narrower strip,

$$-a < \operatorname{Im} k = k_2 < \min(-c, b). \tag{7.4.10}$$

Writing Eq. (7.4.5) as

$$(1 - \lambda \hat{K}(k))\hat{\phi}_-(k) = \hat{f}_-(k) - \hat{\psi}_+(k), \tag{7.4.11}$$

we are content to obtain *one particular solution* of Eq. (7.4.1). We factorize $1 - \lambda\hat{K}(k)$ as before,

$$1 - \lambda\hat{K}(k) = \frac{Y_-(k)}{Y_+(k)}. \tag{7.4.12}$$

Thus we have from Eqs. (7.4.11) and (7.4.12),

$$Y_-(k)\hat{\phi}_-(k) = Y_+(k)\hat{f}_-(k) - Y_+(k)\hat{\psi}_+(k), \tag{7.4.13}$$

where $Y_-(k)\hat{\phi}_-(k)$ is analytic in the lower half-plane and $Y_+(k)\hat{\psi}_+(k)$ is analytic in the upper half-plane. We split $Y_+(k)\hat{f}_-(k)$ into a sum of two functions, one analytic in the upper half-plane and the other analytic in the lower half-plane,

$$Y_+(k)\hat{f}_-(k) = (Y_+(k)\hat{f}_-(k))_+ + (Y_+(k)\hat{f}_-(k))_-.$$

In order to do this, we must construct $Y_+(k)$ such that

$$Y_+(k)\hat{f}_-(k) \to 0 \quad \text{as} \quad k \to \infty,$$

or,

$$Y_+(k) \to \text{constant} \quad \text{as} \quad k \to \infty. \tag{7.4.14}$$

Suppose $\hat{F}(k)$ is analytic inside the strip,

$$-a < \operatorname{Im} k = k_2 < \min(-c, b). \tag{7.4.15}$$

By choosing the contour C inside the strip as in Figure 7.15, we can then apply the Cauchy integral formula.

$$\hat{F}(k) = \frac{1}{2\pi i}\int_C \frac{\hat{F}(\zeta)}{\zeta - k}\,d\zeta = \frac{1}{2\pi i}\int_{C_1} \frac{\hat{F}(\zeta)}{\zeta - k}\,d\zeta - \frac{1}{2\pi i}\int_{-C_2} \frac{\hat{F}(\zeta)}{\zeta - k}\,d\zeta. \tag{7.4.16}$$

By the same argument as in the previous section, we identify

$$\hat{F}_-(k) = -\frac{1}{2\pi i}\int_{-C_2} \frac{\hat{F}(\zeta)}{\zeta - k}\,d\zeta, \tag{7.4.17}$$

$$\hat{F}_+(k) = \frac{1}{2\pi i}\int_{C_1} \frac{\hat{F}(\zeta)}{\zeta - k}\,d\zeta. \tag{7.4.18}$$

Thus, under the assumption that the $Y_+(k)$ satisfy the above-stipulated condition (7.4.14), we obtain

$$(Y_+(k)\hat{f}_-(k))_- = -\frac{1}{2\pi i}\int_{-C_2} \left[\frac{Y_+(\zeta)\hat{f}_-(\zeta)}{(\zeta - k)}\right] d\zeta, \tag{7.4.19}$$

$$(Y_+(k)\hat{f}_-(k))_+ = \frac{1}{2\pi i}\int_{C_1} \left[\frac{Y_+(\zeta)\hat{f}_-(\zeta)}{(\zeta - k)}\right] d\zeta. \tag{7.4.20}$$

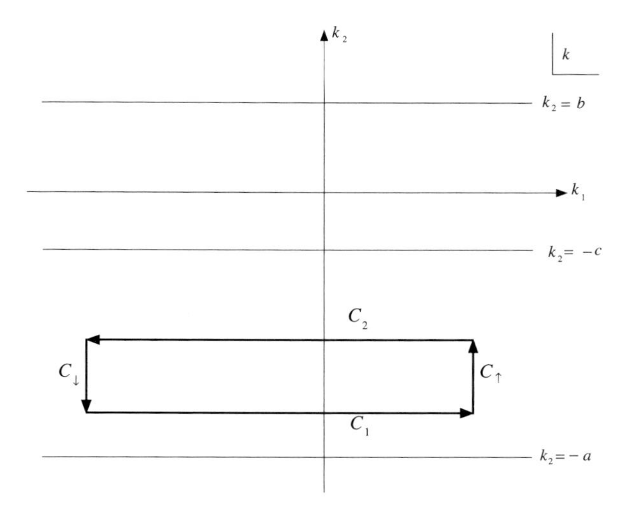

Fig. 7.15: Region of analyticity and the integration contour C for $\hat{F}(k)$ inside the strip, $-a < \operatorname{Im} k < \min(-c, b)$.

Then we write

$$Y_-(k)\hat{\phi}_-(k) - (Y_+(k)\hat{f}_-(k))_- = (Y_+(k)\hat{f}_-(k))_+ - Y_+(k)\hat{\psi}_+(k) \equiv G(k), \quad (7.4.21)$$

where $G(k)$ is entire in k. If we are looking for the most general homogeneous solutions, we set $\hat{f}_-(k) \equiv 0$ and determine the most general form of the entire function $G(k)$. Now we are just looking for *one particular solution to the inhomogeneous equation*, so that we set $G(k) = 0$. Then we have

$$\hat{\phi}_-(k) = \frac{(Y_+(k)\hat{f}_-(k))_-}{Y_-(k)}, \quad (7.4.22)$$

$$\hat{\psi}_+(k) = \frac{(Y_+(k)\hat{f}_-(k))_+}{Y_+(k)}. \quad (7.4.23)$$

Choices of $Y_\pm(k)$ for an inhomogeneous solution are different from those for an homogeneous solution. In view of Eqs. (7.4.22) and (7.4.23), we require that

(1) $1/Y_-(k)$ is analytic in the lower half-plane,

$$\operatorname{Im} k < c', \quad -a < c' < b, \quad (7.4.24)$$

and $1/Y_+(k)$ is analytic in the upper half-plane,

$$\operatorname{Im} k > c'', \quad -a < c'' < b. \quad (7.4.25)$$

(2)

$$\hat{\phi}_-(k) \to 0, \quad \hat{\psi}_+(k) \to 0 \quad \text{as} \quad k \to \infty. \quad (7.4.26)$$

According to this requirement, $Y_-(k)$ for an inhomogeneous solution can have a pole in the lower half-plane. This is all right because then $1/Y_-(k)$ has a zero in the lower half-plane. Once requirements (1) and (2) are satisfied, $\hat{\phi}_-(k)$ and $\hat{\psi}_+(k)$, given by Eq. (7.4.22) and (7.4.23), are analytic in the respective half-plane. Then we construct the following expression from Eqs. (7.4.22) and (7.4.23).

$$[1 - \lambda\hat{K}(k)]\hat{\phi}_-(k) = \hat{f}_-(k) - \hat{\psi}_+(k). \tag{7.4.27}$$

Inverting for $x \geq 0$, we obtain

$$\phi(x) - \lambda\int_0^{+\infty} K(x-y)\phi(y)\,dy = f(x), \quad x \geq 0.$$

Thus $\hat{\phi}_-(k)$ and $\hat{\psi}_+(k)$ derived in Eqs. (7.4.22) and (7.4.23) under the requirements (1) and (2) do provide a particular solution to Eq. (7.4.1).

Case 1. Index $\nu = 0$.

When the index ν of $1 - \lambda\hat{K}(k)$ with respect to the line $\operatorname{Im} k = A$ is equal to *zero*, no nontrivial homogeneous solution exists. From Eqs. (7.3.22) and (7.3.23), we have

$$1 - \lambda\hat{K}(k) = \frac{Y_-(k)}{Y_+(k)}, \tag{7.4.28}$$

where

$$Y_-(k) = \exp\left(-\frac{1}{2\pi i}\int_{-C_2} d\zeta \frac{\ln[1 - \lambda\hat{K}(\zeta)]}{\zeta - k}\right), \qquad \operatorname{Im} k \leq A, \tag{7.4.29}$$

$$Y_+(k) = \exp\left(-\frac{1}{2\pi i}\int_{C_1} d\zeta \frac{\ln[1 - \lambda\hat{K}(\zeta)]}{\zeta - k}\right), \qquad \operatorname{Im} k > A, \tag{7.4.30}$$

and

$$Y_\pm(k) \to 1 \quad \text{as} \quad |k| \to \infty. \tag{7.4.31}$$

Since $Y_+(k)$ ($Y_-(k)$) has no zeros in the upper half-plane (the lower half-plane), $1/Y_+(k)$ ($1/Y_-(k)$) is analytic in the upper half-plane (the lower half-plane). Then we have

$$Y_+(k)\hat{f}_-(k) \to 0 \quad \text{as} \quad k \to \infty,$$

so that $Y_+(k)\hat{f}_-(k)$ can be split up into the $+$ part and the $-$ part as in Eqs. (7.4.19) and (7.4.20).

Using $Y_\pm(k)$ given for the $\nu \equiv 0$ case, Eqs. (7.4.29) and (7.4.30), we construct a *resolvent kernel* $H(x, y)$. Since $1/Y_-(k)$ is analytic in the lower plane and approaches 1 as $k \to \infty$, we define $y_-(x)$ by

$$\frac{1}{Y_-(k)} - 1 \equiv \int_0^{+\infty} dx\, e^{-ikx} y_-(x), \tag{7.4.32}$$

where the left-hand side is analytic in the lower half-plane and vanishes as $k \to \infty$. Inverting Eq. (7.4.32) for $y_-(x)$, we have

$$y_-(x) = \int_{-\infty+iA}^{+\infty+iA} \frac{dk}{2\pi} e^{ikx} \left[\frac{1}{Y_-(k)} - 1 \right] \quad \text{for} \quad x \geq 0, \tag{7.4.33}$$

$$y_-(x) = 0 \quad \text{for} \quad x < 0. \tag{7.4.34}$$

Similarly we define $y_+(x)$ by

$$Y_+(k) - 1 \equiv \int_{-\infty}^{0} dx\, e^{-ikx} y_+(x), \tag{7.4.35}$$

where the left-hand side is analytic in the upper half-plane and vanishes as $k \to \infty$. Inverting Eq. (7.4.35) for $y_+(x)$, we have

$$y_+(x) = \int_{-\infty+iA}^{+\infty+iA} \frac{dk}{2\pi} e^{ikx} [Y_+(k) - 1] \quad \text{for} \quad x < 0, \tag{7.4.36}$$

$$y_+(x) = 0 \quad \text{for} \quad x \geq 0. \tag{7.4.37}$$

We define $\hat{y}_\pm(k)$ by

$$\frac{1}{Y_-(k)} \equiv 1 + \int_0^{+\infty} dx\, e^{-ikx} y_-(x) \equiv 1 + \hat{y}_-(k), \tag{7.4.38}$$

$$Y_+(k) \equiv 1 + \int_{-\infty}^{0} dx\, e^{-ikx} y_+(x) \equiv 1 + \hat{y}_+(k). \tag{7.4.39}$$

Then $\hat{\phi}_-(k)$ given by Eq. (7.4.23) becomes

$$\begin{aligned} \hat{\phi}_-(k) &= \frac{1}{Y_-(k)} (Y_+(k)\hat{f}_-(k))_- = (1 + \hat{y}_-(k))(\hat{f}_-(k) + \hat{y}_+(k)\hat{f}_-(k))_- \\ &= \hat{f}_-(k) + \hat{y}_-(k)\hat{f}_-(k) + (\hat{y}_+(k)\hat{f}_-(k))_- + \hat{y}_-(k)(\hat{y}_+(k)\hat{f}_-(k))_-. \end{aligned} \tag{7.4.40}$$

Inverting Eq. (7.4.40) for $x > 0$,

$$\begin{aligned} \phi(x) &= f(x) + \int_0^{+\infty} y_-(x-y) f(y)\, dy + \int_0^{+\infty} y_+(x-y) f(y)\, dy \\ &\quad + \int_0^{+\infty} y_-(x-z)\, dz \int_0^{+\infty} y_+(z-y) f(y)\, dy \\ &= f(x) + \int_0^{+\infty} H(x,y) f(y)\, dy, \qquad x \geq 0, \end{aligned} \tag{7.4.41}$$

where

$$H(x,y) \equiv y_-(x-y) + y_+(x-y) + \int_0^{+\infty} y_-(x-z) y_+(z-y)\, dz. \tag{7.4.42}$$

It is noted that the existence of the resolvent kernel $H(x, y)$ given above is solely due to the analyticity of $1/Y_-(k)$ in the lower half-plane and that of $Y_+(k)$ in the upper half-plane. Thus, when the index $\nu = 0$, we have a *unique solution, solely consisting of a single particular solution* to Eq. (7.4.11).

Case 2. Index $\nu > 0$.

When the index ν is *positive*, we have ν independent homogeneous solutions given by Eq. (7.3.40). We observed in Section 7.3 that

$$Y_\pm(k) \to k^\nu \quad \text{as} \quad k \to \infty, \tag{7.4.43}$$

where $Y_\pm(k)$ are given by Eqs. (7.3.34) and (7.3.35). On the other hand, in solving for a particular solution, we want $Y_\pm(k)$ to be such that:

(1) $1/Y_-(k)$ $(Y_+(k))$ is analytic in the lower half-plane (the upper half-plane),

$$Y_\pm(k) \to 1 \quad \text{as} \quad |k| \to \infty. \tag{7.4.44}$$

We construct $W_\pm(k)$ as

$$W_\pm(k) = \frac{Y_\pm(k)}{\prod_{j=1}^{\nu}(k - p_l(j))}, \qquad \operatorname{Im} p_l(j) \le A, \quad 1 \le j \le \nu, \tag{7.4.45}$$

where the locations of the $p_l(j)$ are quite arbitrary as long as $\operatorname{Im} p_l(j) \le A$. We notice that the $W_\pm(k)$ satisfy requirements (**1**) and (**2**):

(2)

$$\frac{1}{W_-(k)} = \frac{\prod_{j=1}^{\nu}(k - p_l(j))}{Y_-(k)}$$

is analytic in the lower half-plane, while

$$W_+(k) = \frac{Y_+(k)}{\prod_{j=1}^{\nu}(k - p_l(j))}$$

is analytic in the upper half-plane;

(3)

$$W_\pm(k) \to 1 \quad \text{as} \quad |k| \to \infty. \tag{7.4.46}$$

Thus we use $W_\pm(k)$, Eq. (7.4.45), instead of $Y_\pm(k)$, in the construction of the resolvent $H(x, y)$.

Case 3. Index $\nu < 0$.

When index ν is *negative*, we have no nontrivial homogeneous solution. From Eqs. (7.3.46) and (7.3.47), we have

$$Y_{\pm}(k) \to \frac{1}{k^{|\nu|}} \quad \text{as} \quad |k| \to \infty. \tag{7.4.47}$$

Then we have

$$\frac{1}{Y_{-}(k)} \to k^{|\nu|} \quad \text{as} \quad |k| \to \infty, \tag{7.4.48}$$

while

$$Y_{+}(k)\hat{f}_{-}(k) \to 0 \quad \text{as} \quad |k| \to \infty. \tag{7.4.49}$$

By Liouville's theorem, $\hat{\phi}_{-}(k)$ can grow, at most, as fast as $k^{|\nu|-1}$ as $|k| \to \infty$,

$$\hat{\phi}_{-}(k) = \frac{(Y_{+}(k)\hat{f}_{-}(k))_{-}}{Y_{-}(k)} \sim k^{|\nu|-1} \quad \text{as} \quad |k| \to \infty. \tag{7.4.50}$$

In general, we have

$$\hat{\phi}_{-}(k) \nrightarrow 0 \quad \text{as} \quad |k| \to \infty,$$

so that a particular solution to the inhomogeneous problem may not exist. There are some *exceptions* to this. We analyze $(Y_{+}(k)\hat{f}_{-}(k))_{-}$ more carefully. We know

$$(Y_{+}(k)\hat{f}_{-}(k))_{-} = -\frac{1}{2\pi i}\int_{-C_2} \frac{Y_{+}(\zeta)\hat{f}_{-}(\zeta)}{\zeta - k}\, d\zeta. \tag{7.4.51}$$

Expanding $1/(\zeta - k)$ in power series of ζ/k,

$$\frac{1}{(\zeta - k)} = -\left(\frac{1}{k}\right)\left(1 + \frac{\zeta}{k} + \frac{\zeta^2}{k^2} + \cdots + \frac{\zeta^{|\nu|-1}}{k^{|\nu|-1}} + \cdots\right), \quad \left|\frac{\zeta}{k}\right| < 1,$$

we write

$$\begin{aligned}(Y_{+}(k)\hat{f}_{-}(k))_{-} &= \frac{1}{2\pi i}\int_{-C_2} \frac{1}{k}\sum_{j=0}^{\infty}\left(\frac{\zeta}{k}\right)^{j} Y_{+}(\zeta)\hat{f}_{-}(\zeta)\, d\zeta \\ &= \frac{1}{k}\sum_{j=0}^{\infty}\frac{1}{k^j}\frac{1}{2\pi i}\int_{-C_2} \zeta^j Y_{+}(\zeta)\hat{f}_{-}(\zeta)\, d\zeta.\end{aligned} \tag{7.4.52}$$

In view of Eqs. (7.4.22) and (7.4.52), we realize that

$$\hat{\phi}_-(k) = \frac{(Y_+(k)\hat{f}_-(k))_-}{Y_-(k)} \to 0 \quad \text{as} \quad |k| \to \infty, \tag{7.4.53}$$

if and only if

$$\frac{1}{2\pi i}\int_{-C_2} \zeta^j Y_+(\zeta)\hat{f}_-(\zeta)\,d\zeta = 0, \quad j = 0, \cdots, |\nu| - 1. \tag{7.4.54}$$

If this condition is satisfied, we get from the $j = |\nu|$ term onwards,

$$\hat{\phi}_-(k) \to \frac{C}{k^{1+|\nu|}} \quad \text{as} \quad |k| \to \infty, \tag{7.4.55}$$

so that $\hat{\phi}_-(k)$ can be inverted for $\phi(x)$, which is the *unique solution* to the inhomogeneous problem. To understand this solvability condition (7.4.54), we first recall the *Parseval identity*,

$$\int_{-\infty}^{+\infty} dk\, \hat{h}(k)\hat{g}(-k) = 2\pi \int_{-\infty}^{+\infty} h(y)g(y)\,dy, \tag{7.4.56}$$

where $\hat{h}(k)$ and $\hat{g}(k)$ are the Fourier transforms of $h(y)$ and $g(y)$, respectively. Then we consider the *homogeneous adjoint problem*. Recall that for a real kernel,

$$K^{\text{adj}}(x, y) = K(y, x).$$

Thus, corresponding to the original homogeneous problem,

$$\phi(x) = \lambda \int_0^{+\infty} K(x - y)\phi(y)\,dy,$$

there exists the homogeneous adjoint problem,

$$\phi^{\text{adj}}(x) = \lambda \int_0^{+\infty} K(y - x)\phi^{\text{adj}}(y)\,dy, \tag{7.4.57}$$

whose translation kernel is related to the original one by

$$K^{\text{adj}}(\xi) = K(-\xi). \tag{7.4.58}$$

Now, when we take the Fourier transform of the homogeneous adjoint problem, we find

$$\left[1 - \lambda\hat{K}(-k)\right]\hat{\phi}_-^{\text{adj}}(k) = -\hat{\psi}_+^{\text{adj}}(k), \tag{7.4.59}$$

where the only difference from the original equation is the sign of k inside $\hat{K}(-k)$. However, since $1-\lambda\hat{K}(-k)$ is just the reflection of $1-\lambda\hat{K}(k)$ through the origin, a zero of $1-\lambda\hat{K}(k)$ in the upper half-plane corresponds to a zero of $1 - \lambda\hat{K}(-k)$ in the lower half-plane, etc. Thus, when the original $1-\lambda\hat{K}(k)$ has a negative index $\nu < 0$ with respect to a line, $\text{Im}\, k = k_2 = A$, the homogeneous adjoint problem $1-\lambda\hat{K}(-k)$ has a positive index $|\nu|$ relative to the line, $\mathrm{Im}\,k = k_2 = -A$. So, in that case, although the original problem may have no solutions, the homogeneous adjoint problem has $|\nu|$ independent solutions. Now $1-\lambda\hat{K}(k)$ was found to have the decomposition,

$$1-\lambda\hat{K}(k) = \frac{Y_-(k)}{Y_+(k)},$$

with

$$Y_\pm(k) \to \frac{1}{k^{|\nu|}} \quad \text{as} \quad k\to\infty,$$

we conclude that

$$1-\lambda\hat{K}(-k) = \frac{Y_-(-k)}{Y_+(-k)} = \frac{Y_-^{\mathrm{adj}}(k)}{Y_+^{\mathrm{adj}}(k)}, \tag{7.4.60}$$

from which, we recognize that

$$\begin{cases} Y_-^{\mathrm{adj}}(k) = \dfrac{1}{Y_+(-k)} & \text{is analytic in the lower half-plane,} \quad \to k^{|\nu|}, \\ Y_+^{\mathrm{adj}}(k) = \dfrac{1}{Y_-(-k)} & \text{is analytic in the upper half-plane,} \quad \to k^{|\nu|}. \end{cases}$$

Thus the homogeneous adjoint problem reads

$$Y_-^{\mathrm{adj}}(k)\hat{\phi}_-^{\mathrm{adj}}(k) = -Y_+^{\mathrm{adj}}(k)\psi_+^{\mathrm{adj}}(k) \equiv G(k), \tag{7.4.61}$$

with

$$G(k) = C_0 + C_1 k + \cdots + C_{|\nu|-1}k^{|\nu|-1}. \tag{7.4.62}$$

We then know that the $|\nu|$ independent solutions to the homogeneous adjoint problem are of the form,

$$\hat{\phi}_-^{\mathrm{adj}}(k) = \left\{ \frac{1}{Y_-^{\mathrm{adj}}(k)}, \frac{k}{Y_-^{\mathrm{adj}}(k)}, \cdots, \frac{k^{|\nu|-1}}{Y_-^{\mathrm{adj}}(k)} \right\},$$

which is equivalent to

$$\hat{\phi}_-^{\mathrm{adj}}(k) = \{Y_+(-k), kY_+(-k), \cdots, k^{|\nu|-1}Y_+(-k)\}. \tag{7.4.63}$$

Therefore, to within a constant factor, which is irrelevant, we can write the solvability condition (7.4.54) as

$$\int_{-C_2} \zeta^j Y_+(\zeta)\hat{f}_-(\zeta)\,d\zeta = 0, \qquad j = 0, 1, \cdots, |\nu|-1, \tag{7.4.64}$$

which is equivalent to

$$\int_{-C_2} \hat{\phi}^{\mathrm{adj}}_{-(j)}(-\zeta)\hat{f}_-(\zeta)\,d\zeta = 0, \qquad j = 0, 1, \cdots, |\nu| - 1, \tag{7.4.65}$$

where

$$\hat{\phi}^{\mathrm{adj}}_{-(j)}(\zeta) \equiv \zeta^j Y_+(-\zeta), \qquad j = 0, 1, \cdots, |\nu| - 1. \tag{7.4.66}$$

By the Parseval identity (7.4.56), the *solvability condition* (7.4.65) can now be written as

$$\int_0^{+\infty} \phi^{\mathrm{adj}}_{(j)}(x)f(x)dx = 0, \qquad j = 0, 1, \cdots, |\nu| - 1, \tag{7.4.67}$$

where

$$\phi^{\mathrm{adj}}_{(j)}(x) = \frac{1}{2\pi}\int_{-\infty-iA}^{+\infty-iA} e^{i\zeta x}\hat{\phi}^{\mathrm{adj}}_{-(j)}(\zeta)\,d\zeta, \quad j = 0, 1, \cdots, |\nu| - 1, \quad x \geq 0. \tag{7.4.68}$$

Namely, *if and only if the inhomogeneous term $f(x)$ is orthogonal to all of the homogeneous solutions $\phi^{adj}(x)$ of the homogeneous adjoint problem, the inhomogeneous equation* (7.4.1) *has a unique solution, when the index ν is negative.*

Summary of Wiener–Hopf integral equation

$$\phi(x) = \lambda\int_0^\infty K(x-y)\phi(y)\,dy, \qquad x \geq 0,$$

$$\phi^{\mathrm{adj}}(x) = \lambda\int_0^\infty K(y-x)\phi^{\mathrm{adj}}(y)\,dy, \qquad x \geq 0,$$

$$\phi(x) = f(x) + \lambda\int_0^\infty K(x-y)\phi(y)\,dy, \qquad x \geq 0.$$

1. Index $\nu = 0$:

   The homogeneous problem and its homogeneous adjoint problem have no solutions.

   The inhomogeneous problem has a unique solution.

2. Index $\nu > 0$:

   The homogeneous problem has ν independent solutions and its homogeneous adjoint problem has no solutions.

   The inhomogeneous problem has nonunique solutions (but there are no solvability conditions).

3. Index $\nu < 0$:

   The homogeneous problem has no solutions, and its homogeneous adjoint problem has $|\nu|$ independent solutions.

   The inhomogeneous problem has a unique solution, if and only if the inhomogeneous term is orthogonal to all $|\nu|$ independent solutions to the homogeneous adjoint problem.

7.5 Toeplitz Matrix and Wiener–Hopf Sum Equation

In this section, we consider the application of the Wiener–Hopf method to the *infinite system of the inhomogeneous linear algebraic equation*,

$$M\vec{X} = \vec{f}, \tag{7.5.1}$$

or,

$$\sum_{m} M_{nm} X_m = f_n, \tag{7.5.2}$$

where the coefficient matrix M is *real* and has the *Toeplitz structure*,

$$M_{nm} = M_{n-m}. \tag{7.5.3}$$

We solve Eq. (7.5.1) for two cases.

Case A. Infinite Toeplitz matrix.

Let M be an infinite matrix. Then the *system of the infinite inhomogeneous linear algebraic equation* (7.5.1) becomes

$$\sum_{m=-\infty}^{\infty} M_{n-m} X_m = f_n, \qquad -\infty < n < \infty. \tag{7.5.4}$$

We look for the solution $\{X_m\}_{m=-\infty}^{+\infty}$ assuming the *uniform convergence* of $\{X_m\}_{m=-\infty}^{+\infty}$ and $\{M_m\}_{m=-\infty}^{+\infty}$,

$$\sum_{m=-\infty}^{\infty} |X_m| < \infty, \quad \text{and} \quad \sum_{m=-\infty}^{\infty} |M_m| < \infty. \tag{7.5.5}$$

Multiplying ξ^n on Eq. (7.5.4), and summing over n, we have

$$\sum_{n,m} \xi^n M_{n-m} X_m = \sum_{n} \xi^n f_n. \tag{7.5.6}$$

Assuming uniform convergence, Eq. (7.5.5), the left-hand side of Eq. (7.5.6) can be expressed as

$$\sum_{n,m} \xi^n M_{n-m} X_m = \sum_{n} \xi^{n-m} M_{n-m} \sum_{m} \xi^m X_m = M(\xi) X(\xi),$$

with the interchange of the order of the summations, where $X(\xi)$ and $M(\xi)$ are defined by

$$X(\xi) \equiv \sum_{n=-\infty}^{\infty} X_n \xi^n, \tag{7.5.7}$$

$$M(\xi) \equiv \sum_{n=-\infty}^{\infty} M_n \xi^n. \tag{7.5.8}$$

We also define

$$f(\xi) \equiv \sum_{n=-\infty}^{\infty} f_n \xi^n. \tag{7.5.9}$$

Thus Eq. (7.5.6) takes the following form,

$$M(\xi)X(\xi) = f(\xi). \tag{7.5.10}$$

We assume the following bounds on M_n and f_n,

$$|M_n| = \begin{cases} O(a^{-|n|}) & \text{as} \quad n \to -\infty, \\ O(b^{-n}) & \text{as} \quad n \to +\infty, \end{cases} \qquad a, b > 0, \tag{7.5.11a}$$

$$|f_n| = \begin{cases} O(c^{-|n|}) & \text{as} \quad n \to -\infty, \\ O(d^{-n}) & \text{as} \quad n \to +\infty, \end{cases} \qquad c, d > 0. \tag{7.5.11b}$$

Then $M(\xi)$ is analytic in the annulus in the complex ξ plane,

$$\frac{1}{a} < |\xi| < b, \tag{7.5.12a}$$

provided that $1 < ab$, and $f(\xi)$ is analytic in the annulus in the complex ξ plane,

$$\frac{1}{c} < |\xi| < d, \tag{7.5.12b}$$

provided that $1 < cd$. Hence we obtain

$$X(\xi) = \frac{f(\xi)}{M(\xi)} = \sum_{n=-\infty}^{\infty} X_n \xi^n \quad \text{provided} \quad M(\xi) \neq 0. \tag{7.5.13}$$

$X(\xi)$ is analytic in the annulus

$$\max\left(\frac{1}{a}, \frac{1}{c}\right) < |\xi| < \min(b, d). \tag{7.5.12c}$$

By the *Fourier series inversion formula* on the *unit circle*, we obtain

$$X_n = \frac{1}{2\pi} \int_0^{2\pi} d\theta \exp[-in\theta] X(\exp[i\theta]) = \frac{1}{2\pi i} \oint_{|\xi|=1} d\xi \, \xi^{-n-1} X(\xi), \tag{7.5.14}$$

with

$$M(\xi) \neq 0 \quad \text{for} \quad |\xi| = 1, \tag{7.5.15}$$

which solves Eq. (7.5.4).

Next, consider the *eigenvalue problem* of the following form,

$$\sum_{m=-\infty}^{\infty} M_{n-m} X_m = \mu X_n. \tag{7.5.16}$$

We try

$$X_m = \xi^m. \tag{7.5.17}$$

Then we obtain

$$M(\xi) = \mu. \tag{7.5.18}$$

The roots of Eq. (7.5.18) provide the solutions to Eq. (7.5.16). For $\mu = 0$, we obtain

$$X_m = (\xi_0)^m, \tag{7.5.19}$$

where ξ_0 is a zero of $M(\xi)$.

Case B. Semi-infinite Toeplitz matrix.

Consider now the *system of the semi-infinite inhomogeneous linear algebraic equations*,

$$\sum_{m=0}^{\infty} M_{n-m} X_m = f_n, \quad n \geq 0, \tag{7.5.20}$$

which is the *inhomogeneous Wiener–Hopf sum equation*.

We let

$$M(\xi) \equiv \sum_{n=-\infty}^{\infty} M_n \xi^n, \qquad 0 \leq \arg \xi \leq 2\pi, \tag{7.5.21}$$

be such that

$$M(\exp[i\theta]) \neq 0, \quad \text{for} \qquad 0 \leq \theta \leq 2\pi.$$

We assume that

$$\sum_{n=0}^{\infty} |f_n| < \infty, \tag{7.5.22}$$

and

$$|X_m| < (1+\varepsilon)^{-m}, \quad m \geq 0, \quad \varepsilon > 0. \tag{7.5.23}$$

We define

$$y_n \equiv \sum_{m=0}^{\infty} M_{n-m} X_m \quad \text{for} \quad n \leq -1, \quad \text{and} \quad y_n \equiv 0 \quad \text{for} \quad n \geq 0, \tag{7.5.24a}$$

$$f_n \equiv 0 \quad \text{for} \quad n \leq -1, \tag{7.5.24b}$$

$$\bar{X}(\xi) \equiv \sum_{n=0}^{\infty} X_n \xi^n, \tag{7.5.25}$$

$$\bar{f}(\xi) \equiv \sum_{n=0}^{\infty} f_n \xi^n, \tag{7.5.26}$$

$$Y(\xi) \equiv \sum_{n=-\infty}^{-1} y_n \xi^n. \tag{7.5.27}$$

We look for the solution which satisfies

$$\sum_{n=0}^{\infty} |X_n| < \infty. \tag{7.5.28}$$

Then Eq. (7.5.20) is rewritten as

$$\sum_{m=-\infty}^{\infty} M_{n-m} X_m = f_n + y_n \quad \text{for} \quad -\infty < n < \infty. \tag{7.5.20b}$$

We note that

$$\sum_{n=-\infty}^{\infty} |y_n| = \sum_{n=-\infty}^{\infty} \left| \sum_{m=0}^{\infty} M_{n-m} X_m \right| < \sum_{n=-\infty}^{\infty} |M_n| \sum_{m=0}^{\infty} |X_m| < \infty,$$

where changing the order of the summation is justified since the final expression is finite. Multiplying $\exp[in\theta]$ on both sides of Eq. (7.5.20b) and summing over n, we obtain

$$M(\xi)\bar{X}(\xi) = \bar{f}(\xi) + Y(\xi). \tag{7.5.29}$$

Homogeneous problem : We set

$$f_n = 0 \quad \text{or} \quad \bar{f}(\xi) = 0.$$

The problem we will solve first is

$$M(\xi)\bar{X}(\xi) = Y(\xi), \tag{7.5.30}$$

where

$$\begin{aligned} &\bar{X}(\xi) \quad \text{analytic for} \quad |\xi| < 1 + \varepsilon, \\ &Y(\xi) \quad \text{analytic for} \quad |\xi| > 1. \end{aligned} \tag{7.5.31}$$

We define the *index* ν of $M(\xi)$ in the counter-clockwise direction by

$$\nu \equiv \frac{1}{2\pi i} \ln[M(\exp[i\theta])] \bigg|_{\theta=0}^{\theta=2\pi}. \tag{7.5.32}$$

Suppose that $M(\xi)$ has been factorized into the following form,

$$M(\xi) = N_{\text{in}}(\xi)/N_{\text{out}}(\xi), \tag{7.5.33a}$$

where

$$\begin{cases} N_{\text{in}}(\xi) & \text{analytic for} \quad |\xi| < 1+\varepsilon, \quad \text{and continuous for} \quad |\xi| \leq 1, \\ N_{\text{out}}(\xi) & \text{analytic for} \quad |\xi| > 1, \quad \text{and continuous for} \quad |\xi| \geq 1. \end{cases} \tag{7.5.33b}$$

Then Eq. (7.5.30) is rewritten as

$$N_{\text{in}}(\xi)\bar{X}(\xi) = N_{\text{out}}(\xi)Y(\xi) \equiv G(\xi), \tag{7.5.34}$$

where $G(\xi)$ is entire in the complex ξ plane. The form of $G(\xi)$ is now examined.

Case 1. the **index** $\nu = 0$.

By the now familiar formula, Eq. (7.3.25), we have

$$\oint \frac{\ln[M(\xi')]}{\xi' - \xi}\frac{d\xi'}{2\pi i} = \left(\oint_{C_1} - \oint_{C_2}\right)\left[\frac{\ln[M(\xi')]}{\xi' - \xi}\frac{d\xi'}{2\pi i}\right], \tag{7.5.35}$$

where the integration contours, C_1 and C_2, are displayed in Figure 7.16.

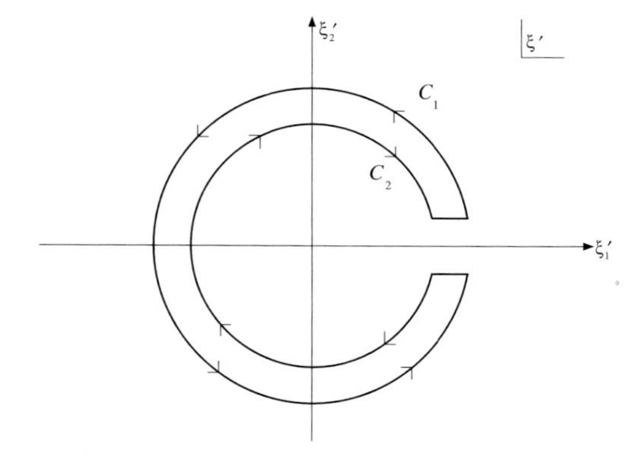

Fig. 7.16: Integration contour of $\ln[M(\xi')]$ when the index $\nu = 0$.

Thus we have

$$N_{\text{in}}(\xi) = \exp\left[\oint_{C_1} \frac{\ln[M(\xi')]}{\xi' - \xi}\frac{d\xi'}{2\pi i}\right] \to 1 \quad \text{as} \quad |\xi| \to \infty, \tag{7.5.36a}$$

$$N_{\text{out}}(\xi) = \exp\left[\oint_{C_2} \frac{\ln[M(\xi')]}{\xi' - \xi}\frac{d\xi'}{2\pi i}\right] \to 1 \quad \text{as} \quad |\xi| \to \infty. \tag{7.5.36b}$$

We find, by Liouville's theorem,

$$G(\xi) = 0, \tag{7.5.37}$$

and hence we have *no nontrivial homogeneous solution* when $\nu = 0$.

Case 2. the *index* $\nu > 0$ (*positive integer*).

We construct the *object with the index zero*, $M(\xi)/\xi^{\nu}$, and obtain

$$M(\xi) = \xi^{\nu} \exp\left(\oint_{C_1} - \oint_{C_2}\right)\left[\frac{\ln[M(\xi')/(\xi')^{\nu}]}{\xi' - \xi}\frac{d\xi'}{2\pi i}\right], \tag{7.5.38}$$

from which, we obtain

$$N_{\text{in}}(\xi) = \xi^{\nu} \exp\left[\oint_{C_1} \frac{\ln[M(\xi')/(\xi')^{\nu}]}{\xi' - \xi}\frac{d\xi'}{2\pi i}\right] \to \xi^{\nu} \quad \text{as} \quad |\xi| \to \infty, \tag{7.5.39a}$$

$$N_{\text{out}}(\xi) = \exp\left[\oint_{C_2} \frac{\ln[M(\xi')/(\xi')^{\nu}]}{\xi' - \xi}\frac{d\xi'}{2\pi i}\right] \to 1 \quad \text{as} \quad |\xi| \to \infty. \tag{7.5.39b}$$

By Liouville's theorem, $G(\xi)$ cannot grow as fast as ξ^{ν}. Hence we have

$$G(\xi) = \sum_{m=0}^{\nu-1} G_m \xi^m, \tag{7.5.40}$$

where th G_m are ν arbitrary constants. Thus $\bar{X}(\xi)$ is given by

$$\bar{X}(\xi) = \sum_{m=0}^{\nu-1} \frac{G_m \xi^m}{N_{\text{in}}(\xi)}, \tag{7.5.41a}$$

from which, we obtain ν *independent homogeneous solutions* X_n by the *Fourier series inversion formula* on the *unit circle*,

$$X_n = \frac{1}{2\pi}\int_0^{2\pi} d\theta \exp[-in\theta] X(\exp[i\theta]) = \frac{1}{2\pi i}\oint_{|\xi|=1} d\xi\, \xi^{-n-1} X(\xi). \tag{7.5.41b}$$

Case 3. the *index* $\nu < 0$ (*negative integer*).

We construct the *object with the index zero*, $M(\xi)\xi^{|\nu|}$, and obtain

$$M(\xi) = \frac{1}{\xi^{|\nu|}} \exp\left(\oint_{C_1} - \oint_{C_2}\right)\left[\frac{\ln[M(\xi')(\xi')^{|\nu|}]}{\xi' - \xi}\frac{d\xi'}{2\pi i}\right], \tag{7.5.42}$$

from which, we obtain

$$N_{\text{in}}(\xi) = \frac{1}{\xi^{|\nu|}} \exp\left[\oint_{C_1} \frac{\ln[M(\xi')(\xi')^{|\nu|}]}{\xi' - \xi}\frac{d\xi'}{2\pi i}\right] \to \frac{1}{\xi^{|\nu|}} \quad \text{as} \quad |\xi| \to \infty, \tag{7.5.43a}$$

$$N_{\text{out}}(\xi) = \exp\left[\oint_{C_2} \frac{\ln[M(\xi')(\xi')^{|\nu|}]}{\xi' - \xi}\frac{d\xi'}{2\pi i}\right] \to 1 \quad \text{as} \quad |\xi| \to \infty. \tag{7.5.43b}$$

By Liouville's theorem, we have

$$G(\xi) = 0, \tag{7.5.44}$$

and hence we have *no nontrivial solution*.

Inhomogeneous problem: We restate the inhomogeneous problem below,

$$M(\xi)\bar{X}(\xi) = \bar{f}(\xi) + Y(\xi), \tag{7.5.45}$$

where we assume that $\bar{f}(\xi)$ is analytic for $|\xi| < 1 + \varepsilon$. Factoring $M(\xi)$ as before,

$$M(\xi) = N_{\text{in}}(\xi)/N_{\text{out}}(\xi), \tag{7.5.46}$$

and multiplying $N_{\text{out}}(\xi)$ on both sides of Eq. (7.5.45), we have

$$\bar{X}(\xi)N_{\text{in}}(\xi) = \bar{f}(\xi)N_{\text{out}}(\xi) + Y(\xi)N_{\text{out}}(\xi), \tag{7.5.47a}$$

or, splitting $\bar{f}(\xi)N_{\text{out}}(\xi)$ into a sum of the *in* function and the *out* function,

$$\bar{X}(\xi)N_{\text{in}}(\xi) - [\bar{f}(\xi)N_{\text{out}}(\xi)]_{\text{in}} = [\bar{f}(\xi)N_{\text{out}}(\xi)]_{\text{out}} + Y(\xi)N_{\text{out}}(\xi) \equiv F(\xi), \tag{7.5.47b}$$

where $F(\xi)$ is entire in the complex ξ plane. Since we wish to obtain *one particular solution* to Eq. (7.5.45), in Eq. (7.5.47b) we set

$$F(\xi) = 0, \tag{7.5.48}$$

resulting in the particular solution,

$$\bar{X}_{\text{part}}(\xi) = [\bar{f}(\xi)N_{\text{out}}(\xi)]_{\text{in}}/N_{\text{in}}(\xi), \tag{7.5.49a}$$

$$Y(\xi) = -[\bar{f}(\xi)N_{\text{out}}(\xi)]_{\text{out}}/N_{\text{out}}(\xi). \tag{7.5.49b}$$

The fact that Eqs. (7.5.49a) and (7.5.49b) satisfy Eq. (7.5.47a) can be easily demonstrated. From Eq. (7.5.49a), the particular solution, $X_{n,\text{part}}$, can be obtained by the *Fourier series inversion formula* on the *unit circle*.

We note that in writing Eq. (7.5.47b), the following property of $N_{\text{out}}(\xi)$ is essential,

$$N_{\text{out}}(\xi) \to 1 \quad \text{as} \quad |\xi| \to \infty. \tag{7.5.50}$$

Case 1. The *index* $\nu = 0$.

Since the homogeneous problem has no nontrivial solution, the *unique particular solution*, $X_{n,\text{part}}$, is obtained for the inhomogeneous problem.

Case 2. The *index* $\nu > 0$ (*positive integer*).

In this case, since the homogeneous problem has ν independent solutions, the solution to the inhomogeneous problem is *not unique*.

Case 3. The *index* $\nu < 0$ (*negative integer*).

In this case, consider the *homogeneous adjoint problem*,

$$\sum_{m=0}^{\infty} M_{m-n} X_m^{\text{adj}} = 0. \tag{7.5.51}$$

Its M function, $M^{\text{adj}}(\xi)$, is defined by

$$M^{\text{adj}}(\xi) \equiv \sum_{n=-\infty}^{+\infty} M_{-n}\xi^{n} = \sum_{n=-\infty}^{+\infty} M_{n}\xi^{-n} = M\left(\frac{1}{\xi}\right). \tag{7.5.52}$$

The index ν^{adj} of $M^{\text{adj}}(\xi)$ is defined by

$$\begin{aligned}\nu^{\text{adj}} &\equiv \frac{1}{2\pi i}\ln[M^{\text{adj}}(\exp[i\theta])]\Big|_{\theta=0}^{\theta=2\pi} = \frac{1}{2\pi i}\ln[M(\exp[-i\theta])]\Big|_{\theta=0}^{\theta=2\pi} \\ &= \frac{1}{2\pi i}\ln[\{M(\exp[i\theta])\}^{*}]\Big|_{\theta=0}^{\theta=2\pi} = -\frac{1}{2\pi i}\ln[M(\exp[i\theta])]\Big|_{\theta=0}^{\theta=2\pi} = -\nu.\end{aligned} \tag{7.5.53}$$

The factorization of $M^{\text{adj}}(\xi)$ is carried out as in the case of $M(\xi)$, with the result,

$$M^{\text{adj}}(\xi) = N_{\text{in}}^{\text{adj}}(\xi)/N_{\text{out}}^{\text{adj}}(\xi) = M(1/\xi) = N_{\text{in}}(1/\xi)/N_{\text{out}}(1/\xi). \tag{7.5.54}$$

From this, we recognize that

$$\begin{cases} N_{\text{in}}^{\text{adj}}(\xi) = N_{\text{out}}^{-1}(1/\xi) & \text{analytic in} \quad |\xi| < 1, \quad \text{and continuous for} \quad |\xi| \le 1, \\ N_{\text{out}}^{\text{adj}}(\xi) = N_{\text{in}}^{-1}(1/\xi) & \text{analytic in} \quad |\xi| > 1, \quad \text{and continuous for} \quad |\xi| \ge 1. \end{cases} \tag{7.5.55}$$

Then, in this case, the homogeneous adjoint problem has $|\nu|$ independent solutions,

$$X_{m}^{\text{adj}(j)} \quad j = 1, \cdots, |\nu|, \quad m \ge 0. \tag{7.5.56}$$

Using an argument similar to the derivation of the solvability condition for the inhomogeneous Wiener–Hopf integral equation of the second kind, discussed in Section 7.4, noting Eq. (7.5.50), we obtain the *solvability condition for the inhomogeneous Wiener–Hopf sum equation* as follows:

$$\sum_{m=0}^{\infty} f_{m} X_{m}^{\text{adj}(j)} = 0, \quad j = 1, \cdots, |\nu|. \tag{7.5.57}$$

Thus, *if and only if the solvability condition* (7.5.57) *is satisfied, i.e., the inhomogeneous term* f_m *is orthogonal to all the* $|\nu|$ *independent solutions* $X_{m}^{adj(j)}$ *to the homogeneous adjoint problem* (7.5.51), then the inhomogeneous Wiener–Hopf sum equation has the *unique solution*, $X_{n,\text{part}}$.

From this analysis of the *inhomogeneous Wiener–Hopf sum equation*, we find that the problem at hand is the *discrete analogue of the inhomogeneous Wiener–Hopf integral equation of the second kind, not of the first kind*, despite its formal appearance.

For an interesting application of the Wiener–Hopf sum equation to the phase transition of the two-dimensional Ising model, the reader is referred to the article by T.T. Wu, cited in the bibliography.

For another interesting application of the Wiener–Hopf sum equation to the Yagi–Uda semi-infinite arrays, the reader is referred to the articles by W. Wasylkiwskyj and A.L. VanKoughnett, cited in the bibliography.

The Cauchy integral formula used in this section should actually be Pollard's theorem which is the generalization of the Cauchy integral formula. We avoided the mathematical technicalities in the presentation of the Wiener–Hopf sum equation.

As for the mathematical details related to the Wiener–Hopf sum equation, Liouville's theorem, the Wiener–Lévy theorem, and Pollard's theorem, we refer the reader to Chapter IX of the book by B. McCoy and T.T. Wu, cited in the bibliography.

Summary of the Wiener–Hopf sum equation

$$\sum_{m=0}^{\infty} M_{n-m} X_m = f_n, \qquad n \geq 0,$$

$$\sum_{m=0}^{\infty} M_{m-n} X_m^{\text{adj}} = 0, \qquad n \geq 0.$$

1) Index $\nu = 0$.

   The homogeneous problem has no nontrivial solution.

   The homogeneous adjoint problem has no nontrivial solution.

   The inhomogeneous problem has a unique solution.

2) Index $\nu > 0$.

   The homogeneous problem has ν independent nontrivial solutions.

   The homogeneous adjoint problem has no nontrivial solution.

   The inhomogeneous problem has non-unique solutions.

3) Index $\nu < 0$.

   The homogeneous problem has no nontrivial solution.

   The homogeneous adjoint problem has $|\nu|$ independent nontrivial solutions.

   The inhomogeneous problem has a unique solution, if and only if the inhomogeneous term is orthogonal to all $|\nu|$ independent solutions to the homogeneous adjoint problem.

7.6 Wiener–Hopf Integral Equation of the First Kind and Dual Integral Equations

In this section, we re-examine the *mixed boundary value problem* considered in Section 7.1 with some generality and show its equivalence to the *Wiener–Hopf integral equation of the first kind* and to the *dual integral equations*. This demonstration of equivalence by no means constitutes a solution to the original problem; rather it provides a hint for solving the Wiener–Hopf integral equation of the first kind, and the dual integral equations, by the methods we developed in Sections 7.1 and 7.3.

❑ **Example 7.5.** Solve the mixed boundary value problem of two-dimensional Laplace equation in the half-plane:

$$\left(\frac{\partial^2}{\partial x^2}+\frac{\partial^2}{\partial y^2}\right)\phi(x,y)=0, \qquad y\geq 0, \tag{7.6.1}$$

with the boundary conditions specified on the x axis,

$$\phi(x,0)=f(x), \qquad x\geq 0, \tag{7.6.2a}$$

$$\phi_y(x,0)=g(x), \qquad x<0, \tag{7.6.2b}$$

$$\phi(x,y)\to 0 \quad \text{as} \quad x^2+y^2\to\infty. \tag{7.6.2c}$$

Solution. We write

$$\phi(x,y)=\lim_{\varepsilon\to 0+}\int_{-\infty}^{+\infty}\frac{dk}{2\pi}e^{ikx-\sqrt{k^2+\varepsilon^2}y}\hat{\phi}(k), \qquad y\geq 0. \tag{7.6.3}$$

Setting $y=0$ in Eq. (7.6.3),

$$\int_{-\infty}^{+\infty}\frac{dk}{2\pi}e^{ikx}\hat{\phi}(k)=\begin{cases} f(x), & x\geq 0,\\ \phi(x,0), & x<0.\end{cases} \tag{7.6.4}$$

Thus we have

$$\hat{\phi}(k)=\int_{-\infty}^{+\infty}dx\,e^{-ikx}\phi(x,0)=\hat{\phi}_+(k)+\hat{f}_-(k), \tag{7.6.5}$$

where

$$\hat{\phi}_+(k)=\int_{-\infty}^{0}dx\,e^{-ikx}\phi(x,0), \tag{7.6.6}$$

$$\hat{f}_-(k)=\int_{0}^{+\infty}dx\,e^{-ikx}f(x). \tag{7.6.7}$$

We know that $\hat{\phi}_+(k)$ ($\hat{f}_-(k)$) is analytic in the upper half-plane (the lower half-plane). Differentiating Eq. (7.6.3) with respect to y, and setting $y=0$, we have

$$\int_{-\infty}^{+\infty}\frac{dk}{2\pi}e^{ikx}(-\sqrt{k^2+\varepsilon^2})\hat{\phi}(k)=\begin{cases} \phi_y(x,0), & x\geq 0,\\ g(x), & x<0.\end{cases} \tag{7.6.8}$$

Then, by inversion, we obtain

$$(-\sqrt{k^2+\varepsilon^2})\hat{\phi}(k)=\hat{g}_+(k)+\hat{\psi}_-(k), \tag{7.6.9}$$

where

$$\hat{g}_{+}(k) = \int_{-\infty}^{0} dx\, e^{-ikx} g(x), \tag{7.6.10}$$

$$\hat{\psi}_{-}(k) = \int_{0}^{+\infty} dx\, e^{-ikx} \phi_y(x,0). \tag{7.6.11}$$

As before, we know that $\hat{g}_{+}(k)$ ($\hat{\psi}_{-}(k)$) is analytic in the upper half-plane (the lower half-plane).

Eliminating $\hat{\phi}(k)$ from Eqs. (7.6.5) and (7.6.9), we obtain

$$\hat{\phi}_{+}(k) + \hat{f}_{-}(k) = \left(-\frac{1}{\sqrt{k^2+\varepsilon^2}}\right)\hat{g}_{+}(k) + \left(-\frac{1}{\sqrt{k^2+\varepsilon^2}}\right)\hat{\psi}_{-}(k). \tag{7.6.12}$$

Inverting Eq. (7.6.12) for $x > 0$, we obtain

$$\int_{-\infty}^{+\infty} \frac{dk}{2\pi} e^{ikx} \left(\hat{\phi}_{+}(k) + \hat{f}_{-}(k) + \frac{1}{\sqrt{k^2+\varepsilon^2}}\hat{g}_{+}(k)\right) = \int_{-\infty}^{+\infty} \frac{dk}{2\pi} e^{ikx} \left(-\frac{1}{\sqrt{k^2+\varepsilon^2}}\right)\hat{\psi}_{-}(k), \qquad x > 0, \tag{7.6.13}$$

where

$$\int_{-\infty}^{+\infty} \frac{dk}{2\pi} e^{ikx} \hat{\phi}_{+}(k) = 0, \quad \text{for} \quad x > 0, \tag{7.6.14}$$

because $\hat{\phi}_{+}(k)$ is analytic in the upper half-plane and the contour of the integration is closed in the upper half-plane for $x > 0$. The remaining terms on the left-hand side of Eq. (7.6.13) are identified as

$$\int_{-\infty}^{+\infty} \frac{dk}{2\pi} e^{ikx} \hat{f}_{-}(k) = f(x), \qquad x > 0, \tag{7.6.15}$$

$$\int_{-\infty}^{+\infty} \frac{dk}{2\pi} e^{ikx} \frac{1}{\sqrt{k^2+\varepsilon^2}}\hat{g}_{+}(k) \equiv G(x), \qquad x > 0. \tag{7.6.16}$$

The right-hand side of Eq. (7.6.13) is identified as

$$\int_{-\infty}^{+\infty} \frac{dk}{2\pi} e^{ikx} \left(-\frac{1}{\sqrt{k^2+\varepsilon^2}}\right)\hat{\psi}_{-}(k) = \int_{0}^{+\infty} \psi(y) K(x-y)\, dy, \quad x > 0, \tag{7.6.17}$$

where $\psi(x)$ and $K(x)$ are defined by

$$\psi(x) \equiv \phi_y(x,0), \quad x > 0, \tag{7.6.18}$$

$$K(x) \equiv \int_{-\infty}^{+\infty} \frac{dk}{2\pi} e^{ikx} \left(-\frac{1}{\sqrt{k^2+\varepsilon^2}}\right). \tag{7.6.19}$$

Thus we obtain the integral equation for $\psi(x)$ from Eq. (7.6.13),

$$\int_0^{+\infty} K(x-y)\psi(y)\,dy = f(x) + G(x), \quad x > 0. \tag{7.6.20}$$

This is the integral equation with a translational kernel of semi-infinite range and is called the *Wiener–Hopf integral equation of the first kind.* As noted earlier, this reduction of the mixed boundary value problem, Eqs. (7.6.1) through (7.6.2c), to the Wiener–Hopf integral equation of the first kind, by no means constitutes a solution to the original mixed boundary value problem.

In order to solve the Wiener–Hopf integral equation of the first kind

$$\int_0^{+\infty} K(x-y)\psi(y)\,dy = F(x), \quad x \geq 0, \tag{7.6.21}$$

we work backwards. Equation (7.6.21) is reduced to the form of Eq. (7.6.12). Defining the left-hand side of Eq. (7.6.21) for $x < 0$ by

$$\int_0^{+\infty} K(x-y)\psi(y)\,dy = H(x), \quad x < 0, \tag{7.6.22}$$

we consider the Fourier transforms of Eqs. (7.6.21) and (7.6.22),

$$\int_0^{+\infty} dx\, e^{-ikx} \int_0^{+\infty} dy K(x-y)\psi(y) = \int_0^{+\infty} dx\, e^{-ikx} F(x) \equiv \hat{F}_-(k), \tag{7.6.23a}$$

$$\int_{-\infty}^0 dx\, e^{-ikx} \int_0^{+\infty} dy K(x-y)\psi(y) = \int_{-\infty}^0 dx\, e^{-ikx} H(x) \equiv \hat{H}_+(k), \tag{7.6.23b}$$

where $\hat{F}_-(k)$ $(\hat{H}_+(k))$ is analytic in the lower half-plane (the upper half-plane). Adding Eqs. (7.6.23a) and (7.6.23b) together, we obtain

$$\int_0^{+\infty} dy e^{-iky}\psi(y) \int_{-\infty}^{+\infty} dx\, e^{-ik(x-y)} K(x-y) = \hat{F}_-(k) + \hat{H}_+(k).$$

Hence we have

$$\hat{\psi}_-(k)\hat{K}(k) = \hat{F}_-(k) + \hat{H}_+(k), \tag{7.6.24}$$

where $\hat{\psi}_-(k)$ and $\hat{K}(k)$, respectively, are defined by

$$\hat{\psi}_-(k) \equiv \int_0^{+\infty} e^{-ikx}\psi(x)dx, \tag{7.6.25}$$

$$\hat{K}(k) \equiv \int_{-\infty}^{+\infty} e^{-ikx} K(x)dx. \tag{7.6.26}$$

From Eq. (7.6.24), we have

$$\hat{\psi}_-(k) = \left(\frac{1}{\hat{K}(k)}\right)(\hat{F}_-(k) + \hat{H}_+(k)). \tag{7.6.27}$$

Carrying out the sum-splitting on the right-hand side of Eq. (7.6.27) either by inspection or by the general method discussed in Section 7.3, we can obtain $\hat{\psi}_-(k)$ as in Section 7.1.

Returning to Example 7.5, we note that the mixed boundary value problem we examined belongs to the general class of the equation,

$$\hat{\phi}_+(k) + \hat{f}_-(k) = \hat{K}(k)(\hat{g}_+(k) + \hat{\psi}_-(k)). \tag{7.6.28}$$

If we directly invert for $\hat{\psi}_-(k)$ for $x > 0$ in Eq. (7.6.28), we obtain the Wiener–Hopf integral equation of the first kind (7.6.20). Instead, we may write Eq. (7.6.28) as a pair of equations,

$$\Phi(k) = \hat{g}_+(k) + \hat{\psi}_-(k), \tag{7.6.29a}$$

$$\hat{K}(k)\Phi(k) = \hat{\phi}_+(k) + \hat{f}_-(k). \tag{7.6.29b}$$

Inverting Eqs. (7.6.29a) and (7.6.29b) for $x < 0$ and $x \geq 0$ respectively, we find a pair of integral equations for $\Phi(k)$ of the following form,

$$\int_{-\infty}^{+\infty} \frac{dk}{2\pi} e^{ikx}\Phi(k) = g(x), \qquad x < 0, \tag{7.6.30a}$$

$$\int_{-\infty}^{+\infty} \frac{dk}{2\pi} e^{ikx}\hat{K}(k)\Phi(k) = f(x), \qquad x \geq 0. \tag{7.6.30b}$$

Such a pair of integral equations, one holding in some range of the independent variable and the other holding in the complementary range, are called the *dual integral equations*. This pair is equivalent to the mixed boundary value problem, Eqs. (7.6.1) through (7.6.2c). A solution to the dual integral equations is again provided by the methods we developed in Sections 7.1 and 7.3.

7.7 Problems for Chapter 7

7.1. (Due to H. C.) Solve the Sommerfeld diffraction problem in two dimensions, with the boundary condition,

$$\phi_x(x, 0) = 0 \quad \text{for} \quad x < 0.$$

7.2. (Due to H. C.) Solve the half-line problem,

$$\left(\frac{\partial^2}{\partial x^2} + \frac{\partial^2}{\partial y^2} - p^2\right)\phi(x, y) = 0 \quad \text{with} \quad \phi(x, 0) = e^x \quad \text{for} \quad x \leq 0,$$

and

$$\phi(x, y) \to 0 \quad \text{as} \quad x^2 + y^2 \to \infty.$$

It is assumed that $\phi(x, y)$ and $\phi_y(x, y)$ are continuous except on the half-line $y = 0$ with $x \leq 0$.

7.3. (Due to D. M.) Solve the boundary value problem,

$$\left(\frac{\partial^2}{\partial x^2} + \frac{\partial^2}{\partial y^2} - p^2\right)\phi(x,y) = 0 \quad \text{with} \quad \phi_y(x,0) = e^{i\alpha x} \quad \text{for} \quad x \geq 0,$$

and

$$\phi(x,y) \to 0 \quad \text{as} \quad \sqrt{x^2+y^2} \to \infty,$$

by using the Wiener–Hopf method. In this problem, the point (x,y) lies in the region stated in the previous problem. Note that the Sommerfeld radiation condition is now replaced by the usual condition of zero limit. Compare your answer with the previous one.

7.4. (Due to H. C.) Solve

$$\nabla^2\phi(x,y) = 0,$$

with a cut on the positive x axis, subject to the boundary conditions,

$$\phi(x,0) = e^{-ax} \quad \text{for} \quad x \geq 0,$$

$$\phi(x,y) \to 0 \quad \text{as} \quad x^2+y^2 \to \infty.$$

7.5. (Due to H. C.) Solve

$$\nabla^2\phi(x,y) = 0, \qquad 0 < y < 1,$$

subject to the boundary conditions,

$$\phi(x,0) = 0 \quad \text{for} \quad x \geq 0,$$

$$\phi(x,1) = e^{-x} \quad \text{for} \quad x \geq 0,$$

and

$$\phi_y(x,1) = 0 \quad \text{for} \quad x < 0.$$

7.6. (Due to H. C.) Solve

$$\nabla^2\phi(x,y) = 0, \qquad 0 < y < 1,$$

subject to the boundary conditions,

$$\phi_y(x,0) = 0 \quad \text{for} \quad -\infty < x < \infty, \quad \phi_y(x,1) = 0 \quad \text{for} \quad x < 0,$$

and

$$\phi(x,1) = e^{-x} \quad \text{for} \quad x \geq 0.$$

7.7. (Due to H. C.) Solve

$$\left(\frac{\partial^2}{\partial x^2} + 2\frac{\partial}{\partial x} + \frac{\partial^2}{\partial y^2}\right)\phi(x,y) = \phi(x,y), \qquad y > 0,$$

with the boundary conditions,

$$\phi(x,0) = e^{-x} \quad \text{for} \quad x > 0, \quad \phi_y(x,0) = 0 \quad \text{for} \quad x < 0,$$

and

$$\phi(x,y) \to 0 \quad \text{as} \quad x^2 + y^2 \to \infty.$$

7.8. (Due to H. C.) Solve

$$\phi(x) = \lambda \int_0^{+\infty} K(x-y)\phi(y)\,dy, \qquad x \geq 0,$$

with

$$K(x) \equiv \int_{-\infty}^{+\infty} e^{-ikx} \frac{1}{\sqrt{k^2+1}} \frac{dk}{2\pi} \quad \text{and} \quad \lambda > 1.$$

Find also the resolvent $H(x,y)$ of this kernel.

7.9. (Due to H. C.) Solve

$$\phi(x) = \lambda \int_0^{+\infty} e^{-(x-y)^2} \phi(y)\,dy, \qquad 0 \leq x < \infty.$$

7.10. Solve

$$\phi(x) = \frac{\lambda}{2} \int_0^{+\infty} E_1(|x-y|)\phi(y)\,dy, \quad x \geq 0, \qquad 0 < \lambda \leq 1,$$

with

$$E_1(x) \equiv \int_x^{+\infty} \left(\frac{e^{-\zeta}}{\zeta}\right) d\zeta.$$

7.11. (Due to H. C.) Consider the eigenvalue equation,

$$\phi(x) = \lambda \int_0^{+\infty} K(x-y)\phi(y)\,dy \quad \text{where} \quad K(x) = x^2 e^{-x^2}.$$

a) What is the behavior of $\phi(x)$ so that the integral above is convergent?

b) What is the behavior of $\psi(x)$ as $x \to -\infty$ (where $\psi(x)$ is the integral above for $x < 0$)? What is the region of analyticity for $\hat{\psi}(k)$?

c) Find $\hat{K}(k)$. What is the region of analyticity for $\hat{K}(k)$?

d) It is required that $\phi(x)$ does not blow up faster than a polynomial of x as $x \to \infty$. Find the spectrum of λ and the number of independent eigenfunctions for each eigenvalue λ.

7.12. Solve

$$\phi(x) = e^{-|x|} + \lambda \int_0^{+\infty} e^{-|x-y|}\phi(y)\,dy, \qquad x \geq 0.$$

Hint:

$$\int_{-\infty}^{\infty} dx\, e^{ikx} e^{-|x|} = \frac{2}{k^{2.} + 1}.$$

7.13. (Due to H. C.) Solve

$$\phi(x) = \cosh\frac{x}{2} + \lambda \int_0^{+\infty} e^{-|x-y|}\phi(y)\,dy, \qquad x \geq 0.$$

Hint:

$$\int_{-\infty}^{+\infty} \frac{e^{ikx}}{\cosh x} dx = \frac{\pi}{\cosh\left(\frac{\pi k}{2}\right)}.$$

7.14. (Due to H. C.) Solve

$$\phi(x) = 1 + \lambda \int_0^1 \frac{1}{x + x'}\phi(x')dx', \qquad 0 \leq x \leq 1.$$

Hint: Perform the change of variables from x and x' to t and t',

$$x = \exp(-t) \quad \text{and} \quad x' = \exp(-t') \quad \text{with} \quad t, t' \in [0, +\infty).$$

7.15. Solve

$$\phi(x) = \lambda \int_0^{+\infty} \frac{1}{\alpha^2 + (x-y)^2}\phi(y)\,dy + f(x), \quad x \geq 0, \quad \alpha > 0.$$

7.16. Solve

$$T_{n+1}(z) + T_{n-1}(z) = 2zT_n(z), \quad n \geq 1, \quad -1 \leq z \leq 1,$$

with

$$T_0(z) = 1, \quad \text{and} \quad T_1(z) = z.$$

Hint: Factorize the $M(\xi)$ function by inspection.

7.17. Solve

$$U_{n+1}(z) + U_{n-1}(z) = 2zU_n(z), \quad n \geq 1, \quad -1 \leq z \leq 1,$$

with

$$U_0(z) = 0, \quad \text{and} \quad U_1(z) = \sqrt{1 - z^2}.$$

7.18. Solve

$$\sum_{k=0}^{\infty} \exp[i\rho\,|j-k|]\xi_k - \lambda\xi_j = q^j, \quad j = 0, 1, 2, \cdots,$$

with

$$\operatorname{Im}\rho > 0, \quad \text{and} \quad |q| < 1.$$

7.19. (Due to H. C.) Solve the inhomogeneous Wiener–Hopf sum equation which originates from the two-dimensional Ising model,

$$\sum_{m=0}^{\infty} M_{n-m} X_m = f_n, \quad n \geq 0,$$

with

$$M(\xi) = \sum_{n=-\infty}^{\infty} M_n \xi^n \equiv \sqrt{\frac{(1-\alpha_1\xi)(1-\alpha_2\xi^{-1})}{(1-\alpha_1\xi^{-1})(1-\alpha_2\xi)}}, \quad \text{and} \quad f_n = \delta_{n0}.$$

Consider the following five cases,

a) $\alpha_1 < 1 < \alpha_2$,

b) $\alpha_1 < \alpha_2 < 1$,

c) $\alpha_1 < \alpha_2 = 1$,

d) $1 < \alpha_1 < \alpha_2$,

e) $\alpha_1 = \alpha_2$.

Hint: Factorize the $M(\xi)$ function by inspection for the above five cases and determine the functions, $N_{\text{in}}(\xi)$ and $N_{\text{out}}(\xi)$.

7.20. Solve the Wiener–Hopf integral equation of the first kind,

$$\int_0^{+\infty} \frac{1}{2\pi} K_0(\alpha\,|x-y|)\phi(y)\,dy = 1, \qquad x \geq 0,$$

where the kernel is given by

$$K_0(x) \equiv \int_0^{+\infty} \frac{\cos kx}{\sqrt{k^2+1}}\,dk = \frac{1}{2}\int_{-\infty}^{+\infty} \frac{e^{ikx}}{\sqrt{k^2+1}}\,dk.$$

7.21. (Due to D. M.) Solve the Wiener–Hopf integral equation of the first kind,

$$\int_0^{\infty} K(x-y)\phi(y)\,dy = 1, \qquad x \geq 0,$$

where the kernel is given by

$$K(x) = |x| \exp[-|x|].$$

7.22. Solve the Wiener–Hopf integral equation of the first kind,

$$\int_0^{+\infty} K(z-\varsigma)\phi(\varsigma)\,d\varsigma = 0, \qquad z \geq 0,$$

$$K(z) \equiv \frac{1}{2}[H_0^{(1)}(k|z|) + H_0^{(1)}(k\sqrt{d^2+z^2})],$$

where $H_0^{(1)}(k|z|)$ is the zeroth-order Hankel function of the first kind.

7.23. Solve the Wiener–Hopf integral equation of the first kind,

$$\int_0^{+\infty} K(z-\varsigma)\phi(\varsigma)\,d\varsigma = 0, \qquad z \geq 0,$$

$$K(z) \equiv \frac{1}{2}[H_0^{(1)}(k|z|) - H_0^{(1)}(k\sqrt{d^2+z^2})],$$

where $H_0^{(1)}(k|z|)$ is the zeroth-order Hankel function of the first kind.

7.24. Solve the integro-differential equation,

$$\left(\frac{\partial^2}{\partial z^2} + k^2\right) \int_0^{+\infty} K(z-\varsigma)\phi(\varsigma)\,d\varsigma = 0, \quad z \geq 0,$$

$$K(z) \equiv \frac{1}{2}\left[H_0^{(1)}(k|z|) + H_0^{(1)}(k\sqrt{d^2+z^2})\right],$$

where $H_0^{(1)}(k|z|)$ is the zeroth-order Hankel function of the first kind.

7.25. Solve the integro-differential equation,

$$\left(\frac{\partial^2}{\partial z^2} + k^2\right) \int_0^{+\infty} K(z-\varsigma)\phi(\varsigma)\,d\varsigma = 0, \quad z \geq 0,$$

$$K(z) \equiv \frac{1}{2}\left[H_0^{(1)}(k|z|) - H_0^{(1)}(k\sqrt{d^2+z^2})\right],$$

where $H_0^{(1)}(k|z|)$ is the zeroth-order Hankel function of the first kind.

Hint: for Problems 7.23 through 7.26:

The zeroth-order Hankel function of the first kind $H_0^{(1)}(kD)$ is given by

$$H_0^{(1)}(kD) = \frac{1}{\pi i}\int_{-\infty}^{\infty} \frac{\exp\left[ik\sqrt{D^2+\xi^2}\right]}{\sqrt{D^2+\xi^2}}d\xi, \quad D > 0,$$

and hence its Fourier transform is given by

$$\frac{1}{2}\int_{-\infty}^{\infty} H_0^{(1)}(k\sqrt{D^2+z^2})\exp[i\omega z]dz = \frac{\exp[iv(\omega)D]}{v(\omega)}, \quad v(\omega) = \sqrt{k^2-\omega^2},$$

$$\operatorname{Im} v(\omega) > 0.$$

The problems are thus reduced to factorizing the following functions,

$$\psi(\omega) \equiv 1 + \exp[iv(\omega)d] = \psi_+(\omega)\psi_-(\omega),$$

$$\varphi(\omega) \equiv 1 - \exp[iv(\omega)d] = \varphi_+(\omega)\varphi_-(\omega),$$

where $\psi_+(\omega)$ ($\varphi_+(\omega)$) is analytic and has no zeroes in the upper half-plane, $\operatorname{Im}\omega \geq 0$, and $\psi_-(\omega)$ ($\varphi_-(\omega)$) is analytic and has no zeroes in the lower half-plane, $\operatorname{Im}\omega \leq 0$.

For the integro-differential equations, the differential operator

$$\frac{\partial^2}{\partial z^2} + k^2$$

can be brought inside the integral symbol and we obtain the extra factor,

$$v^2(\omega) = k^2 - \omega^2,$$

for the Fourier transforms, multiplying onto the functions to be factorized. The functions to be factorized are given by

$$\tilde{K}(\omega)_{\text{Prob. 7.23}} = \frac{\psi(\omega)}{v(\omega)}, \quad \tilde{K}(\omega)_{\text{Prob. 7.24}} = \frac{\varphi(\omega)}{v(\omega)},$$

$$v(\omega)\tilde{K}(\omega)_{\text{Prob. 7.25}} = v(\omega)\psi(\omega), \quad v(\omega)\tilde{K}(\omega)_{\text{Prob. 7.26}} = v(\omega)\varphi(\omega).$$

7.26. Solve the Wiener–Hopf integral equation of the first kind,

$$\int_0^{+\infty} K(z-\varsigma)\phi(\varsigma)\,d\varsigma = 0, \qquad z \geq 0,$$

$$K(z) \equiv \frac{a}{2}\int_{-\infty}^{\infty} J_1(v(\omega)a)H_1^{(1)}(v(\omega)a)\exp[i\omega z]\,d\omega,$$

with

$$v(\omega) = \sqrt{k^2-\omega^2},$$

where $J_1(va)$ is the 1st order Bessel function of the first kind and $H_1^{(1)}(va)$ is the first-order Hankel function of the first kind.

7.27. Solve the integro-differential equation,

$$\left(\frac{\partial^2}{\partial z^2} + k^2\right) \int_0^{+\infty} K(z - \varsigma)\phi(\varsigma)\, d\varsigma = 0, \qquad z \geq 0,$$

$$K(z) \equiv \frac{a}{2} \int_{-\infty}^{\infty} J_0(v(\omega)a) H_0^{(1)}(v(\omega)a) \exp[i\omega z]\, d\omega,$$

with

$$v(\omega) = \sqrt{k^2 - \omega^2},$$

where $J_0(va)$ is the 0th order Bessel function of the first kind and $H_0^{(1)}(va)$ is the zeroth-order Hankel function of the first kind.

Hint: for Problems 7.27 and 7.28:

The functions to be factorized are

$$\tilde{K}(\omega)_{\text{Prob. 7.27}} = \pi a J_1(va) H_1^{(1)}(va), \quad v^2 \tilde{K}(\omega)_{\text{Prob. 7.28}} = \pi a v^2 J_0(va) H_0^{(1)}(va).$$

The factorization procedures are identical to those in the previous problems.

For the details of the factorizations for Problems 7.23 through 7.28, we refer the reader to the following monograph.

Weinstein, L.A.: "*The theory of diffraction and the factorization method*", Golem Press, (1969). Chapters 1 and 2.

7.28. Solve the dual integral equations of the following form,

$$\int_0^{\infty} y f(y) J_n(yx)\, dy = x^n \quad \text{for} \quad 0 \leq x < 1,$$

$$\int_0^{\infty} f(y) J_n(yx)\, dy = 0 \quad \text{for} \quad 1 \leq x < \infty,$$

where n is the non-negative integer and $J_n(yx)$ is the nth-order Bessel function of the first kind.

Hint: Jackson, J.D.:"*Classical Electrodynamics* ", 3rd edition, John Wiley & Sons, New York, (1999). Section 3.13.

7.29. Solve the dual integral equations of the following form,

$$\int_0^{\infty} f(y) J_n(yx)\, dy = x^n \quad \text{for} \quad 0 \leq x < 1,$$

$$\int_0^{\infty} y f(y) J_n(yx)\, dy = 0 \quad \text{for} \quad 1 \leq x < \infty,$$

where n is the non-negative integer and $J_n(yx)$ is the nth-order Bessel function of the first kind.

Hint: Jackson, J.D.:"*Classical Electrodynamics* ", 3rd edition, John Wiley & Sons, New York, (1999). Section 5.13.

7.30. Solve the dual integral equations of the following form,

$$\int_0^\infty y^\alpha f(y) J_\mu(yx)\, dy = g(x) \quad \text{for} \quad 0 \le x < 1,$$

$$\int_0^\infty f(y) J_\mu(yx)\, dy = 0 \quad \text{for} \quad 1 \le x < \infty.$$

Hint: Kondo, J.: "*Integral Equations*", Kodansha Ltd., Tokyo, (1991). p412.

8 Nonlinear Integral Equations

8.1 Nonlinear Integral Equation of Volterra type

In Chapter 3, the integral equations of Volterra type, all of which are *linear*, are examined. We applied the Laplace transform technique for a translation kernel. As an application of the *Laplace transform technique*, we can solve a *nonlinear Volterra integral equation of convolution type*:

$$\phi(x) = f(x) + \lambda \int_0^x \phi(y)\phi(x-y)\,dy. \tag{8.1.1}$$

Taking the Laplace transform of Eq. (8.1.1), we obtain

$$\bar{\phi}(s) = \bar{f}(s) + \lambda[\bar{\phi}(s)]^2. \tag{8.1.2}$$

Hence we have

$$\bar{\phi}(s) = \frac{1 \pm (1 - 4\lambda \bar{f}(s))^{1/2}}{2\lambda}.$$

We assume that $\bar{f}(s) \to 0$ as $\mathrm{Re}\, s \to \infty$, and require that $\bar{\phi}(s) \to 0$ as $\mathrm{Re}\, s \to \infty$. Then only one of the two solutions survives. Specifically it is

$$\bar{\phi}(s) = \frac{1 - (1 - 4\lambda \bar{f}(s))^{1/2}}{2\lambda}, \tag{8.1.3a}$$

and

$$\begin{aligned} \phi(x) &= L^{-1}\left(\frac{1 - (1 - 4\lambda \bar{f}(s))^{1/2}}{2\lambda}\right) \\ &= \int_{\gamma - i\infty}^{\gamma + i\infty} \frac{ds}{2\pi i} e^{sx} \frac{1 - (1 - 4\lambda \bar{f}(s))^{1/2}}{2\lambda}, \end{aligned} \tag{8.1.3b}$$

where the inversion path is to the right of all singularities of the integrand. We examine two specific cases.

❑ **Example 8.1.** $f(x) = 0$.

Solution. In this case, Eq. (8.1.3b) gives

$$\phi(x) = 0.$$

Thus there is no nontrivial solution.

Applied Mathematics in Theoretical Physics. Michio Masujima

ISBN: 3-527-40534-8

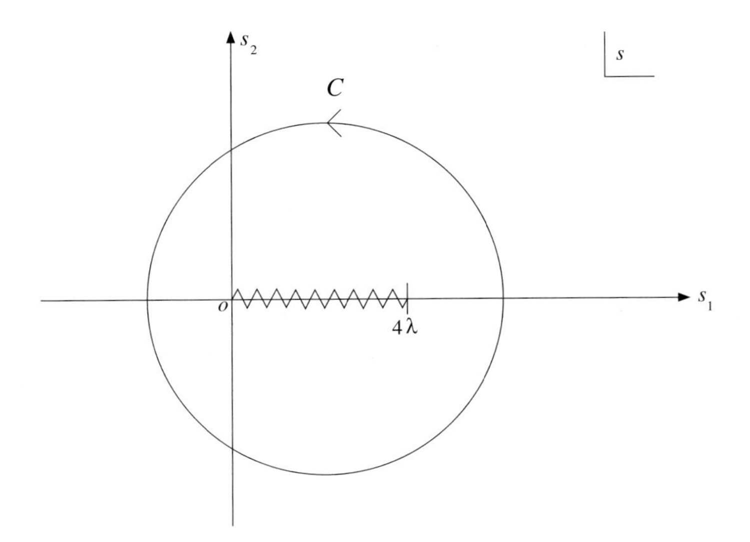

Fig. 8.1: Branch cut of the integrand of Eq. (8.1.5a) from $s = 0$ to $s = 4\lambda$.

❑ **Example 8.2.** $f(x) = 1$.

Solution. In this case,

$$\bar{f}(s) = \frac{1}{s}, \tag{8.1.4}$$

and Eq. (8.1.3b) gives

$$\phi(x) = \frac{1}{2\lambda} \int_{\gamma - i\infty}^{\gamma + i\infty} \frac{ds}{2\pi i} e^{sx} \left[1 - \sqrt{\frac{s - 4\lambda}{s}} \right]. \tag{8.1.5a}$$

The integrand has a branch cut from $s = 0$ to $s = 4\lambda$ as in Figure 8.1.

Let $\lambda > 0$, then the branch cut is as illustrated in Figure 8.2. By deforming the contour, we get

$$\phi(x) = \frac{1}{2\lambda} \oint_C \frac{ds}{2\pi i} e^{sx} \left(1 - \sqrt{\frac{s - 4\lambda}{s}} \right) = -\frac{1}{2\lambda} \oint_C \frac{ds}{2\pi i} e^{sx} \sqrt{\frac{s - 4\lambda}{s}}, \tag{8.1.5b}$$

where C is the contour wrapped around the branch cut, as shown in Figure 8.2. By evaluating the values of the integrand on the two sides of the branch cut, we get

$$\phi(x) = \frac{1}{2\pi\lambda} \int_0^{4\lambda} ds\, e^{sx} \sqrt{\frac{4\lambda - s}{s}} = \frac{2}{\pi} \int_0^1 dt\, e^{4\lambda t x} \sqrt{\frac{1 - t}{t}}, \tag{8.1.6}$$

where we have made the change of variable,

$$s = 4\lambda t, \quad 0 \le t \le 1.$$

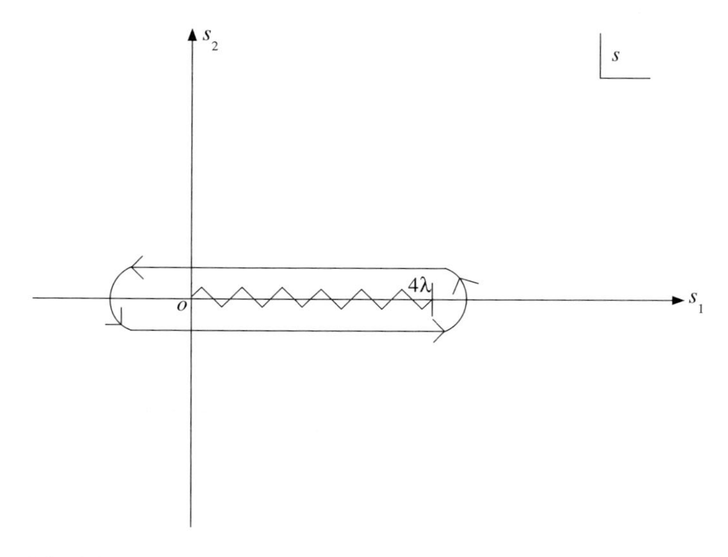

Fig. 8.2: The contour of integration C wrapping around the branch cut of Figure 8.1 for $\lambda > 0$.

The integral in Eq. (8.1.6) can be explicitly evaluated.

$$\phi(x) = \frac{2}{\pi} \sum_{n=0}^{\infty} \frac{(4\lambda x)^n}{n!} \int_0^1 dt\, t^n \sqrt{\frac{1-t}{t}} = \frac{2}{\pi} \sum_{n=0}^{\infty} \frac{(4\lambda x)^n}{n!} \frac{\Gamma\left(n+\frac{1}{2}\right)\Gamma\left(\frac{3}{2}\right)}{\Gamma(n+2)}, \tag{8.1.7}$$

where we have made use of the formula,

$$\int_0^1 dt\, t^{n-1}(1-t)^{m-1} = \frac{\Gamma(n)\Gamma(m)}{\Gamma(n++m)}.$$

Now, the *confluent hypergeometric function* is given by

$$F(a;c;z) = 1 + \frac{a}{c}\frac{z}{1!} + \frac{a}{c}\frac{a+1}{c+1}\frac{z^2}{2!} + \cdots = \sum_{n=0}^{\infty} \frac{\Gamma(c)}{\Gamma(a)} \frac{\Gamma(a+n)}{\Gamma(c+n)} \frac{z^n}{n!}. \tag{8.1.8}$$

From Eqs. (8.1.7) and (8.1.8), we find that

$$\phi(x) = F\left(\frac{1}{2}; 2; 4\lambda x\right) \tag{8.1.9}$$

satisfies the nonlinear integral equation,

$$\phi(x) = 1 + \lambda \int_0^x \phi(x-y)\phi(y)\, dy. \tag{8.1.10}$$

Although the above is proved only for $\lambda > 0$, we may verify that it is also true for $\lambda < 0$ by repeating the same argument. Alternatively, we may prove this in the following way. Let us substitute Eq. (8.1.9) into Eq. (8.1.10). Since $F\left(\frac{1}{2};2;4\lambda x\right)$ is an entire function of λ, each side of the resulting equation is also an entire function of λ. Since this equation is satisfied for $\lambda > 0$, it must be satisfied for all λ by analytic continuation. Thus the integral equation (8.1.10) has the unique solution given by Eq. (8.1.9), for all values of λ.

In closing this section, we classify the nonlinear integral equations of Volterra type in the following manner:

(1) The kernel part is nonlinear,

$$\phi(x) - \int_a^x H(x, y, \phi(y))\, dy = f(x). \tag{VN.1}$$

(2) The particular part is nonlinear,

$$G(\phi(x)) - \int_a^x K(x, y)\phi(y)\, dy = f(x). \tag{VN.2}$$

(3) Both parts are nonlinear,

$$G(\phi(x)) - \int_a^x H(x, y, \phi(y))\, dy = f(x). \tag{VN.3}$$

(4) The nonlinear Volterra integral equation of the first kind,

$$\int_a^x H(x, y, \phi(y))\, dy = f(x). \tag{VN.4}$$

(5) The homogeneous nonlinear Volterra integral equation of the first kind, where the kernel part is nonlinear,

$$\phi(x) = \int_a^x H(x, y, \phi(y))\, dy. \tag{VN.5}$$

(6) The homogeneous nonlinear Volterra integral equation of the first kind where the particular part is nonlinear,

$$G(\phi(x)) = \int_a^x K(x, y)\phi(y)\, dy. \tag{VN.6}$$

(7) Homogeneous nonlinear Volterra integral equation of the first kind where the both parts are nonlinear,

$$G(\phi(x)) = \int_a^x H(x, y, \phi(y))\, dy. \tag{VN.7}$$

8.2 Nonlinear Integral Equation of Fredholm Type

In this section, we give a brief discussion of the *nonlinear integral equation of Fredholm type*. Let us recall that the linear algebraic equation

$$\phi = f + \lambda K\phi \tag{8.2.1}$$

has the solution

$$\phi = (1 - \lambda K)^{-1} f. \tag{8.2.2}$$

In particular, the solution of Eq. (8.2.1) exists and is unique as long as the corresponding homogeneous equation

$$\phi = \lambda K\phi \tag{8.2.3}$$

has no nontrivial solutions.

Nonlinear equations behave quite differently. Consider, for example, the nonlinear algebraic equation obtained from Eq. (8.2.1) by replacing K with ϕ,

$$\phi = f + \lambda\phi^2. \tag{8.2.4a}$$

The solutions of Eq. (8.2.4a) are,

$$\phi = \frac{1 \pm \sqrt{1 - 4\lambda f}}{2\lambda}. \tag{8.2.4b}$$

We first observe that the solution is not unique. Indeed, if we require the solutions to be real, then Eq. (8.2.4a) has two solutions if

$$1 - 4\lambda f > 0, \tag{8.2.5}$$

and no solution if

$$1 - 4\lambda f < 0. \tag{8.2.6}$$

Thus the number of solutions changes from 2 to 0 as the value of λ passes $1/4f$. The point $\lambda = 1/4f$ is called a *bifurcation point* of Eq. (8.2.4a). Note that at the bifurcation point, Eq. (8.2.4a) has only one solution.

We also observe from Eq. (8.2.4b) that another special point for Eq. (8.2.4a) is $\lambda = 0$. At this point, one of the two solutions is infinite. Since the number of solutions remains to be two as the value of λ passes $\lambda = 0$, the point $\lambda = 0$ is not a bifurcation point. We shall call it a *singular point*.

Consider now the equation

$$\phi = \lambda\phi^2, \tag{8.2.7}$$

obtained from Eq. (8.2.4a) by setting $f = 0$. This equation always has the nontrivial solution

$$\phi = \frac{1}{\lambda} \tag{8.2.8}$$

provided that

$$\lambda \neq 0. \tag{8.2.9}$$

There is no connection between the existence of the solutions for Eq. (8.2.4a) and the absence of the solutions for Eq. (8.2.4a) with $f = 0$, quite unlike the case of linear algebraic equations.

Nonlinear integral equations share these properties. This is evident in the following examples.

❑ **Example 8.3.** Solve

$$\phi(x) = 1 + \lambda \int_0^1 \phi^2(y)\, dy. \tag{8.2.10}$$

Solution. The right-hand side of Eq. (8.2.10) is independent of x. Thus $\phi(x)$ is constant. Let

$$\phi(x) = a.$$

Then Eq. (8.2.10) becomes

$$a = 1 + \lambda a^2. \tag{8.2.11}$$

Equation (8.2.11) is just Eq. (8.2.4.a) with $f = 1$. Thus

$$\phi(x) = \frac{1 \pm \sqrt{1 - 4\lambda}}{2\lambda}. \tag{8.2.12}$$

There are two real solutions for $\lambda < 1/4$, and no real solutions for $\lambda > 1/4$. Thus $\lambda = 1/4$ is a bifurcation point.

❑ **Example 8.4.** Solve

$$\phi(x) = 1 + \lambda \int_0^1 \phi^3(y)\, dy. \tag{8.2.13}$$

Solution. The right-hand side of Eq. (8.2.13) is independent of x. Thus $\phi(x)$ is constant. Letting

$$\phi(x) = a,$$

we get

$$a = 1 + \lambda a^3. \tag{8.2.14a}$$

Equation (8.2.14a) is cubic, and hence has three solutions. Not all of these solutions are real. Let us rewrite Eq. (8.2.14a) as

$$\frac{a - 1}{\lambda} = a^3, \tag{8.2.14b}$$

and plot $(a-1)/\lambda$ as well as a^3 in the figure. The points of intersection between these two curves are the solutions of Eq. (8.2.14a). For λ negative, there is obviously only one real root, while for λ positive and very small, there are three real roots (two positive roots and one negative root). Thus $\lambda = 0$ is a bifurcation point. For λ large and positive, there is again only one real root. The change of numbers of roots can be shown to occur at $\lambda = 4/27$, which is another bifurcation point.

We may generalize the above considerations to

$$\phi(x) = c + \lambda \int_0^1 K(\phi(y))\, dy. \tag{8.2.15a}$$

The right-hand side of Eq. (8.2.15a) is independent of x. Thus $\phi(x)$ is constant. Letting

$$\phi(x) = a,$$

we get

$$\frac{(a-c)}{\lambda} = K(a). \tag{8.2.15b}$$

The roots of the equation above can be graphically obtained by plotting $K(a)$ and $(a-c)/\lambda$. Obviously, with a proper choice of $K(a)$, the number of solutions as well as the number of bifurcation points may take any value. For example, if

$$K(\phi) = \phi \sin \pi\phi, \tag{8.2.16}$$

there are infinitely many solutions as long as $|\lambda| < 1$. As another example, for

$$K(\phi) = \frac{\sin \phi}{(\phi^2+1)}, \tag{8.2.17}$$

there are infinitely many bifurcation points.

In summary, we have found the following conclusions for nonlinear integral equations.

(1) There may be more than one solution.

(2) There may be one or more bifurcation points.

(3) There is no significant relationship between the integral equation with $f \neq 0$ and the one obtained from it by setting $f = 0$.

The above considerations for simple examples may be extended to more general cases. Consider the integral equation

$$\phi(x) = f(x) + \int_0^1 K(x, y, \phi(x), \phi(y))\, dy. \tag{8.2.18}$$

If K is separable, i.e.,

$$K(x, y, \phi(x), \phi(y)) = g(x, \phi(x))h(y, \phi(y)), \tag{8.2.19}$$

then the integral equation (8.2.18) is solved by

$$\phi(x) = f(x) + ag(x, \phi(x)), \tag{8.2.20}$$

with

$$a = \int_0^1 h(x, \phi(x))\, dx. \tag{8.2.21}$$

We may solve Eq. (8.2.20) for $\phi(x)$ and express $\phi(x)$ as a function of x and a. There may be more than one solution. Substituting any one of these solutions into Eq. (8.2.21), we may obtain an equation for a. Thus the nonlinear integral equation (8.2.18) is equivalent to one or more nonlinear algebraic equations for a.

Similarly, if K is a sum of the separable terms,

$$K(x, y, \phi(x), \phi(y)) = \sum_{n=1}^{N} g_n(x, \phi(x)) h_n(y, \phi(y)), \tag{8.2.22}$$

then the integral equation is equivalent to one or more systems of N coupled nonlinear algebraic equations.

In closing this section, we classify the nonlinear integral equations of Fredholm type in the following manner:

(1) The kernel part is nonlinear,

$$\phi(x) - \int_a^b H(x, y, \phi(y))\, dy = f(x). \tag{FN.1}$$

(2) The particular part is nonlinear,

$$G(\phi(x)) - \int_a^b K(x, y)\phi(y)\, dy = f(x). \tag{FN.2}$$

(3) Both parts are nonlinear,

$$G(\phi(x)) - \int_a^b H(x, y, \phi(y))\, dy = f(x). \tag{FN.3}$$

(4) The nonlinear Fredholm integral equation of the first kind,

$$\int_a^b H(x, y, \phi(y))\, dy = f(x). \tag{FN.4}$$

(5) Homogeneous nonlinear Fredholm integral equation of the first kind where the kernel part is nonlinear,

$$\phi(x) = \int_a^b H(x, y, \phi(y))\, dy. \tag{FN.5}$$

(6) The homogeneous nonlinear Fredholm integral equation of the first kind where the particular part is nonlinear,

$$G(\phi(x)) = \int_a^b K(x, y)\phi(y)\, dy. \tag{FN.6}$$

(7) The homogeneous nonlinear Fredholm integral equation of the first kind where both parts are nonlinear,

$$G(\phi(x)) = \int_a^b H(x, y, \phi(y))\, dy. \tag{FN.7}$$

8.3 Nonlinear Integral Equation of Hammerstein type

The inhomogeneous term $f(x)$ in Eq. (8.2.18) is not particularly meaningful. This is because we may define

$$\psi(x) \equiv \phi(x) - f(x), \tag{8.3.1}$$

then Eq. (8.2.18) is of the form

$$\psi(x) = \int_0^1 K(x, y, \psi(x) + f(x), \psi(y) + f(y))\, dy. \tag{8.3.2}$$

The *nonlinear integral equation of Hammerstein type* is a special case of Eq. (8.3.2),

$$\psi(x) = \int_0^1 K(x, y) f(y, \psi(y))\, dy. \tag{8.3.3}$$

For the remainder of this section, we discuss this latter equation, (8.3.3). We shall show that if f *uniformly* satisfies a *Lipschitz condition* of the form

$$|f(y, u_1) - f(y, u_2)| < C(y)\, |u_1 - u_2|\,, \tag{8.3.4}$$

and if

$$\int_0^1 A(y)C^2(y)\, dy = M^2 < 1, \tag{8.3.5}$$

then the solution of Eq. (8.3.3) is unique and can be obtained by iteration. The function $A(x)$ in Eq. (8.3.5) is given by

$$A(x) = \int_0^1 K^2(x, y)\, dy. \tag{8.3.6}$$

We begin by setting

$$\psi_0(x) = 0 \tag{8.3.7}$$

and

$$\psi_n(x) = \int_0^1 K(x,y) f(y, \psi_{n-1}(y))\, dy. \tag{8.3.8}$$

If

$$f(y,0) = 0, \tag{8.3.9}$$

then Eq. (8.3.3) is solved by

$$\psi(x) = 0. \tag{8.3.10}$$

If

$$f(y,0) \neq 0, \tag{8.3.11}$$

we have

$$\psi_1^2(x) \leq A(x) \int_0^1 f^2(y,0)\, dy = A(x) \left\| f \right\|^2 , \tag{8.3.12}$$

where

$$\left\| f \right\|^2 \equiv \int_0^1 f^2(y,0)\, dy. \tag{8.3.13}$$

Also, as a consequence of the Lipschitz condition (8.3.4),

$$|\psi_n(x) - \psi_{n-1}(x)| < \int_0^1 |K(x,y)|\, C(y)\, |\psi_{n-1}(y) - \psi_{n-2}(y)|\, dy.$$

Thus

$$[\psi_n(x) - \psi_{n-1}(x)]^2 \leq A(x) \int_0^1 C^2(y) [\psi_{n-1}(y) - \psi_{n-2}(y)]^2\, dy. \tag{8.3.14}$$

From Eqs. (8.3.4) and (8.3.5), we get

$$[\psi_2(x) - \psi_1(x)]^2 \leq A(x) \left\| f \right\|^2 M^2.$$

By induction,

$$[\psi_n(x) - \psi_{n-1}(x)]^2 \leq A(x) \left\| f \right\|^2 (M^2)^{n-1}. \tag{8.3.15}$$

Thus the series

$$\psi_1(x) + [\psi_2(x) - \psi_1(x)] + [\psi_3(x) - \psi_2(x)] + \cdots$$

is convergent, due to Eq. (8.3.5).

The proof of uniqueness will be left to the reader.

* * *

The most typical nonlinear integral equations in quantum field theory and quantum statistical mechanics are Schwinger–Dyson equations, which are actually the coupled nonlinear integro-differential equations. Schwinger–Dyson equations in quantum field theory and quantum statistical mechanics can be solved iteratively. These topics are discussed in Sections 10.3 and 10.4 of Chapter 10.

8.4 Problems for Chapter 8

8.1. (Due to H. C.) Prove the uniqueness of the solution to the nonlinear integral equation of Hammerstein type,

$$\psi(x) = \int_0^1 K(x,y) f(y, \psi(y))\, dy.$$

8.2. (Due to H. C.) Consider

$$\frac{d^2}{dt^2} x(t) + x(t) = \frac{1}{\pi^2} x^2(t), \quad t > 0,$$

with the initial conditions,

$$x(0) = 0, \quad \text{and} \quad \left.\frac{dx}{dt}\right|_{t=0} = 1.$$

a) Transform this nonlinear ordinary differential equation into an integral equation.

b) Obtain an approximate solution, accurate to a few percent.

8.3. (Due to H. C.) Consider

$$\left(\frac{\partial^2}{\partial x^2} + \frac{\partial^2}{\partial y^2} \right) \phi(x,y) = \frac{1}{\pi^2} \phi^2(x,y), \quad x^2 + y^2 < 1,$$

with

$$\phi(x,y) = 1 \quad \text{on} \quad x^2 + y^2 = 1.$$

a) Construct a Green's function satisfying

$$\left(\frac{\partial^2}{\partial x^2} + \frac{\partial^2}{\partial y^2} \right) G(x,y;x',y') = \delta(x-x')\delta(y-y'),$$

with the boundary condition,

$$G(x,y;x',y') = 0 \quad \text{on} \quad x^2 + y^2 = 1.$$

Prove that

$$G(x,y;x',y') = G(x',y';x,y).$$

Hint: To construct the Green's function $G(x,y;x',y')$ which vanishes on the unit circle, use the method of images.

b) Transform the nonlinear partial differential equation above to an integral equation, and obtain an approximate solution accurate to a few percent.

8.4. (Due to H. C.)

a) Discuss a phase transition of $\rho(\theta)$ which is given by the nonlinear integral equation,

$$\rho(\theta) = Z^{-1} \exp\left[\beta \int_0^{2\pi} \cos(\theta - \phi)\rho(\phi)\, d\phi\right], \quad \beta = \frac{J}{k_B T}, \quad 0 \leq \theta \leq 2\pi,$$

where Z is the normalization constant such that $\rho(\theta)$ is normalized to unity,

$$\int_0^{2\pi} \rho(\theta)\, d\theta = 1.$$

b) Determine the nature of the phase transition.

Hint: You may need the following special functions,

$$I_0(z) = \sum_{m=0}^{\infty} \frac{1}{(m!)^2} \left(\frac{z}{2}\right)^{2m},$$

$$I_1(z) = \sum_{m=0}^{\infty} \frac{1}{m!(m+1)!} \left(\frac{z}{2}\right)^{2m+1},$$

and generally

$$I_n(z) = \frac{1}{\pi} \int_0^{\pi} e^{z \cos\theta} (\cos n\theta)\, d\theta.$$

For $|z| \to 0$, we have

$$I_0(z) \to 1,$$

$$I_1(z) \to \frac{z}{2}.$$

8.5. Solve the nonlinear integral equation of Fredholm type,

$$\phi(x) - \int_0^{0.1} xy\left[1 + \phi^2(y)\right] dy = x + 1.$$

8.6. Solve the nonlinear integral equation of Fredholm type,

$$\phi(x) - 60 \int_0^1 xy\phi^2(y)\, dy = 1 + 20x - x^2.$$

8.7. Solve the nonlinear integral equation of Fredholm type,

$$\phi(x) - \lambda \int_0^1 xy\left[1 + \phi^2(y)\right] dy = x + 1.$$

8.8. Solve the nonlinear integral equation of Fredholm type,

$$\phi^2(x) - \int_0^1 xy\phi(y)\,dy = 4 + 10x + 9x^2.$$

8.9. Solve the nonlinear integral equation of Fredholm type,

$$\phi^2(x) - \lambda \int_0^1 xy\phi(y)\,dy = 1 + x^2.$$

8.10. Solve the nonlinear integral equation of Fredholm type,

$$\phi^2(x) + \int_0^2 \phi^3(y)\,dy = 1 - 2x + x^2.$$

8.11. Solve the nonlinear integral equation of Fredholm type,

$$\phi^2(x) = \frac{5}{2} \int_0^1 xy\phi(y)\,dy.$$

Hint for Problems 8.5 through 8.11: The integrals are, at most, linear in x.

8.12. Solve the nonlinear integral equation of Volterra type,

$$\phi^2(x) - \int_0^x (x-y)\phi(y)\,dy = 1 + 3x + \frac{1}{2}x^2 - \frac{1}{2}x^3.$$

8.13. Solve the nonlinear integral equation of Volterra type,

$$\phi^2(x) + \int_0^x \sin(x-y)\phi(y)\,dy = \exp[x].$$

8.14. Solve the nonlinear integral equation of Volterra type,

$$\phi(x) - \int_0^x (x-y)^2\phi^2(y)\,dy = x.$$

8.15. Solve the nonlinear integral equation of Volterra type,

$$\phi^2(x) - \int_0^x (x-y)^3\phi^3(y)\,dy = 1 + x^2.$$

8.16. Solve the nonlinear integral equation of Volterra type,

$$2\phi(x) - \int_0^x \phi(x-y)\phi(y)\,dy = \sin x.$$

Hint for Problems 8.12 through 8.16: Take the Laplace transform of the given nonlinear integral equations of Volterra type.

8.17. Solve the nonlinear integral equation,

$$\phi(x) - \lambda \int_0^1 \phi^2(y)\, dy = 1.$$

In particular, identify the bifurcation points of this equation. What are the non-trivial solutions of the corresponding homogeneous equations?

9 Calculus of Variations: Fundamentals

9.1 Historical Background

The calculus of variations was first found in the late $17^{\text{th.}}$ century soon after calculus was invented. The main people involved were Newton, the two Bernoulli brothers, Euler, Lagrange, Legendre and Jacobi.

Isaac Newton (1642–1727) formulated the fundamental laws of motion. The fundamental quantities of motion were established as momentum and force. Newton's laws of motion state:

1. In the inertial frame, every body remains at rest or in uniform motion unless acted on by a force $\vec{F}$. The condition $\vec{F} = \vec{0}$ implies a constant velocity $\vec{v}$ and a constant momentum $\vec{p} = m\vec{v}$.

2. In the inertial frame, the application of force $\vec{F}$ alters the momentum $\vec{p}$ by an amount specified by

$$\vec{F} = \frac{d}{dt}\vec{p}. \tag{9.1.1}$$

3. To each action of a force, there is an equal and opposite action of another force. Thus if $\vec{F}_{21}$ is the force exerted on particle 1 by particle 2, then

$$\vec{F}_{21} = -\vec{F}_{12}, \tag{9.1.2}$$

and these forces act along the line separating the particles.

Contrary to the common belief that Newton discovered the gravitational force by observing that the apple dropped from the tree at Trinity College, he actually deduced Newton's laws of motion from the careful analysis of Kepler's laws. He also invented the calculus, named methodus fluxionum in 1666, about 10 years ahead of Leibniz. In 1687, Newton published his "*Philosophiae naturalis principia mathematica*", often called "*Principia*". It consists of three parts: Newton's laws of motion, the laws of the gravitational force, and the laws of motion of the planets.

The Bernoulli brothers, Jacques (1654–1705) and Jean (1667–1748), came from a family of mathematicians in Switzerland. They solved the problem of *Brachistochrone*. They established the principle of virtual work as a general principle of statics with which all problems of equilibrium could be solved. Remarkably, they also compared the motion of a particle in a given field of force with that of light in an optically heterogeneous medium and tried to

Applied Mathematics in Theoretical Physics. Michio Masujima

ISBN: 3-527-40534-8

provide a mechanical theory of the refractive index. The Bernoulli brothers were the forerunners of the theory of Hamilton which has shown that the principle of least action in classical mechanics and Fermat's principle of shortest time in geometrical optics, are strikingly analogous to each other. They used the notation, g, for the gravitational acceleration for the first time.

Leonhard Euler (1707–1783) grew up under the influence of the Bernoulli family in Switzerland. He made an extensive contribution to the development of calculus after Leibniz, and initiated the calculus of variations. He started the systematic study of isoperimetric problems. He also contributed in an essential way to the variational treatment of classical mechanics, providing the Euler equation

$$\frac{\partial f}{\partial y} - \frac{d}{dx}\left(\frac{\partial f}{\partial y'}\right) = 0, \tag{9.1.3}$$

for the extremization problem

$$\delta I = \delta \int_{x_1}^{x_2} f(x, y, y')\, dx = 0, \tag{9.1.4}$$

with

$$\delta y(x_1) = \delta y(x_2) = 0. \tag{9.1.5}$$

Joseph Louis Lagrange (1736–1813) provided the solution to the isoperimetric problems by the method presently known as the method of Lagrange multipliers, quite independently of Euler. He started the whole field of the calculus of variations. He also introduced the notion of generalized coordinates, $\{q_r(t)\}_{r=1}^{f}$, into classical mechanics and completely reduced the mechanical problem to that of the differential equations now known as the Lagrange equations of motion,

$$\frac{d}{dt}\left(\frac{\partial L(q_s, \dot{q}_s, t)}{\partial \dot{q}_r}\right) - \frac{\partial L(q_s, \dot{q}_s, t)}{\partial q_r} = 0, \quad r = 1, \cdots, f, \quad \text{with} \quad \dot{q}_r \equiv \frac{d}{dt} q_r, \tag{9.1.6}$$

with the Lagrangian $L(q_r(t), \dot{q}_r(t), t)$ appropriately chosen in terms of the kinetic energy and the potential energy. He successfully converted classical mechanics into analytical mechanics using the variational principle. He also carried out research on the Fermat problem, the general treatment of the theory of ordinary differential equations, and the theory of elliptic functions.

Adrien Marie Legendre (1752–1833) announced his research on the form of the planet in 1784. In his article, the Legendre polynomials were used for the first time. He provided the Legendre test in the maximization–minimization problem of the calculus of variations, among his numerous and diverse contributions to mathematics. As one of his major accomplishments, his classification of elliptic integrals into three types stated in "*Exercices de calcul intégral*", published in 1811, should be mentioned. In 1794, he published "*Éléments de géométrie, avec notes*". He further developed the method of transformations for thermodynamics which are currently known as the Legendre transformations and are used even today in quantum field theory.

William Rowan Hamilton (1805–1865) started his research on optics around 1823 and introduced the notion of the characteristic function. His results formed the basis of the later development of the concept of the eikonal in optics. He also succeeded in transforming the Lagrange equations of motion (of the second order) into a set of differential equations of the first order with twice as many variables, by the introduction of the momenta $\{p_r(t)\}_{r=1}^{f}$ canonically conjugate to the generalized coordinates $\{q_r(t)\}_{r=1}^{f}$ by

$$p_r(t) = \frac{\partial L(q_s, \dot{q}_s, t)}{\partial \dot{q}_r}, \qquad r = 1, \ldots, f. \tag{9.1.7}$$

His equations are known as Hamilton's canonical equations of motion:

$$\begin{aligned} \frac{d}{dt} q_r(t) &= \frac{\partial H(q_s(t), p_s(t), t)}{\partial p_r(t)}, \\ \frac{d}{dt} p_r(t) &= -\frac{\partial H(q_s(t), p_s(t), t)}{\partial q_r(t)}, \qquad r = 1, \ldots, f. \end{aligned} \tag{9.1.8}$$

He formulated classical mechanics in terms of the principle of least action. The variational principles formulated by Euler and Lagrange apply only to conservative systems. He also recognized that the principle of least action in classical mechanics and Fermat's principle of shortest time in geometrical optics are strikingly analogous, permitting the interpretation of the optical phenomena in terms of mechanical terms and vice versa. He was one step short of discovering wave mechanics by analogy with wave optics as early as 1834, although he did not have any experimentally compelling reason to take such a step. On the other hand, by 1924, L. de Broglie and E. Schrödinger had sufficient experimentally compelling reasons to take this step.

Carl Gustav Jacob Jacobi (1804–1851), in 1824, quickly recognized the importance of the work of Hamilton. He realized that Hamilton was using just one particular choice of a set of the variables $\{q_r(t)\}_{r=1}^{f}$ and $\{p_r(t)\}_{r=1}^{f}$ to describe the mechanical system and carried out the research on the canonical transformation theory with the Legendre transformation. He duly arrived at what is now known as the Hamilton–Jacobi equation. His research on the canonical transformation theory is summarized in "*Vorlesungen über Dynamik*", published in 1866. He formulated his version of the principle of least action for the time-independent case and provided the Jacobi test in the maximization–minimization problem of the calculus of variations. In 1827, he introduced the elliptic functions as the inverse functions of the elliptic integrals.

From our discussions, we may be led to the conclusion that the calculus of variations is the completed subject of the 19$^{\text{th.}}$ century. We note, however, that from the 1940s to 1950s, there was a resurgence of the action principle for the systemization of quantum field theory.

Feynman's action principle and Schwinger's action principle are the main subject matter. In contemporary particle physics, if we begin with the Lagrangian density of the system under consideration, the extremization of the action functional is still employed as the starting point of the discussion (See Chapter 10). Furthermore, the Legendre transformation is used in the computation of the effective potential in quantum field theory.

We now define the problem of the calculus of variations. Suppose that we have an unknown function $y(x)$ of the independent variable x which satisfies some condition C. We construct the functional $I[y]$ which involves the unknown function $y(x)$ and its derivatives. We now want to determine the unknown function $y(x)$ which extremizes the functional $I[y]$ under the infinitesimal variation $\delta y(x)$ of $y(x)$ subject to the condition C. A simple example is the extremization of the following functional:

$$I[y] = \int_{x_1}^{x_2} L(x, y, y')\, dx, \quad C : y(x_1) = y(x_2) = 0. \tag{9.1.9}$$

The problem is reduced to solving the Euler equation, which we discuss later. This problem and its solution constitute the problem of the calculus of variations.

Many basic principles of physics can be cast in the form of the calculus of variations. Most of the problems in classical mechanics and classical field theory are of this form, with certain generalizations, and the following replacements:
for classical mechanics, we replace

$$\begin{cases} x & \text{with} \quad t, \\ y & \text{with} \quad q(t), \\ y' & \text{with} \quad \dot{q}(t) \equiv \dfrac{dq(t)}{dt}, \end{cases} \tag{9.1.10}$$

and for classical field theory, we replace

$$\begin{cases} x & \text{with} \quad (t, \vec{r}), \\ y & \text{with} \quad \psi(t, \vec{r}), \\ y' & \text{with} \quad \left(\dfrac{\partial \psi(t, \vec{r})}{\partial t}, \vec{\nabla}\psi(t, \vec{r})\right). \end{cases} \tag{9.1.11}$$

On the other hand, when we want to solve some differential equation, subject to the condition C, we may be able to reduce the problem of solving the original differential equation to that of the extremization of the functional $I[y]$ subject to the condition C, provided that the Euler equation of the extremization problem coincides with the original differential equation we want to solve. With this reduction, we can obtain an approximate solution of the original differential equation.

The problem of *Brachistochrone*, to be defined later, the isoperimetric problem, to be defined later, and the problem of finding the shape of the soap membrane with the minimum surface area, are the classic problems of the calculus of variations.

9.2 Examples

We list examples of the problems to be solved.

❑ **Example 9.1. The shortest distance between two points is a straight line: Minimize**

$$I = \int_{x_1}^{x_2} \sqrt{1 + (y')^2}\, dx. \tag{9.2.1}$$

❑ **Example 9.2. The largest area enclosed by an arc of fixed length is a circle: Minimize**

$$I = \int y\, dx, \tag{9.2.2}$$

subject to the condition that

$$\int \sqrt{1 + (y')^2}\, dx \quad \text{fixed.} \tag{9.2.3}$$

❑ **Example 9.3. Catenary. The surface formed by two circular wires dipped in a soap solution: Minimize**

$$\int_{x_1}^{x_2} y\sqrt{1 + (y')^2}\, dx. \tag{9.2.4}$$

❑ **Example 9.4. Brachistochrone. Determine a path down which a particle falls under gravity in the shortest time: Minimize**

$$\int \sqrt{\frac{1 + (y')^2}{y}}\, dx. \tag{9.2.5}$$

❑ **Example 9.5. Hamilton's Action Principle in Classical Mechanics: Minimize the *action integral I* defined by**

$$I \equiv \int_{t_1}^{t_2} L(q, \dot{q})\, dt \quad \text{with} \quad q(t_1) \text{ and } q(t_2) \quad \text{fixed,} \tag{9.2.6}$$

where $L(q(t), \dot{q}(t))$ is the *Lagrangian* of the mechanical system.

The last example is responsible for beginning the whole field of the calculus of variations.

9.3 Euler Equation

We shall derive the fundamental equation for the calculus of variations, the Euler equation.

We shall extremize

$$I = \int_{x_1}^{x_2} f(x, y, y')\, dx, \tag{9.3.1}$$

with the endpoints

$$y(x_1), \quad y(x_2) \quad \text{fixed.} \tag{9.3.2}$$

We consider a small variation in $y(x)$ of the following form,

$$y(x) \to y(x) + \varepsilon\nu(x), \tag{9.3.3a}$$

$$y'(x) \to y'(x) + \varepsilon\nu'(x), \tag{9.3.3b}$$

with

$$\nu(x_1) = \nu(x_2) = 0, \quad \varepsilon = \text{positive infinitesimal.} \tag{9.3.3c}$$

Then the variation in I is given by

$$\begin{aligned} \delta I &= \int_{x_1}^{x_2} \left(\frac{\partial f}{\partial y}\varepsilon\nu(x) + \frac{\partial f}{\partial y'}\varepsilon\nu'(x) \right) dx \\ &= \frac{\partial f}{\partial y'}\varepsilon\nu(x) \bigg|_{x=x_1}^{x=x_2} + \int_{x_1}^{x_2} \left(\frac{\partial f}{\partial y} - \frac{d}{dx}\left(\frac{\partial f}{\partial y'} \right) \right) \varepsilon\nu(x)\, dx \\ &= \int_{x_1}^{x_2} \left(\frac{\partial f}{\partial y} - \frac{d}{dx}\left(\frac{\partial f}{\partial y'} \right) \right) \varepsilon\nu(x)\, dx = 0, \end{aligned} \tag{9.3.4}$$

for $\nu(x)$ arbitrary other than the condition (9.3.3c).

We set

$$J(x) = \frac{\partial f}{\partial y} - \frac{d}{dx}\left(\frac{\partial f}{\partial y'} \right). \tag{9.3.5}$$

We suppose that $J(x)$ is positive in the interval $[x_1, x_2]$,

$$J(x) > 0 \quad \text{for} \quad x \in [x_1, x_2]. \tag{9.3.6}$$

Then, by choosing

$$\nu(x) > 0 \quad \text{for} \quad x \in [x_1, x_2], \tag{9.3.7}$$

we can make δI positive,

$$\delta I > 0. \tag{9.3.8}$$

We now suppose that $J(x)$ is negative in the interval $[x_1, x_2]$,

$$J(x) < 0 \quad \text{for} \quad x \in [x_1, x_2]. \tag{9.3.9}$$

Then, by choosing

$$\nu(x) < 0 \quad \text{for} \quad x \in [x_1, x_2], \tag{9.3.10}$$

we can make δI positive,

$$\delta I > 0. \tag{9.3.11}$$

We lastly suppose that $J(x)$ alternates its sign in the interval $[x_1, x_2]$. Then by choosing

$$\nu(x) \gtrless 0 \quad \text{wherever} \quad J(x) \gtrless 0, \tag{9.3.12}$$

we can make δI positive,

$$\delta I > 0. \tag{9.3.13}$$

Thus, in order to have Eq. (9.3.4) for $\nu(x)$ arbitrary, together with the condition (9.3.3c), we must have $J(x)$ identically equal to zero,

$$\frac{\partial f}{\partial y} - \frac{d}{dx}\left(\frac{\partial f}{\partial y'}\right) = 0, \tag{9.3.14}$$

which is known as the *Euler equation.*

We now illustrate the Euler equation by solving some of the examples listed above.

❑ **Example 9.1. The shortest distance between two points.**

In this case, f is given by

$$f = \sqrt{1 + (y')^2}. \tag{9.3.15a}$$

The Euler equation simply gives

$$\frac{y'}{\sqrt{1 + (y')^2}} = \text{constant}, \Rightarrow y' = c. \tag{9.3.15b}$$

In general, if $f = f(y')$, independent of x and y, then the Euler equation always gives

$$y' = \text{constant}. \tag{9.3.16}$$

❑ **Example 9.4. The brachistochrone problem.**

In this case, f is given by

$$f = \sqrt{\frac{1 + (y')^2}{y}}. \tag{9.3.17}$$

The Euler equation gives

$$-\frac{1}{2}\sqrt{\frac{1 + (y')^2}{y^3}} - \frac{d}{dx}\frac{y'}{\sqrt{(1 + (y')^2)y}} = 0.$$

This appears somewhat difficult to solve. However, there is a simple way which is applicable to many cases.

Suppose

$$f = f(y, y'), \quad \text{independent of } x, \tag{9.3.18a}$$

then

$$\frac{d}{dx} = y'\frac{\partial}{\partial y} + y''\frac{\partial}{\partial y'} + \frac{\partial}{\partial x}$$

where the last term is absent when acting on $f = f(y, y')$. Thus

$$\frac{d}{dx}f = y'\frac{\partial f}{\partial y} + y''\frac{\partial f}{\partial y'}.$$

Making use of the Euler equation on the first term of the right-hand side, we have

$$\frac{d}{dx}f = y'\frac{d}{dx}\left(\frac{\partial f}{\partial y'}\right) + \left(\frac{d}{dx}y'\right)\frac{\partial f}{\partial y'} = \frac{d}{dx}\left(y'\frac{\partial f}{\partial y'}\right),$$

i.e.,

$$\frac{d}{dx}\left(f - y'\frac{\partial f}{\partial y'}\right) = 0.$$

Hence we obtain

$$f - y'\frac{\partial f}{\partial y'} = \text{constant}. \tag{9.3.18b}$$

Returning to the *Brachistochrone problem*, we have

$$\sqrt{\frac{1 + (y')^2}{y}} - y'\frac{y'}{\sqrt{(1 + (y')^2)y}} = \text{constant},$$

or,

$$y(1 + (y')^2) = 2R.$$

Solving for y', we obtain

$$\frac{dy}{dx} = y' = \sqrt{\frac{2R - y}{y}}. \tag{9.3.19}$$

Hence we have

$$\int dy \sqrt{\frac{y}{2R - y}} = x.$$

We set

$$y = 2R\sin^2\left(\frac{\theta}{2}\right) = R(1 - \cos\theta). \tag{9.3.20a}$$

Then we easily get

$$x = 2R\int \sin^2\left(\frac{\theta}{2}\right)\,d\theta = R(\theta - \sin\theta). \tag{9.3.20b}$$

Equations (9.3.20a) and (9.3.20b) are the *parametric equations for a cycloid*, the curve traced by a point on the rim of a wheel rolling on the x axis. The shape of a cycloid is displayed in Figure 9.1.

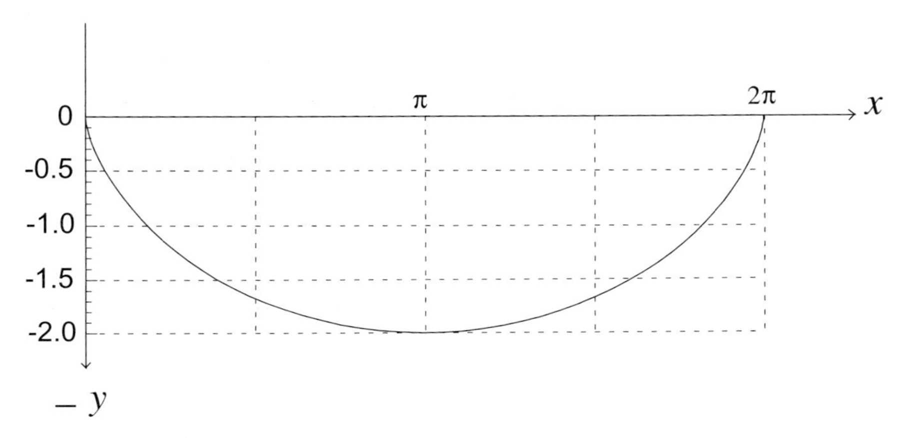

Fig. 9.1: The curve traced by a point on the rim of a wheel rolling on the x axis.

We state several facts for **Example 9.4:**

1. The solution of the fastest fall is not a straight line. It is a cycloid with infinite initial slope.

2. There exists a unique solution. In our parametric representation of a cycloid, the range of θ is implicitly assumed to be $0 \le \theta \le 2\pi$. Setting $\theta = 0$, we find that the starting point is chosen to be at the origin,

$$(x_1, y_1) = (0, 0). \tag{9.3.21}$$

The question is that, given the endpoint (x_2, y_2), can we uniquely determine a radius of the wheel R? We put $y_2 = R(1 - \cos\theta_0)$, and $x_2 = R(\theta_0 - \sin\theta_0)$, or,

$$\frac{1-\cos\theta_0}{\theta_0 - \sin\theta_0} = \frac{y_2}{x_2}, \quad 2R = \frac{2y_2}{1-\cos\theta_0} = \frac{y_2}{\sin^2(\theta_0/2)},$$

which has a unique solution in the range $0 < \theta_0 < \pi$.

3. The shortest time of descent is

$$T = \int_0^{x_2} dx\sqrt{\frac{1+(y')^2}{y}} = 2\sqrt{2R}\int_0^{\theta_0/2} d\theta = \sqrt{y_2}\frac{\theta_0}{\sin(\theta_0/2)}. \tag{9.3.22}$$

❑ **Example 9.5. Hamilton's Action Principle in Classical Mechanics.**

Consider the infinitesimal variation $\delta q(t)$ of $q(t)$, vanishing at $t = t_1$ and $t = t_2$,

$$\delta q(t_1) = \delta q(t_2) = 0. \tag{9.3.23}$$

Then Hamilton's Action Principle demands that

$$\begin{aligned}
\delta I &= \delta \int_{t_1}^{t_2} L(q(t), \dot{q}(t), t)\, dt = \int_{t_1}^{t_2} \left(\delta q(t) \frac{\partial L}{\partial q(t)} + \delta \dot{q}(t) \frac{\partial L}{\partial \dot{q}(t)} \right) dt \\
&= \int_{t_1}^{t_2} dt \left(\delta q(t) \frac{\partial L}{\partial q(t)} + \left(\frac{d}{dt} \delta q(t) \right) \frac{\partial L}{\partial \dot{q}(t)} \right) \\
&= \left[\delta q(t) \frac{\partial L}{\partial \dot{q}(t)} \right]_{t=t_1}^{t=t_2} + \int_{t_1}^{t_2} dt \delta q(t) \left(\frac{\partial L}{\partial q(t)} - \frac{d}{dt} \left(\frac{\partial L}{\partial \dot{q}(t)} \right) \right) \\
&= \int_{t_1}^{t_2} dt \delta q(t) \left(\frac{\partial L}{\partial q(t)} - \frac{d}{dt} \left(\frac{\partial L}{\partial \dot{q}(t)} \right) \right) = 0,
\end{aligned} \tag{9.3.24}$$

where $\delta q(t)$ is arbitrary other than the condition (9.3.23). From this, we obtain the *Lagrange equation of motion*,

$$\frac{d}{dt} \left(\frac{\partial L}{\partial \dot{q}(t)} \right) - \frac{\partial L}{\partial q(t)} = 0, \tag{9.3.25}$$

which is nothing but the Euler equation (9.3.14), with the identification,

$$t \Rightarrow x, \quad q(t) \Rightarrow y(x), \quad L(q(t), \dot{q}(t), t) \Rightarrow f(x, y, y').$$

When the Lagrangian $L(q(t), \dot{q}(t), t)$ does not depend on t explicitly, the following quantity,

$$\dot{q}(t) \frac{\partial L}{\partial \dot{q}(t)} - L(q(t), \dot{q}(t)) \equiv E, \tag{9.3.26}$$

is a constant of motion and is called the *energy integral*, which is simply Eq. (9.3.18b). Solving the energy integral for $\dot{q}(t)$, we can obtain the *differential equation* for $q(t)$.

9.4 Generalization of the Basic Problems

❑ **Example 9.6. Free endpoint: y_2 arbitrary.**

An example is to consider, in the *Brachistochrone problem*, the dependence of the shortest time of fall as a function of y_2. The question is: What is the height of fall y_2 which, for a given x_2, minimizes this time of fall? We may, of course, start by taking the expression for the shortest time of fall:

$$T = \sqrt{y_2} \frac{\theta_0}{\sin(\theta_0/2)} = \sqrt{2x_2} \frac{\theta_0}{\sqrt{\theta_0 - \sin\theta_0}},$$

which, for a given x_2, has a minimum at $\theta_0 = \pi$, where $T = \sqrt{2\pi x_2}$. We shall, however, give a treatment for the general problem of free endpoint.

To extremize

$$I = \int_{x_1}^{x_2} f(x, y, y')\, dx, \tag{9.4.1}$$

with

$$y(x_1) = y_1, \quad \text{and} \quad y_2 \text{ arbitrary}, \tag{9.4.2}$$

we require

$$\delta I = \left. \frac{\partial f}{\partial y'} \varepsilon \nu(x) \right|_{x=x_2} + \varepsilon \int_{x_1}^{x_2} \left[\frac{\partial f}{\partial y} - \frac{d}{dx}\left(\frac{\partial f}{\partial y'} \right) \right] \nu(x)\, dx = 0. \tag{9.4.3}$$

By choosing $\nu(x_2) = 0$, we get the Euler equation. Next we choose $\nu(x_2) \neq 0$ and obtain in addition,

$$\left. \frac{\partial f}{\partial y'} \right|_{x=x_2} = 0. \tag{9.4.4}$$

Note that y_2 is determined by these equations.

For the *Brachistochrone problem* of arbitrary y_2, we get

$$\left. \frac{\partial f}{\partial y'} \right|_{x=x_2} = \frac{y'}{\sqrt{(1 + (y')^2)y}} = 0 \Rightarrow y' = 0.$$

Thus $\theta_0 = \pi$, and $y_2 = x_2(\frac{2}{\pi})$, as obtained previously.

❑ **Example 9.7. Endpoint on the curve $y = g(x)$:**

An example is to find the shortest time of descent to a curve. Note that, in this problem, neither x_2 nor y_2 are given. They are to be determined and related by $y_2 = g(x_2)$.

Suppose that $y = y(x)$ is the solution. This means that if we make a variation

$$y(x) \rightarrow y(x) + \varepsilon \nu(x), \tag{9.4.5}$$

which intersects the curve at $(x_2 + \Delta x_2, y_2 + \Delta y_2)$, then $y(x_2 + \Delta x_2) + \varepsilon \nu(x_2 + \Delta x_2) = g(x_2 + \Delta x_2)$, or,

$$\varepsilon \nu(x_2) = (g'(x_2) - y'(x_2))\Delta x_2, \tag{9.4.6}$$

and that the variation δI vanishes:

$$\begin{aligned} \delta I &= \int_{x_1}^{x_2 + \Delta x_2} f(x, y + \varepsilon \nu, y' + \varepsilon \nu')\, dx - \int_{x_1}^{x_2} f(x, y, y')\, dx \\ &\simeq \left[f(x, y, y')\Delta x_2 + \frac{\partial f}{\partial y'} \varepsilon \nu(x) \right]_{x=x_2} + \int_{x_1}^{x_2} \left(\frac{\partial f}{\partial y} - \frac{d}{dx}\frac{\partial f}{\partial y'} \right) \varepsilon \nu(x)\, dx = 0. \end{aligned}$$

Thus, in addition to the Euler equation, we have

$$\left[f(x,y,y')\Delta x_2 + \frac{\partial f}{\partial y'}\varepsilon\nu(x)\right]_{x=x_2} = 0,$$

or, using Eq. (9.4.6), we obtain

$$\left[f(x,y,y') + \frac{\partial f}{\partial y'}(g' - y')\right]_{x=x_2} = 0. \tag{9.4.7}$$

Applying the above equation to the *Brachistochrone problem*, we get

$$y'g' = -1.$$

This means that the path of fastest descent intersects the curve $y = g(x)$ at a right angle.

❑ **Example 9.8. The isoperimetric problem**: Find $y(x)$ which extremizes

$$I = \int_{x_1}^{x_2} f(x,y,y')\,dx \tag{9.4.8}$$

while keeping

$$J = \int_{x_1}^{x_2} F(x,y,y')\,dx \quad \text{fixed.} \tag{9.4.9}$$

An example is the classic problem of finding the maximum area enclosed by a curve of fixed length.

Let $y(x)$ be the solution. This means that if we make a variation of y which does not change the value of J, the variation of I must vanish. Since J cannot change, this variation is not of the form $\varepsilon\nu(x)$, with $\nu(x)$ arbitrary. Instead, we must put

$$y(x) \to y(x) + \varepsilon_1\nu_1(x) + \varepsilon_2\nu_2(x), \tag{9.4.10}$$

where ε_1 and ε_2 are so chosen that

$$\delta J = \int_{x_1}^{x_2} \left(\frac{\partial F}{\partial y} - \frac{d}{dx}\frac{\partial F}{\partial y'}\right)(\varepsilon_1\nu_1(x) + \varepsilon_2\nu_2(x))\,dx = 0. \tag{9.4.11}$$

For these kinds of variations, $y(x)$ extremizes I:

$$\delta I = \int_{x_1}^{x_2} \left(\frac{\partial f}{\partial y} - \frac{d}{dx}\frac{\partial f}{\partial y'}\right)(\varepsilon_1\nu_1(x) + \varepsilon_2\nu_2(x))\,dx = 0. \tag{9.4.12}$$

Eliminating ε_2, we get

$$\varepsilon_1 \int_{x_1}^{x_2} \left[\frac{\partial}{\partial y}(f + \lambda F) - \frac{d}{dx}\frac{\partial}{\partial y'}(f + \lambda F)\right]\nu_1(x)\,dx = 0, \tag{9.4.13a}$$

where

$$\lambda = \frac{\left[\int_{x_1}^{x_2} \left(\frac{\partial f}{\partial y} - \frac{d}{dx}\frac{\partial f}{\partial y'}\right) \nu_2(x)\, dx\right]}{\left[\int_{x_1}^{x_2} \left(\frac{\partial F}{\partial y} - \frac{d}{dx}\frac{\partial F}{\partial y'}\right) \nu_2(x)\, dx\right]}. \tag{9.4.13b}$$

Thus $(f + \lambda F)$ satisfies the Euler equation. The λ is determined by solving the Euler equation, substituting y into the integral for J, and requiring that J takes the prescribed value.

❑ **Example 9.9. The integral involves more than one function:**
Extremize

$$I = \int_{x_1}^{x_2} f(x, y, y', z, z')\, dx, \tag{9.4.14}$$

with y and z taking prescribed values at the endpoints. By varying y and z successively, we get

$$\frac{\partial f}{\partial y} - \frac{d}{dx}\frac{\partial f}{\partial y'} = 0, \tag{9.4.15a}$$

and

$$\frac{\partial f}{\partial z} - \frac{d}{dx}\frac{\partial f}{\partial z'} = 0. \tag{9.4.15b}$$

❑ **Example 9.10. The integral involves y'':**

$$I = \int_{x_1}^{x_2} f(x, y, y', y'')\, dx \tag{9.4.16}$$

with y and y' taking prescribed values at the endpoints. The Euler equation is

$$\frac{\partial f}{\partial y} - \frac{d}{dx}\frac{\partial f}{\partial y'} + \frac{d^2}{dx^2}\frac{\partial f}{\partial y''} = 0. \tag{9.4.17}$$

❑ **Example 9.11. The integral is the multi-dimensional:**

$$I = \int dxdt\, f(x, t, y, y_x, y_t), \tag{9.4.18}$$

with y taking the prescribed values at the boundary. The Euler equation is

$$\frac{\partial f}{\partial y} - \frac{d}{dx}\frac{\partial f}{\partial y_x} - \frac{d}{dt}\frac{\partial f}{\partial y_t} = 0. \tag{9.4.19}$$

9.5 More Examples

❑ **Example 9.12. Catenary:**

(a) **The shape of a chain hanging on two pegs**: The gravitational potential of a chain of uniform density is proportional to

$$I = \int_{x_1}^{x_2} y\sqrt{1+(y')^2}\,dx. \tag{9.5.1a}$$

The equilibrium position of the chain minimizes I, subject to the condition that the length of the chain is fixed,

$$J = \int_{x_1}^{x_2} \sqrt{1+(y')^2}\,dx \quad \text{fixed.} \tag{9.5.1b}$$

Thus we extremize $I + \lambda J$ and obtain

$$\frac{(y+\lambda)}{\sqrt{1+(y')^2}} = \alpha \quad \text{constant,}$$

or,

$$\int \frac{dy}{\sqrt{(y+\lambda)^2/\alpha^2 - 1}} = \int dx. \tag{9.5.2}$$

We put

$$\frac{y+\lambda}{\alpha} = \cosh\theta, \tag{9.5.3}$$

then

$$x - \beta = \alpha\theta, \tag{9.5.4}$$

and the shape of chain is given by

$$y = \alpha\cosh\left(\frac{x-\beta}{\alpha}\right) - \lambda. \tag{9.5.5}$$

The constants, α β and λ, are determined by the two boundary conditions and the requirement that J is equal to the length of the chain.

Let us consider the case

$$y(-L) = y(L) = 0. \tag{9.5.6}$$

Then the boundary conditions give

$$\beta = 0, \quad \lambda = \alpha\cosh\frac{L}{\alpha}. \tag{9.5.7}$$

The condition that J is constant gives

$$\frac{\alpha}{L}\sinh\frac{L}{\alpha}=\frac{l}{L}, \tag{9.5.8}$$

where $2l$ is the length of the chain. It is easily shown that, for $l \geq L$, a unique solution is obtained.

(b) **A soap film formed by two circular wires:** The surface of the soap film takes a minimum area as a result of surface tension. Thus we minimize

$$I=\int_{x_1}^{x_2} y\sqrt{1+(y')^2}\,dx, \tag{9.5.9}$$

obtaining as in (a),

$$y=\alpha\cosh\left(\frac{x-\beta}{\alpha}\right), \tag{9.5.10}$$

where α and β are constants of integration, to be determined from the boundary conditions at $x=\pm L$.

Let us consider the special case in which the two circular wires are of equal radius R. Then

$$y(-L)=y(L)=R. \tag{9.5.11}$$

We easily find that $\beta=0$, and that α is determined by the equation

$$\frac{R}{L}=\frac{\alpha}{L}\cosh\frac{L}{\alpha}, \tag{9.5.12}$$

which has zero, one, or two solutions depending on the ratio R/L. In order to decide if any of these solutions actually minimizes I, we must study the *second variation*, which we shall discuss in Section 9.7.

❑ **Example 9.5. Hamilton's Action Principle in Classical Mechanics.**

Let the Lagrangian $L(q(t),\dot{q}(t))$ be defined by

$$L(q(t),\dot{q}(t))\equiv T(q(t),\dot{q}(t))-V(q(t),\dot{q}(t)), \tag{9.5.13}$$

where T and V are the *kinetic energy* and the *potential energy* of the mechanical system, respectively. In general, T and V can depend on both $q(t)$ and $\dot{q}(t)$. When T and V are given respectively by

$$T=\frac{1}{2}m\dot{q}(t)^2, \quad V=V(q(t)), \tag{9.5.14}$$

the Lagrange equation of motion (9.3.25) provides us with *Newton's equation of motion*,

$$m\ddot{q}(t)=-\frac{d}{dq(t)}V(q(t)) \quad \text{with} \quad \ddot{q}(t)=\frac{d^2}{dt^2}q(t). \tag{9.5.15}$$

In other words, the extremization of the action integral I given by

$$I = \int_{t_1}^{t_2} \left[\frac{1}{2} m\dot{q}(t)^2 - V(q(t)) \right] dt$$

with $\delta q(t_1) = \delta q(t_2) = 0$, leads us to Newton's equation of motion (9.5.15). With T and V given by Eq. (9.5.14), the energy integral E given by Eq. (9.3.26) assumes the following form,

$$E = \frac{1}{2} m\dot{q}(t)^2 + V(q(t)), \tag{9.5.16}$$

which represents the total mechanical energy of the system, very appropriate for the terminology, "the *energy integral*".

❑ **Example 9.13. Fermat's Principle in Geometrical Optics.**
The path of a light ray between two given points in a medium is the one which minimizes the time of travel. Thus the path is determined from minimizing

$$T = \frac{1}{c} \int_{x_1}^{x_2} dx \sqrt{1 + \left(\frac{dy}{dx}\right)^2 + \left(\frac{dz}{dx}\right)^2} \, n(x, y, z), \tag{9.5.17}$$

where $n(x, y, z)$ is the index of refraction. If n is independent of x, we get

$$\frac{n(y, z)}{\sqrt{1 + \left(\frac{dy}{dx}\right)^2 + \left(\frac{dz}{dx}\right)^2}} = \text{constant}. \tag{9.5.18}$$

From Eq. (9.5.18), we easily derive the *law of reflection* and the *law of refraction* (*Snell's law*).

9.6 Differential Equations, Integral Equations, and Extremization of Integrals

We now consider the inverse problem: If we are to solve a differential or an integral equation, can we formulate the problem in terms of one which extremizes an integral? This will have practical advantages when we try to obtain approximate solutions for differential equations and approximate eigenvalues.

❑ **Example 9.14.** Solve

$$\frac{d}{dx}\left[p(x) \frac{d}{dx} y(x) \right] - q(x)y(x) = 0, \qquad x_1 < x < x_2, \tag{9.6.1}$$

with

$$y(x_1), \quad y(x_2) \quad \text{specified.} \tag{9.6.2}$$

This problem is equivalent to extremizing the integral

$$I = \frac{1}{2} \int_{x_1}^{x_2} \left[p(x)(y'(x))^2 + q(x)(y(x))^2 \right] dx. \tag{9.6.3}$$

❑ **Example 9.15. Solve the Sturm–Liouville eigenvalue problem**

$$\frac{d}{dx}\left[p(x)\frac{d}{dx}y(x)\right] - q(x)y(x) = \lambda r(x)y(x), \qquad x_1 < x < x_2, \tag{9.6.4}$$

with

$$y(x_1) = y(x_2) = 0. \tag{9.6.5}$$

This problem is equivalent to extremizing the integral

$$I = \frac{1}{2}\int_{x_1}^{x_2} \left[p(x)(y'(x))^2 + q(x)(y(x))^2\right] dx, \tag{9.6.6}$$

while keeping

$$J = \frac{1}{2}\int_{x_1}^{x_2} r(x)(y(x))^2\, dx \quad \text{fixed.} \tag{9.6.7}$$

In practice, we find the approximation to the lowest eigenvalue and the corresponding eigenfunction of the Sturm–Liouville eigenvalue problem by minimizing I/J. Note that an eigenfunction, good to first order, yields an eigenvalue which is good to second order.

❑ **Example 9.16.** Solve

$$\frac{d^2}{dx^2}y(x) = -\lambda y(x), \qquad 0 < x < 1, \tag{9.6.8}$$

with

$$y(0) = y(1) = 0. \tag{9.6.9}$$

Solution. We choose the trial function to be the one in Figure 9.2. Then $I = 2 \cdot h^2/\varepsilon$, $J = h^2(1 - \frac{4}{3}\varepsilon)$, and $I/J = 2/\left[\varepsilon(1 - \frac{4}{3}\varepsilon)\right]$, which has a minimum value of $\frac{32}{3}$ at $\varepsilon = \frac{3}{8}$. This is compared with the exact value, $\lambda = \pi^2$. Note that $\lambda = \pi^2$ is a lower bound for I/J. If we choose the trial function to be

$$y(x) = x(1 - x),$$

we get $I/J = 10$. This is accurate to almost one percent.

In order to obtain an accurate estimate of the eigenvalue, it is important to choose a trial function which satisfies the boundary condition and looks qualitatively like the expected solution. For instance, if we are calculating the lowest eigenvalue, it would be unwise to use a trial function which has a zero inside the interval.

Let us now calculate the next eigenvalue. This is done by choosing the trial function $y(x)$ which is orthogonal to the exact lowest eigenfunction $u_0(x)$, i.e.,

$$\int_0^1 y(x)u_0(x)r(x)\, dx = 0, \tag{9.6.10}$$

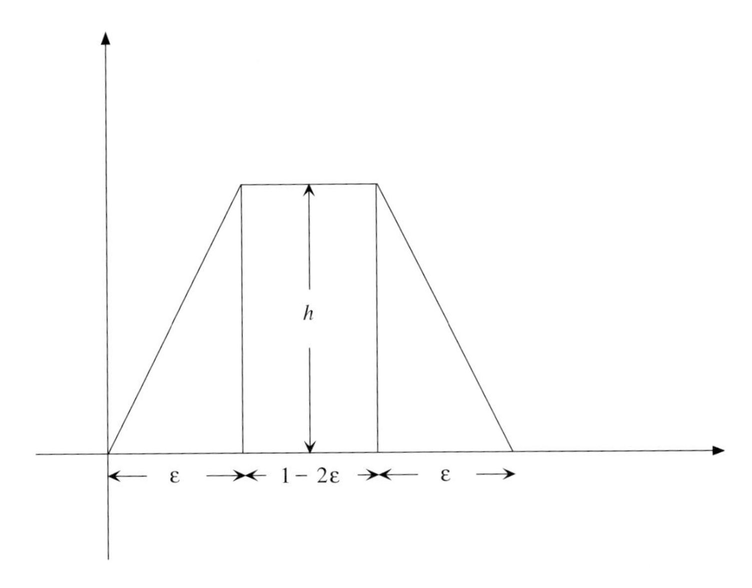

Fig. 9.2: The shape of the trial function for Example 9.16.

and find the minimum value of I/J with respect to some parameters in the trial function. Let us choose the trial function to be

$$y(x) = x(1-x)(1-ax),$$

and then the requirement that it is orthogonal to $x(1-x)$, instead of $u_0(x)$ which is unknown, gives $a = 2$. Note that this trial function has one zero inside the interval $[0, 1]$. This looks qualitatively like the expected solution. For this trial function, we have

$$\frac{I}{J} = 42,$$

and this compares well with the exact value, $4\pi^2$.

❑ **Example 9.17. Solve the Laplace equation**

$$\nabla^2\phi = 0, \tag{9.6.11}$$

with ϕ given at the boundary.

This problem is equivalent to extremizing

$$I = \int (\vec{\nabla}\phi)^2 \, dV. \tag{9.6.12}$$

❑ **Example 9.18. Solve the wave equation**

$$\nabla^2 \phi = k^2 \phi, \tag{9.6.13a}$$

with

$$\phi = 0 \tag{9.6.13b}$$

at the boundary.

This problem is equivalent to extremizing

$$\frac{\displaystyle\int (\vec{\nabla}\phi)^2 \, dV}{\displaystyle\int \phi^2 \, dV}. \tag{9.6.14}$$

❑ **Example 9.19.** Estimate the lowest frequency of a circular drum of radius R.

Solution.

$$k^2 \leq \frac{\displaystyle\int (\vec{\nabla}\phi)^2 \, dV}{\displaystyle\int \phi^2 \, dV}. \tag{9.6.15}$$

Try a rotationally symmetric trial function,

$$\phi(r) = 1 - \frac{r}{R}, \quad 0 \leq r \leq R. \tag{9.6.16}$$

Then we get

$$k^2 \leq \frac{\displaystyle\int_0^R (\phi_r(r))^2 2\pi r \, dr}{\displaystyle\int_0^R (\phi(r))^2 2\pi r \, dr} = \frac{6}{R^2}. \tag{9.6.17}$$

This is compared with the exact value,

$$k^2 = \frac{5.7832}{R^2}. \tag{9.6.18}$$

Note that the numerator on the right-hand side of Eq. (9.6.18) is the square of the smallest zero in magnitude of the zeroth-order Bessel function of the first kind, $J_0(kR)$.

The homogeneous Fredholm integral equations of the second kind for the localized, monochromatic, and highly directive classical current distributions in two and three dimensions can be derived by maximizing the directivity D in the far field while constraining $C = N/T$, where N is the integral of the square of the magnitude of the current density and T is proportional to the total radiated power. The homogeneous Fredholm integral equations of the second kind and the inhomogeneous Fredholm integral equations of the second kind are now derived from the calculus of variations in general terms.

❑ **Example 9.20. Solve the homogeneous Fredholm integral equation of the second kind,**

$$\phi(x) = \lambda \int_0^h K(x,x')\phi(x')\,dx', \tag{9.6.19}$$

where the *square-integrable* kernel is $K(x,x')$, and the projection is unity on $\psi(x)$,

$$\int_0^h \psi(x)\phi(x)\,dx = 1. \tag{9.6.20}$$

This problem is equivalent to extremizing the integral

$$I = \int_0^h \int_0^h \psi(x)K(x,x')\phi(x')\,dxdx', \tag{9.6.21}$$

with respect to $\psi(x)$, while keeping

$$J = \int_0^h \psi(x)\phi(x)\,dx = 1 \quad \text{fixed.} \tag{9.6.22}$$

The extremization of Eq. (9.6.21) with respect to $\phi(x)$, while keeping J fixed, results in the *homogeneous adjoint integral equation* for $\psi(x)$,

$$\psi(x) = \lambda \int_0^h \psi(x')K(x',x)\,dx'. \tag{9.6.23}$$

With the real and symmetric kernel, $K(x,x') = K(x',x)$, the homogeneous integral equations for $\phi(x)$ and $\psi(x)$, Eqs. (9.6.19) and (9.6.23), become identical and Eq. (9.6.20) provides the normalization of $\phi(x)$ and $\psi(x)$ to unity, respectively.

❑ **Example 9.21. Solve the inhomogeneous Fredholm integral equation of the second kind,**

$$\phi(x) - \lambda \int_0^h K(x,x')\phi(x')\,dx' = f(x), \tag{9.6.24}$$

with the *square-integrable* kernel $K(x,x')$, $0 < \|K\|^2 < \infty$.

This problem is equivalent to extremizing the integral

$$I = \int_0^h \left[\left\{ \frac{1}{2}\phi(x) - \lambda \int_0^h K(x,x')\phi(x')\,dx' \right\} \phi(x) + F(x)\frac{d}{dx}\phi(x) \right] dx, \tag{9.6.25}$$

where $F(x)$ is defined by

$$F(x) = \int^x f(x')\,dx'. \tag{9.6.26}$$

9.7 The Second Variation

The *Euler equation* is *necessary* to extremize the integral, but it is *not sufficient*. In order to find out whether the solution of the Euler equation actually extremizes the integral, we must study the *second variation*. This is similar to the case of finding an extremum of a function. To confirm that the point at which the first derivative of a function vanishes is an extremum of the function, we must study the second derivatives.

Consider the extremization of

$$I = \int_{x_1}^{x_2} f(x, y, y')\, dx, \tag{9.7.1}$$

with

$$y(x_1), \quad \text{and} \quad y(x_2) \quad \text{specified.} \tag{9.7.2}$$

Suppose $y(x)$ is such a solution.

Let us consider a *weak variation*,

$$\begin{cases} y(x) & \rightarrow \quad y(x) + \varepsilon\nu(x), \\ y'(x) & \rightarrow \quad y'(x) + \varepsilon\nu'(x), \end{cases} \tag{9.7.3}$$

as opposed to a *strong variation* in which $y'(x)$ is also varied, independent of $\varepsilon\nu'(x)$.

Then

$$I \rightarrow I + \varepsilon I_1 + \frac{1}{2}\varepsilon^2 I_2 + \cdots, \tag{9.7.4}$$

where

$$I_1 = \int_{x_1}^{x_2} \left[\frac{\partial f}{\partial y}\nu(x) + \frac{\partial f}{\partial y'}\nu'(x)\right] dx = \int_{x_1}^{x_2} \left[\frac{\partial f}{\partial y} - \frac{d}{dx}\frac{\partial f}{\partial y'}\right] \nu(x)\, dx, \tag{9.7.5}$$

and

$$I_2 = \int_{x_1}^{x_2} \left[\frac{\partial^2 f}{\partial y^2}\nu^2(x) + 2\frac{\partial^2 f}{\partial y \partial y'}\nu(x)\nu'(x) + \frac{\partial^2 f}{\partial y'^2}\nu'^2(x)\right] dx. \tag{9.7.6a}$$

If $y = y(x)$ indeed minimizes or maximizes I, then I_2 must be positive or negative for all variations $\nu(x)$ vanishing at the endpoints, when the solution of the Euler equation, $y = y(x)$, is substituted into the integrand of I_2.

The integrand of I_2 is a quadratic form of $\nu(x)$ and $\nu'(x)$. Therefore, this integral is always positive if

$$\left(\frac{\partial^2 f}{\partial y \partial y'}\right)^2 - \left(\frac{\partial^2 f}{\partial y^2}\right)\left(\frac{\partial^2 f}{\partial y'^2}\right) < 0, \tag{9.7.7a}$$

and

$$\frac{\partial^2 f}{\partial y'^2} > 0. \tag{9.7.7b}$$

Thus, if the above conditions hold throughout

$$x_1 < x < x_2,$$

I_2 is always positive and $y(x)$ minimizes I. Similar considerations hold, of course, for maximizations. The above conditions are, however, too crude, i.e., are stronger than necessary. This is because $\nu(x)$ and $\nu'(x)$ are not independent.

Let us first state the necessary and sufficient conditions for the solution to the Euler equation to give the *weak minimum*:
Let

$$P(x) \equiv f_{yy}, \quad Q(x) \equiv f_{yy'}, \quad R(x) \equiv f_{y'y'}, \tag{9.7.8}$$

where $P(x)$, $Q(x)$ and $R(x)$ are evaluated at the point which extremizes the integral I defined by

$$I \equiv \int_{x_1}^{x_2} f(x, y, y')\, dx.$$

We express I_2 in terms of $P(x)$, $Q(x)$ and $R(x)$ as

$$I_2 = \int_{x_1}^{x_2} [P(x)\nu^2(x) + 2Q(x)\nu(x)\nu'(x) + R(x)\nu'^2(x)]\, dx. \tag{9.7.6b}$$

Then we have the following conditions for the weak minimum,

	Necessary condition	**Sufficient condition**	
Legendre test	$R(x) \geq 0.$	$R(x) > 0.$	(9.7.9)
Jacobi test	$\xi \geq x_2.$	$\xi > x_2,$ or no such ξ exists.	

where ξ is the *conjugate point* to be defined later. Both conditions must be satisfied for sufficiency and both are necessary separately. Before we go on to prove the above assertion, it is perhaps helpful to have an intuitive understanding of why these two tests are relevant.

Let us consider the case in which $R(x)$ is positive at $x = a$, and negative at $x = b$, where a and b are both between x_1 and x_2. Let us first choose $\nu(x)$ to be the function as in Figure 9.3. Note that $\nu(x)$ is of the order of ε, while $\nu'(x)$ is of the order $\sqrt{\varepsilon}$. If we choose ε to be sufficiently small, I_2 is positive. Next, we consider the same variation located at $x = b$. By the same consideration, I_2 is negative for this variation. Thus $y(x)$ does not extremize I. This shows that the *Legendre test* is a necessary condition.

The relevance of the *Jacobi test* is best illustrated by the problem of finding the shortest path between two points on the surface of a sphere. The solution is obtained by going along the great circle on which these two points lie. There are, however, two paths connecting these two points on the great circle. One of them is truly the shortest path, while the other is neither the shortest nor the longest. Take the circle on the surface of the sphere which passes one of the two points. The other point at which this circle and the great circle intersect lies on one arc of the great circle which is neither the shortest nor the longest arc. This point is the conjugate point ξ of this problem.

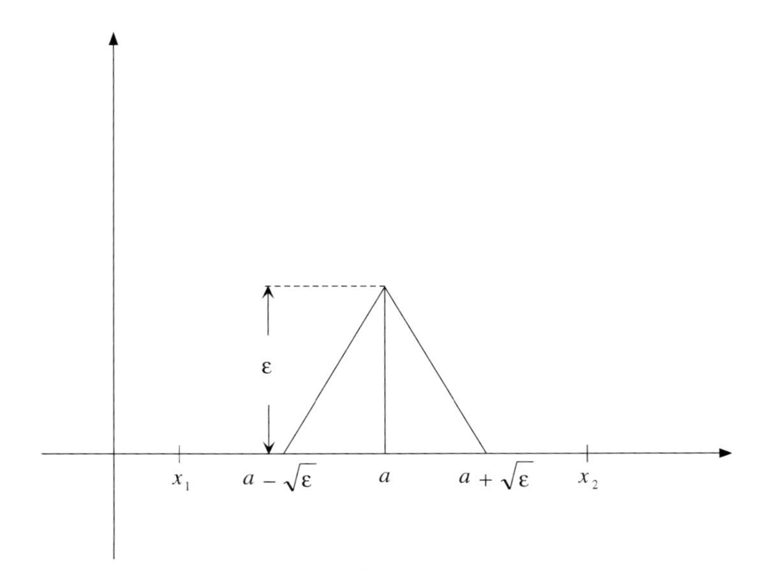

Fig. 9.3: The shape of $\nu(x)$ for the Legendre test.

We discuss the *Legendre test* first. We add to I_2, (9.7.6b), the following term which is zero,

$$\int_{x_1}^{x_2} \frac{d}{dx}(\nu^2(x)\omega(x))\,dx = \int_{x_1}^{x_2} (\nu^2(x)\omega'(x) + 2\nu(x)\nu'(x)\omega(x))\,dx. \tag{9.7.10}$$

Then we have I_2 to be

$$I_2 = \int_{x_1}^{x_2} \Big[\big(P(x) + \omega'(x)\big)\nu^2(x) + 2\big(Q(x) + \omega(x)\big)\nu(x)\nu'(x) + R(x)\nu'^2(x)\Big]\,dx. \tag{9.7.11}$$

We require the integrand to be a *complete square*, i.e.,

$$(Q(x) + \omega(x))^2 = (P(x) + \omega'(x))R(x). \tag{9.7.12}$$

Hence, if we can find $\omega(x)$ satisfying Eq. (9.7.12), we will have

$$I_2 = \int_{x_1}^{x_2} R(x)\left[\nu'(x) + \frac{Q(x) + \omega(x)}{R(x)}\nu(x)\right]^2 dx. \tag{9.7.13}$$

Now, it is not possible to have $R(x) < 0$ in any region between x_1 and x_2 for the minimum. If $R(x) < 0$ in some region, we can solve the differential equation

$$\omega'(x) = -P(x) + \frac{(Q(x) + \omega(x))^2}{R(x)} \tag{9.7.14}$$

for $\omega(x)$ in this region. Restricting the variation $\nu(x)$ such that

$$\begin{cases} \nu(x) \equiv 0, & \text{outside the region where } R(x) < 0, \\ \nu(x) \neq 0, & \text{inside the region where } R(x) < 0, \end{cases}$$

we can have $I_2 < 0$. Thus it is necessary to have $R(x) \geq 0$ for the entire region for the minimum. If

$$P(x)R(x) - Q^2(x) > 0, \qquad P(x) > 0, \tag{9.7.15a}$$

then we have no need to go further to find $\omega(x)$. In many cases, however, we have

$$P(x)R(x) - Q^2(x) \leq 0, \tag{9.7.15b}$$

and so we have to examine further. The differential equation (9.7.14) for $\omega(x)$, can be rewritten as

$$(Q(x) + \omega(x))' = -P(x) + Q'(x) + \frac{(Q(x) + \omega(x))^2}{R(x)}, \tag{9.7.16}$$

which is the *Riccatti differential equation*. Making the *Riccatti substitution*,

$$Q(x) + \omega(x) = -R(x)\frac{u'(x)}{u(x)}, \tag{9.7.17}$$

we obtain the *second-order linear ordinary differential equation for* $u(x)$,

$$\frac{d}{dx}\left[R(x)\frac{d}{dx}u(x)\right] + (Q'(x) - P(x))u(x) = 0. \tag{9.7.18}$$

Expressing everything in Eq. (9.7.13) in terms of $u(x)$, I_2 becomes

$$I_2 = \int_{x_1}^{x_2} \left[R(x)\frac{(\nu'(x)u(x) - u'(x)\nu(x))^2}{u^2(x)}\right] dx. \tag{9.7.19}$$

If we can find any $u(x)$ such that

$$u(x) \neq 0, \qquad x_1 \leq x \leq x_2, \tag{9.7.20}$$

we will have

$$I_2 > 0 \quad \text{for} \quad R(x) > 0, \tag{9.7.21}$$

which is the sufficient condition for the minimum. This completes the derivation of the Legendre test. We further clarify the Legendre test after the discussion of the conjugate point ξ of the Jacobi test.

The ordinary differential equation (9.7.18) for $u(x)$ is related to the Euler equation by the *infinitesimal variation of the initial condition.* In the Euler equation,

$$f_y(x,y,y') - \frac{d}{dx} f_{y'}(x,y,y') = 0 \tag{9.7.22}$$

we make the following infinitesimal variation,

$$y(x) \to y(x) + \varepsilon u(x), \quad \varepsilon = \text{positive infinitesimal.} \tag{9.7.23}$$

Writing out this variation explicitly,

$$f_y(x, y + \varepsilon u, y' + \varepsilon u') - \frac{d}{dx} f_{y'}(x, y + \varepsilon u, y' + \varepsilon u') = 0,$$

we have, using the Euler equation,

$$f_{yy}u + f_{yy'}u' - \frac{d}{dx}(f_{y'y}u + f_{y'y'}u') = 0,$$

or,

$$Pu + Qu' - \frac{d}{dx}(Qu + Ru') = 0, \tag{9.7.24}$$

i.e.,

$$\frac{d}{dx}\left[R(x)\frac{d}{dx}u(x)\right] + (Q'(x) - P(x))u(x) = 0,$$

which is simply the ordinary differential equation (9.7.18). The solution to the differential equation (9.7.18) corresponds to the infinitesimal change in the initial condition. We further note that the differential equation (9.7.18) is of the **self-adjoint form**. Thus its Wronskian $W(u_1(x), u_2(x))$ is given by

$$W(x) \equiv u_1(x)u_2'(x) - u_1'(x)u_2(x) = \frac{C}{R(x)}, \tag{9.7.25}$$

where $u_1(x)$ and $u_2(x)$ are the linearly independent solutions of Eq. (9.7.18) and C in Eq. (9.7.25) is some constant.

We now discuss the *conjugate point* ξ and the *Jacobi test*. We claim that the sufficient condition for the minimum is that $R(x) > 0$ on (x_1, x_2) and that the conjugate point ξ lies outside (x_1, x_2), i.e., both the Legendre test and the Jacobi test are satisfied. Suppose that

$$R(x) > 0 \quad \text{on} \quad (x_1, x_2), \tag{9.7.26}$$

which implies that the Wronskian has the same sign on (x_1, x_2). Suppose that $u_1(x)$ vanishes **only** at $x = \xi_1$ and $x = \xi_2$,

$$u_1(\xi_1) = u_1(\xi_2) = 0, \qquad x_1 < \xi_1 < \xi_2 < x_2. \tag{9.7.27}$$

We claim that $u_2(x)$ must vanish at least once between ξ_1 and ξ_2. The Wronskian W evaluated at $x = \xi_1$ and $x = \xi_2$ is given respectively by

$$W(\xi_1) = -u_1'(\xi_1)u_2(\xi_1), \tag{9.7.28a}$$

and

$$W(\xi_2) = -u_1'(\xi_2)u_2(\xi_2). \tag{9.7.28b}$$

By the continuity of $u_1(x)$ on (x_1, x_2), $u_1'(\xi_1)$ has the opposite sign to $u_1'(\xi_2)$, i.e.,

$$u_1'(\xi_1)u_1'(\xi_2) < 0. \tag{9.7.29}$$

But, we have

$$W(\xi_1)W(\xi_2) = u_1'(\xi_1)u_1'(\xi_2)u_2(\xi_1)u_2(\xi_2) > 0, \tag{9.7.30}$$

from which we conclude that $u_2(\xi_1)$ has the opposite sign to $u_2(\xi_2)$, i.e.,

$$u_2(\xi_1)u_2(\xi_2) < 0. \tag{9.7.31}$$

Hence $u_2(x)$ must vanish at least once between ξ_1 and ξ_2.

Since the $u(x)$ provide the infinitesimal variation of the initial condition, we choose $u_1(x)$ and $u_2(x)$ to be

$$u_1(x) \equiv \frac{\partial}{\partial \alpha} y(x, \alpha, \beta), \qquad u_2(x) \equiv \frac{\partial}{\partial \beta} y(x, \alpha, \beta), \tag{9.7.32}$$

where $y(x, \alpha, \beta)$ is the solution of the Euler equation,

$$f_y - \frac{d}{dx} f_{y'} = 0, \tag{9.7.33a}$$

with the initial conditions,

$$y(x_1) = \alpha, \qquad y(x_2) = \beta, \tag{9.7.33b}$$

and $u_i(x)$ $(i = 1, 2)$ satisfy the differential equation (9.7.18). We now claim that the sufficient condition for the weak minimum is that

$$R(x) > 0, \quad \text{and} \quad \xi > x_2. \tag{9.7.34}$$

We construct $U(x)$ by

$$U(x) \equiv u_1(x)u_2(x_1) - u_1(x_1)u_2(x), \tag{9.7.35}$$

which vanishes at $x = x_1$. We define the conjugate point ξ as the solution of the equation,

$$U(\xi) = 0, \qquad \xi \neq x_1. \tag{9.7.36}$$

The function $U(x)$ represents another infinitesimal change in the solution to the Euler equation which would also pass through $x = x_1$. Since

$$U(x_1) = U(\xi) = 0, \tag{9.7.37}$$

there exists another solution $u(x)$ such that

$$u(x) = 0 \quad \text{for} \quad x \in (x_1, \xi). \tag{9.7.38}$$

We choose x_3 such that

$$x_2 < x_3 < \xi \tag{9.7.39}$$

and another solution $u(x)$ to be

$$u(x) = u_1(x)u_2(x_3) - u_1(x_3)u_2(x), \qquad u(x_3) = 0. \tag{9.7.40}$$

We claim that

$$u(x) \neq 0 \quad \text{on} \quad (x_1, x_2). \tag{9.7.41}$$

Suppose that $u(x) = 0$ in this interval. Then any other solution of the differential equation (9.7.18) must vanish between these points. But, $U(x)$ does not vanish. This completes the derivation of the Jacobi test.

We further clarify the Legendre test and the Jacobi test. We now assume that

$$\xi < x_2, \quad \text{and} \quad R(x) > 0 \quad \text{for} \quad x \in (x_1, x_2), \tag{9.7.42}$$

and show that

$$I_2 < 0 \quad \text{for some } \nu(x) \quad \text{such that} \quad \nu(x_1) = \nu(x_2) = 0. \tag{9.7.43}$$

We choose x_3 such that

$$\xi < x_3 < x_2.$$

We construct the solution $U(x)$ to Eq. (9.7.18) such that

$$U(x_1) = U(\xi) = 0.$$

Also, we construct the solution $\pm v(x)$, independent of $U(x)$, such that

$$v(x_3) = 0,$$

i.e., we choose x_3 such that

$$U(x_3) \neq 0.$$

We choose the sign of $v(x)$ such that the Wronskian $W(U(x), v(x))$ is given by

$$W(U(x), v(x)) = U(x)v'(x) - v(x)U'(x) = \frac{C}{R(x)}, \quad R(x) > 0, \quad C > 0.$$

The function $U(x)-v(x)$ solves the differential equation (9.7.18). It must vanish at least once between x_1 and ξ. We call this point $x=a$, so that

$$U(a)=v(a),$$

and

$$x_1<a<\xi<x_3<x_2.$$

We define $\nu(x)$ to be

$$\nu(x)\equiv\begin{cases}U(x), & \text{for} \quad x\in(x_1,a),\\ v(x), & \text{for} \quad x\in(a,x_3),\\ 0, & \text{for} \quad x\in(x_3,x_2).\end{cases}$$

In I_2, we rewrite the term involving $Q(x)$ as

$$2Q(x)\nu(x)\nu'(x)=Q(x)d(\nu^2(x)),$$

and we perform integration by parts in I_2 to obtain

$$\begin{aligned}I_2 &= \left[Q(x)\nu^2(x)+R(x)\nu'(x)\nu(x)\right]\Big|_{x_1}^{x_2}\\ &\quad -\int_{x_1}^{x_2}\nu(x)\left[R(x)\nu''(x)+R'(x)\nu'(x)+(Q'(x)-P(x))\nu(x)\right]dx.\end{aligned}$$

The integral in the second term on the right-hand side is broken up into three separate integrals, each of which vanishes identically since $\nu(x)$ satisfies the differential equation (9.7.18) in all three regions. The $Q(x)$ term of the integrated part of I_2 also vanishes, i.e.,

$$Q(x)\nu^2(x)\Big|_{x_1}^{a}+Q(x)\nu^2(x)\Big|_{a}^{x_3}+Q(x)\nu^2(x)\Big|_{x_3}^{x_2}=0.$$

We consider now the $R(x)$ term of the integrated part of I_2,

$$\begin{aligned}I_2 &= R(x)\nu'(x)\nu(x)\big|_{a-\varepsilon}-R(x)\nu'(x)\nu(x)\big|_{a+\varepsilon}\\ &= R(a)\left[U'(x)v(x)-v'(x)U(x)\right]_{x=a},\end{aligned}$$

where the continuity of $U(x)$ and $v(x)$ at $x=a$ is used. Thus we have

$$\begin{aligned}I_2 &= R(a)\left[U'(a)v(a)-v'(a)U(a)\right]=-R(a)W(U(a),v(a))\\ &= -R(a)\left[\frac{C}{R(a)}\right]=-C<0,\end{aligned}$$

i.e.,

$$I_2=-C<0.$$

This ends the clarification of both the Legendre and the Jacobi tests.

❑ **Example 9.22. Catenary**. Discuss a solution of soap film sustained between two circular wires.

Solution. The surface area is given by

$$I = \int_{-L}^{+L} 2\pi y \sqrt{1 + y'^2}\, dx.$$

Thus $f(x, y, y')$ is given by

$$f(x, y, y') = y\sqrt{1 + y'^2},$$

and is independent of x. Then we have

$$f - y' f_{y'} = \alpha,$$

where α is an arbitrary integration constant.

After a little algebra, we have

$$\frac{dy}{\sqrt{\frac{y^2}{\alpha^2} - 1}} = \pm dx.$$

We perform a change of variable as follows,

$$y = \alpha \cosh\theta, \quad dy = \alpha \sinh\theta \cdot d\theta.$$

Hence we have

$$\alpha\, d\theta = \pm dx,$$

or,

$$\theta = \pm\left(\frac{x - \beta}{\alpha}\right),$$

i.e.,

$$y = \alpha \cosh\left(\frac{x - \beta}{\alpha}\right),$$

where α and β are arbitrary constants of integration, to be determined from the boundary conditions at $x = \pm L$. We have, as the boundary conditions,

$$R_1 = y(-L) = \alpha \cosh\left(\frac{L + \beta}{\alpha}\right), \quad R_2 = y(+L) = \alpha \cosh\left(\frac{L - \beta}{\alpha}\right).$$

For the sake of simplicity, we assume

$$R_1 = R_2 \equiv R.$$

Then we have

$$\beta = 0, \qquad \frac{R}{L} = \frac{\alpha}{L} \cosh \frac{L}{\alpha},$$

and

$$y = \alpha \cosh \frac{x}{\alpha}.$$

Setting

$$v = \frac{L}{\alpha},$$

the boundary conditions at $x = \pm L$ read as

$$\frac{R}{L} = \frac{\cosh v}{v}.$$

Defining the function $G(v)$ by

$$G(v) \equiv \frac{\cosh v}{v}, \tag{9.7.44}$$

$G(v)$ is plotted in Figure 9.4.

If the geometry of the problem is such that

$$\frac{R}{L} > 1.5089,$$

then there exist *two candidates for the solution* at $v = v_<$ and $v = v_>$, where

$$0 < v_< < 1.1997 < v_>,$$

and if

$$\frac{R}{L} < 1.5089,$$

then there exists *no solution*. If the geometry is such that

$$\frac{R}{L} = 1.5089,$$

then there exists *one candidate for the solution* at

$$v_< = v_> = v = \frac{L}{\alpha} = 1.1997.$$

We apply *two tests for the minimum.*

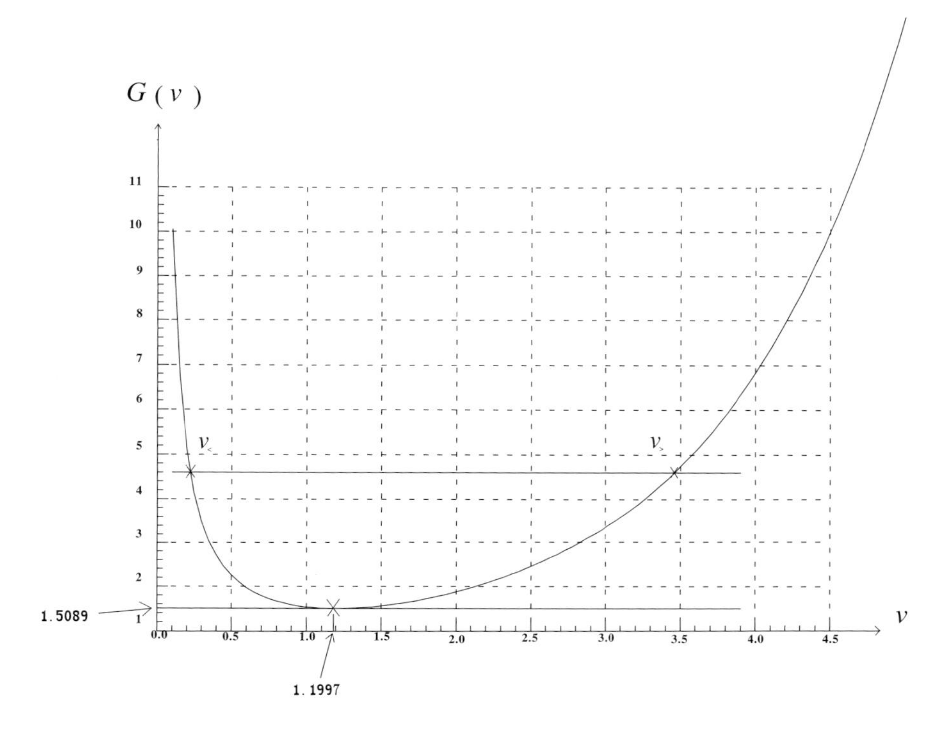

Fig. 9.4: Plot of $G(v)$.

Legendre test:

$$f_{y'y'} = \frac{y}{(1+y'^2)^{3/2}} = R > 0,$$

thus *passing the Legendre test.*

Jacobi test: Since

$$y(x, \alpha, \beta) = \alpha \cosh\left(\frac{x-\beta}{\alpha}\right),$$

we have

$$u_1(x) = \frac{\partial y}{\partial \alpha} = \cosh\left(\frac{x-\beta}{\alpha}\right) - \left(\frac{x-\beta}{\alpha}\right) \sinh\left(\frac{x-\beta}{\alpha}\right),$$

and

$$u_2(x) = -\frac{\partial y}{\partial \beta} = \sinh\left(\frac{x-\beta}{\alpha}\right),$$

where the minus sign for $u_2(x)$ does not matter. We construct a solution $U(x)$ which vanishes at $x = x_1$,

$$U(x) = (\cosh\tilde{v} - \tilde{v}\sinh\tilde{v})\sinh\tilde{v}_1 - \sinh\tilde{v}(\cosh\tilde{v}_1 - \tilde{v}_1\sinh\tilde{v}_1),$$

where

$$\tilde{v} \equiv \frac{x-\beta}{\alpha}, \quad \tilde{v}_1 \equiv \frac{x_1-\beta}{\alpha}.$$

Then we have

$$\frac{U(x)}{\sinh\tilde{v}\sinh\tilde{v}_1} = (\coth\tilde{v} - \tilde{v}) - (\coth\tilde{v}_1 - \tilde{v}_1).$$

Defining the function $F(\tilde{v})$ by

$$F(\tilde{v}) \equiv \coth\tilde{v} - \tilde{v}, \tag{9.7.45}$$

$F(\tilde{v})$ is plotted in Figure 9.5.

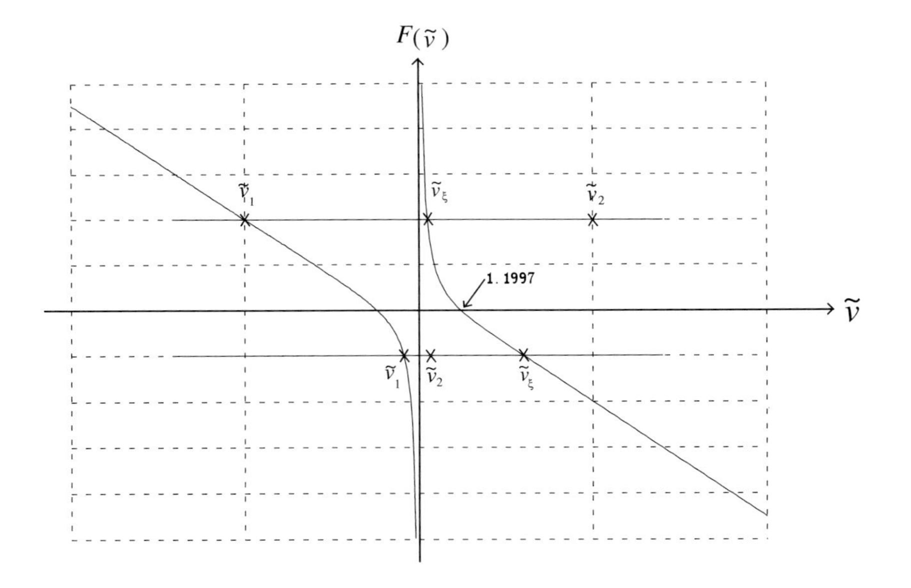

Fig. 9.5: Plot of $F(\tilde{v})$.

We have

$$\frac{U(x)}{\sinh\tilde{v}\sinh\tilde{v}_1} = F(\tilde{v}) - F(\tilde{v}_1).$$

Note that $F(\tilde{v})$ is an odd function of $\tilde{v}$,

$$F(-\tilde{v}) = -F(\tilde{v}).$$

We set

$$\tilde{v}_1 \equiv -\frac{L}{\alpha}, \qquad \tilde{v}_2 \equiv +\frac{L}{\alpha}, \qquad \tilde{v}_\xi \equiv \frac{\xi}{\alpha},$$

where

$$\beta = 0,$$

is used. The equation,

$$U(\xi) = 0, \qquad \xi \neq x_1,$$

which determines the conjugate point ξ, is equivalent to the following equation,

$$F(\tilde{v}_\xi) = F(\tilde{v}_1), \qquad \tilde{v}_\xi \neq \tilde{v}_1.$$

If

$$F(\tilde{v}_1) > 0, \tag{9.7.46}$$

then, from Figure 9.5, we have

$$\tilde{v}_1 < \tilde{v}_\xi < \tilde{v}_2,$$

thus *failing the Jacobi test*. If, on the other hand,

$$F(\tilde{v}_1) < 0, \tag{9.7.47}$$

then, from Figure 9.5, we have

$$\tilde{v}_1 < \tilde{v}_2 < \tilde{v}_\xi,$$

thus *passing the Jacobi test*. The dividing line

$$F(\tilde{v}_1) = 0,$$

corresponds to

$$\tilde{v}_1 < \tilde{v}_2 = \tilde{v}_\xi,$$

and thus we have *one solution* at

$$\tilde{v}_2 = -\tilde{v}_1 = \frac{L}{\alpha} = 1.1997.$$

Having derived the *particular statements of the Jacobi test* as applied to this example, Eq. (9.7.46) and (9.7.47), we now test which of the two candidates for the solution, $v = v_<$ and $v = v_>$, is actually the minimizing solution. Two functions, $G(v)$ and $F(v)$, defined by Eq. (9.7.44) and (9.7.45), are related to each other through

$$\frac{d}{dv}G(v) = -\left(\frac{\sinh v}{v^2}\right)F(v). \tag{9.7.48}$$

At $v = v_<$, we know from Figure 9.4 that

$$\frac{d}{dv}G(v_<) < 0,$$

which implies

$$F(-v_<) = -F(v_<) < 0,$$

so that *one candidate*, $v = v_<$, *passes the Jacobi test* and is the *solution*, whereas at $v = v_>$, we know from Figure 9.4 that

$$\frac{d}{dv}G(v_>) > 0,$$

which implies

$$F(-v_>) = -F(v_>) > 0,$$

so that the *other candidate*, $v = v_>$, *fails the Jacobi test* and is *not the solution.*
When *two candidates*, $v = v_<$ *and* $v = v_>$, *coalesce to a single point*,

$$v = 1.1997,$$

where the first derivative of $G(v)$ vanishes, i.e.,

$$\frac{d}{dv}G(v) = 0,$$

$v_< = v_> = 1.1997$ is the *solution.*

We now consider a *strong variation* and the condition for the *strong minimum.* In the strong variation, since the varied derivatives behave very differently from the original derivatives, we cannot expand the integral I in a Taylor series. Instead, we consider the *Weierstrass E function* defined by

$$E(x, y_0, y_0', p) \equiv f(x, y_0, p) - [f(x, y_0, y_0') + (p - y_0')f_{y'}(x, y_0, y_0')]\,. \tag{9.7.49}$$

Necessary and sufficient conditions for the strong minimum are given by

Necessary condition	Sufficient condition
1) $f_{y'y'}(x, y_0, y_0') \geq 0$, where y_0 is the solution of the Euler equation.	1) $f_{y'y'}(x, y, p) > 0$, for every (x, y) close to (x, y_0), and every finite p.
2) $\xi \geq x_2$.	2) $\xi > x_2$.
3) $E(x, y_0, y_0', p) \geq 0$, for all finite p, and $x \in [x_1, x_2]$	

(9.7.50)

We note that if

$$p \sim y_0',$$

then we have

$$E \sim f_{y'y'}(x, y_0, y_0') \frac{(p - y_0')^2}{2!},$$

just like the *mean value theorem* of the ordinary function $f(x)$, which is given by

$$f(x) - f(x_0) - f'(x_0) = \frac{(x - x_0)^2}{2!} f''(x_0 + \lambda(x - x_0)), \qquad 0 < \lambda < 1.$$

9.8 Weierstrass–Erdmann Corner Relation

In this section, we consider the variational problem with the solutions which are the *piecewise continuous functions with corners*, i.e., the function itself is continuous, but the derivative is not. We maximize the integral

$$I = \int_{x_1}^{x_2} f(x, y, y')\, dx. \tag{9.8.1}$$

We put a point of the corner at $x = a$. We write the solution for $x \leq a$, as $y(x)$ and, for $x > a$, as $Y(x)$. Then we have from the continuity of the solution,

$$y(a) = Y(a). \tag{9.8.2}$$

Now, consider the variation of $y(x)$ and $Y(x)$ of the following forms,

$$y(x) \to y(x) + \varepsilon\nu(x), \quad Y(x) \to Y(x) + \varepsilon V(x), \tag{9.8.3a}$$

and

$$y'(x) \to y'(x) + \varepsilon\nu'(x), \quad Y'(x) \to Y'(x) + \varepsilon V'(x). \tag{9.8.3b}$$

Under these variations, we require I to be stationary,

$$I = \int_{x_1}^{a} f(x, y, y')\, dx + \int_{a}^{x_2} f(x, Y, Y')\, dx. \tag{9.8.4}$$

First, we consider the variation problem with the point of the discontinuity of the derivative at $x = a$ fixed. We have

$$\nu(a) = V(a). \tag{9.8.5}$$

Performing the above variations, we have

$$\begin{aligned}
\delta I &= \int_{x_1}^{a} [f_y \varepsilon\nu + f_{y'}\varepsilon\nu']\, dx + \int_{a}^{x_2} [f_Y \varepsilon V + f_{Y'}\varepsilon V']\, dx \\
&= f_{y'}\varepsilon\nu\Big|_{x=x_1}^{x=a} + \int_{x_1}^{a} \varepsilon\nu \left[f_y - \frac{d}{dx} f_{y'} \right] dx \\
&\quad + f_{Y'}\varepsilon V\Big|_{x=a}^{x=x_2} + \int_{a}^{x_2} \varepsilon V \left[f_Y - \frac{d}{dx} f_{Y'} \right] dx = 0.
\end{aligned}$$

If the variations vanish at both ends ($x = x_1$ and $x = x_2$), i.e.,

$$\nu(x_1) = V(x_2) = 0, \tag{9.8.6}$$

we have the following equations,

$$f_y - \frac{d}{dx} f_{y'} = 0, \qquad x \in [x_1, x_2] \tag{9.8.7}$$

and

$$f_{y'}\big|_{x=a^-} = f_{Y'}\big|_{x=a^+} ,$$

namely

$$f_{y'} \text{ is continuous at } x = a. \tag{9.8.8}$$

Next, we consider the variation problem with the point of the discontinuity of the derivative at $x = a$ varied, i.e.,

$$a \rightarrow a + \Delta a. \tag{9.8.9}$$

The point of the discontinuity becomes shifted, and yet the solutions are continuous at $x = a + \Delta a$, i.e.,

$$y(a) + y'(a)\Delta a + \varepsilon\nu(a) = Y(a) + Y'(a)\Delta a + \varepsilon V(a),$$

or,

$$\left[y'(a) - Y'(a)\right] \Delta a = \varepsilon \left[V(a) - \nu(a)\right] , \tag{9.8.10}$$

which is the condition on Δa, $V(a)$ and $\nu(a)$.

The integral I becomes changed into

$$\begin{aligned} I \rightarrow & \int_{x_1}^{a+\Delta a} f(x, y + \varepsilon\nu, y' + \varepsilon\nu')\, dx + \int_{a+\Delta a}^{x_2} f(x, Y + \varepsilon V, Y' + \varepsilon V')\, dx \\ &= \int_{x_1}^{a} f(x, y + \varepsilon\nu, y' + \varepsilon\nu')\, dx + f(x, y, y')\Big|_{x=a^-} \Delta a \\ &\quad + \int_{a}^{x_2} f(x, Y + \varepsilon V, Y' + \varepsilon V')\, dx - f(x, Y, Y')\Big|_{x=a^+} \Delta a\, . \end{aligned}$$

The integral parts, after the integration by parts, vanish due to the Euler equation in the respective region, and what remain are the integrated parts, i.e.,

$$\begin{aligned} \delta I &= \varepsilon\nu f_{y'}\big|_{x=a^-} - \varepsilon V f_{Y'}\big|_{x=a^+} + (f\big|_{x=a^-} - f\big|_{x=a^+})\, \Delta a \\ &= \varepsilon(\nu - V)\, f_{y'}\big|_{x=a} + (f\big|_{x=a^-} - f\big|_{x=a^+})\, \Delta a = 0, \end{aligned} \tag{9.8.11}$$

where the first term of the second line above follows from the continuity of $f_{y'}$. From the continuity condition (9.8.10) and the expression (9.8.11), by eliminating Δa, we obtain

$$-\left[y'(a) - Y'(a)\right] f_{y'} + (f\big|_{x=a^-} - f\big|_{x=a^+}) = 0,$$

or,

$$f - y'(a)f_{y'}|_{x=a^-} = f - Y'(a)f_{y'}|_{x=a^+},$$

i.e.,

$$f - y'f_{y'} \quad \text{continuous at } x = a. \tag{9.8.12}$$

For the solution of the variation problem with the discontinuous derivative, we have

$$\begin{aligned} & f_y - \frac{d}{dx}f_{y'} = 0, && \text{Euler equation,} \\ & f_{y'} && \text{continuous at } x = a, \\ & f - y'f_{y'} && \text{continuous at } x = a, \end{aligned} \tag{9.8.13}$$

which are called the *Weierstrass–Erdmann corner relation.*

❑ **Example 9.23.** Extremize

$$I = \int_0^1 (y' + 1)^2 y'^2 \, dx$$

with the endpoints fixed as below,

$$y(0) = 2, \quad y(1) = 0.$$

Are there solutions with discontinuous derivatives? Find the minimum value of I.

Solution. We have

$$f(x, y, y') = (y' + 1)^2 y'^2,$$

which is independent of x. Thus

$$f - y'f_{y'} = \text{constant},$$

where $f_{y'}$ is calculated to be

$$f_{y'} = 2(y' + 1)y'(2y' + 1).$$

Hence we have

$$f - y'f_{y'} = -(3y' + 1)(y' + 1)y'^2.$$

1) If we want to have a *continuous* solution alone, we have $y' = a$(constant), from which, we conclude that $y = ax + b$. From the endpoint conditions, the solution is

$$y = 2(-x + 1), \quad y' = -2.$$

Thus the integral I is evaluated to be $I_{\text{cont.}} = 4$.

2) Suppose that y' is *discontinuous* at $x = a$. Setting

$$p = y'_<, \quad p' = y'_>,$$

we have, from the corner relation at $x = a$,

$$\begin{cases} p(p+1)(2p+1) &= p'(p'+1)(2p'+1), \\ (3p+1)(p+1)p^2 &= (3p'+1)(p'+1)p'^2. \end{cases}$$

Setting

$$u = p + p', \quad v = p^2 + pp' + p'^2,$$

we have

$$\begin{cases} 3u + 2v + 1 = 0, \\ 3u(2v - u^2) + u + 4v = 0, \end{cases} \Rightarrow \begin{cases} p = 0, \\ p' = -1, \end{cases} \quad \text{or} \quad \begin{cases} p = -1, \\ p' = 0. \end{cases}$$

Thus the discontinuous solution gives $y' + 1 = 0$, or $y' = 0$. Then we have $I_{\text{disc.}} = 0 < I_{\text{cont.}} = 4$.

In the above example, the solution $y(x)$ itself became discontinuous. Depending on the boundary conditions, the discontinuous solution may not be present.

9.9 Problems for Chapter 9

9.1 (Due to H. C.) Find an approximate value for the lowest eigenvalue E_0 for

$$\left[\frac{\partial^2}{\partial x^2} + \frac{\partial^2}{\partial y^2} - x^2y^2\right]\psi(x,y) = -E\psi(x,y), \qquad -\infty < x, y < \infty,$$

where $\psi(x,y)$ is normalized to unity,

$$\int_{-\infty}^{+\infty}\int_{-\infty}^{+\infty} |\psi(x,y)|^2 \, dxdy = 1.$$

9.2 (Due to H. C.) A light ray in the x-y plane is incident on the lower half-plane ($y \leq 0$) of the medium with an index of refraction $n(x,y)$ given by

$$n(x,y) = n_0(1 + ay), \qquad y \leq 0,$$

where n_0 and a are positive constants.

a) Find the path of a ray passing through $(0,0)$ and $(l,0)$.

b) Find the apparent depth of the object located at $(0,0)$ when viewed from $(l,0)$.

9.3 (Due to H. C.) Estimate the lowest frequency of a circular drum of radius R with a rotationally symmetric trial function,

$$\phi(r) = 1 - \left(\frac{r}{R}\right)^n, \qquad 0 \leq r \leq R.$$

Find the n which gives the best estimate.

9.4 Minimize the integral

$$I \equiv \int_1^2 x^2 (y')^2 \, dx,$$

with the endpoints fixed as below,

$$y(1) = 0, \quad y(2) = 1.$$

Apply the Legendre test and the Jacobi test to determine if your solution is a minimum. Is it a strong minimum or a weak minimum? Are there solutions with discontinuous derivatives?

9.5 Extremize

$$\int_0^1 \frac{(1+y^2)^2}{(y')^2} \, dx,$$

with the endpoints fixed as below,

$$y(0) = 0, \quad y(1) = 1.$$

Is your solution a weak minimum? Is it a strong minimum? Are there solutions with discontinuous derivatives?

9.6 Extremize

$$\int_0^1 (y'^2 - y'^4) \, dx,$$

with the endpoints fixed as below,

$$y(0) = y(1) = 0.$$

Is your solution a weak minimum? Is it a strong minimum? Are there solutions with discontinuous derivatives?

9.7 Extremize

$$\int_0^2 (xy' + y'^2) \, dx,$$

with the endpoints fixed as below,

$$y(0) = 1, \quad y(2) = 0.$$

Is your solution a weak minimum? Is it a strong minimum? Are there solutions with discontinuous derivatives?

9.8 Extremize

$$\int_1^2 \frac{x^3}{y'^2}\,dx,$$

with the endpoints fixed as below,

$$y(1) = 1, \quad y(2) = 4.$$

Is your solution a weak minimum? Is it a strong minimum? Are there solutions with discontinuous derivatives?

9.9 (Due to H. C.) An airplane with speed v_0 flies in a wind of speed a_x. What is the trajectory of the airplane if it is to enclose the greatest area in a given amount of time?

Hint: You may assume that the velocity of the airplane is

$$\frac{dx}{dt}\vec{e}_x + \frac{dy}{dt}\vec{e}_y = (v_x + a_x)\vec{e}_x + v_y\vec{e}_y,$$

with

$$v_x^2 + v_y^2 = v_0^2.$$

9.10 (Due to H. C.) Find the shortest distance between two points on a cylinder. Apply the Jacobi test and the Legendre test to verify if your solution is a minimum.

10 Calculus of Variations: Applications

10.1 Feynman's Action Principle in Quantum Mechanics

In this section, we discuss Feynman's action principle in quantum mechanics. The derivation of Feynman's action principle in quantum mechanics can be found in the monograph by M. Masujima, but we shall be content here with the deduction of the canonical formalism of quantum mechanics from Feynman's action principle for the *nonsingular Lagrangian.*

The operator $\hat{q}(t)$ at all time t forms the complete set of the commuting operators, i.e., the quantum mechanical state vector $|\psi(t)>$ can be expressed in terms of the linear superposition of the complete set of the eigenket $|q,t>$ of the commuting operator $\hat{q}(t)$. Feynman's action principle asserts that the *transformation function* $<q'',t''|\,q',t'>$ is given by

$$< q'',t''|\,q',t' > = \frac{1}{N(\Omega)}\int_{q(t')=q'}^{q(t'')=q''} \mathcal{D}[q(t)] \exp\left[\frac{i}{\hbar}\int_{t'}^{t''} dt\, L(q(t),\dot{q}(t))\right], \quad (10.1.1)$$

$\Omega = \Omega(t'',t') =$ the "space–time region" sandwiched between t' and t''.

We state here three assumptions.

(A-1) In the "space–time region", $\Omega_1 + \Omega_2$, where Ω_1 and Ω_2 are neighboring each other, we have the *principle of superposition* in the following form,

$$\int_{q(t')=q'}^{q(t'')=q''} \mathcal{D}[q(t)] = \int_{t=t'''} dq''' \int_{q_2(t''')=q'''}^{q_2(t'')=q''} \mathcal{D}[q_2(t)] \int_{q_1(t')=q'}^{q_1(t''')=q'''} \mathcal{D}[q_1(t)]. \quad (10.1.2)$$

(A-2) *Functional integration by parts* is allowed. The requisite damping factor which kills the contribution from the functional infinities will be supplied by an "$i\varepsilon$" piece which originates from the wave function of the vacuum at $t = \mp\infty$.

(A-3) We have the following *resolution of identity,*

$$\int |q',t'> dq' < q',t'| = 1. \quad (10.1.3)$$

From the consistency of (A-1), (A-2) and (A-3), the normalization constant $N(\Omega)$ must satisfy the following equation,

$$N(\Omega_1 + \Omega_2) = N(\Omega_1)N(\Omega_2). \tag{10.1.4}$$

Equation (10.1.4) also originates from the additivity of the action functional,

$$I[q;t'',t'] = I[q;t'',t'''] + I[q;t''',t'],$$

where the action functional is given by,

$$I[q;t'',t'] = \int_{t'}^{t''} dtL(q(t),\dot{q}(t)).$$

In Feynman's action principle, the operator $\hat{q}(t)$ is defined by its matrix elements,

$$< q'',t''|\,\hat{q}(t)\,|q',t' > = \frac{1}{N(\Omega)}\int_{q(t')=q'}^{q(t'')=q''} \mathcal{D}[q(t)]q(t)\exp\left[\frac{i}{\hbar}I\,[q;t'',t']\right]. \tag{10.1.5}$$

If t of $\hat{q}(t)$ is on the t''-surface, we have

$$< q'',t''|\,\hat{q}(t'')\,|q',t' > = q'' < q'',t''|\,q',t' >, \tag{10.1.6}$$

and from the assumed completeness of the eigenket $|q',t' >$, we have

$$< q'',t''|\,\hat{q}(t'') = q'' < q'',t''|\,. \tag{10.1.7}$$

Equation (10.1.7) is the defining equation of the eigenbra $< q'',t''|$ of $\hat{q}(t)$.

Consider the variation of the action functional,

$$\begin{aligned}\delta I[q;t'',t'] &= \int_{t'}^{t''} dt\left[\frac{\partial L(q(t),\dot{q}(t))}{\partial q(t)} - \frac{d}{dt}\left(\frac{\partial L(q(t),\dot{q}(t))}{\partial \dot{q}(t)}\right)\right]\delta q(t) \\ &\quad + \int_{t'}^{t''} dt\frac{d}{dt}\left[\frac{\partial L(q(t),\dot{q}(t))}{\partial \dot{q}(t)}\delta q(t)\right].\end{aligned} \tag{10.1.8}$$

We first employ the variation which vanishes at the endpoints, $\delta q(t') = \delta q(t'') = 0$. Then the second term of Eq. (10.1.8) vanishes and we obtain the Euler derivative,

$$\frac{\delta I[q;t'',t']}{\delta q(t)} = \frac{\partial L(q(t),\dot{q}(t))}{\partial q(t)} - \frac{d}{dt}\left(\frac{\partial L(q(t),\dot{q}(t))}{\partial \dot{q}(t)}\right). \tag{10.1.9}$$

We now evaluate the matrix elements of the operator, $\delta I[\hat{q};t'',t']/\delta\hat{q}(t)$, in accordance with Feynman's action principle.

$$\begin{aligned}&\left\langle q'',t''\left|\frac{\delta I[\hat{q};t'',t']}{\delta\hat{q}(t)}\right|q',t'\right\rangle \\ &= \frac{1}{N(\Omega)}\int_{q(t')=q'}^{q(t'')=q''} \mathcal{D}[q(t)]\frac{\delta I[q;t'',t']}{\delta q(t)}\exp\left[\frac{i}{\hbar}I[q;t'',t']\right] \\ &= \frac{1}{N(\Omega)}\int_{q(t')=q'}^{q(t'')=q''} \mathcal{D}[q(t)]\frac{\hbar}{i}\frac{\delta}{\delta q(t)}\exp\left[\frac{i}{\hbar}I[q;t'',t']\right] = 0.\end{aligned} \tag{10.1.10}$$

From the assumed completeness of the eigenket $|q', t' >$, we shall obtain the *Lagrange equation of motion at the operator level*, $\delta I[\hat{q}]/\delta\hat{q}(t) = 0$, i.e., we obtain,

$$\frac{d}{dt}\left(\frac{\partial L(\hat{q}(t), \dot{\hat{q}}(t))}{\partial\dot{\hat{q}}(t)}\right) - \frac{\partial L(\hat{q}(t), \dot{\hat{q}}(t))}{\partial\hat{q}(t)} = 0. \tag{10.1.11}$$

As for the *time-ordered product*, we can obtain the following identity by mathematical induction, starting from $n = 2$, by the repeated use of Eqs. (10.1.2) and (10.1.3).

$$\begin{aligned} \frac{1}{N(\Omega)} \int_{q(t')=q'}^{q(t'')=q''} & \mathcal{D}[q(t)]q(t_1) \cdot \ldots \cdot q(t_n) \exp\left[\frac{i}{\hbar} I[q; t'', t']\right] \\ &= < q'', t''|\, \mathrm{T}[\hat{q}(t_1) \cdot \ldots \cdot \hat{q}(t_n)]\, |q', t' > . \end{aligned} \tag{10.1.12}$$

We define the *momentum operator* $\hat{p}(t)$ as the displacement operator,

$$< q'', t''|\, \hat{p}(t'')\, |q', t' > = \frac{\hbar}{i}\frac{\partial}{\partial q''} < q'', t''|\, q', t' > . \tag{10.1.13}$$

In order to express the right-hand side of Eq. (10.1.13) in the form in which we can use Feynman's action principle, we consider the following variation.

(1) Inside Ω, we take the infinitesimal variation of $q(t)$, $q(t) \to q(t) + \delta q(t)$,

$$\delta q(t') = 0, \quad \delta q(t) = \xi(t), \quad \delta q(t'') = \xi''. \tag{10.1.14}$$

(2) Inside Ω, we assume that the physical system evolves with time t in accordance with the *Lagrange equation of motion*.

The response of the action functional $I[q; t'', t']$ to the variation, (10.1.14), is given by

$$\delta I[q; t'', t'] = \int_{t'}^{t''} dt\, \frac{d}{dt}\left[\frac{\partial L(q(t), \dot{q}(t)}{\partial\dot{q}(t)}\delta q(t)\right] = \frac{\partial L(q(t''), \dot{q}(t''))}{\partial\dot{q}(t'')}\xi''. \tag{10.1.15}$$

Thus we obtain

$$\frac{\delta I[q; t'', t']}{\delta q(t'')} = \frac{\partial L(q(t''), \dot{q}(t''))}{\partial\dot{q}(t'')}. \tag{10.1.16}$$

Using Eq. (10.1.16), the right-hand side of Eq. (10.1.13) can be expressed as,

$$\begin{aligned}
\frac{\hbar}{i}\frac{\partial}{\partial q''} < q'', t''|q', t' >&= \frac{\hbar}{i}\lim_{\xi''\to 0}\frac{< q''+\xi'', t''|\, q', t' > - < q'', t''|\, q', t' >}{\xi''} \\
&= \frac{\hbar}{i}\lim_{\xi''\to 0}\frac{1}{N(\Omega)}\int_{q(t')=q'}^{q(t'')=q''}\mathcal{D}[q(t)]\exp\left[\frac{i}{\hbar}I[q;t'',t']\right] \\
&\quad\times\frac{1}{\xi''}\left\{\exp\left[\frac{i}{\hbar}\Big(I[q+\xi'',t'',t'] - I[q,t'',t']\Big)\right] - 1\right\} \\
&= \lim_{\xi''\to 0}\frac{1}{N(\Omega)}\int_{q(t')=q'}^{q(t'')=q''}\mathcal{D}[q(t)]\exp\left[\frac{i}{\hbar}I[q;t'',t']\right] \\
&\quad\times\frac{1}{\xi''}\left[I[q+\xi'';t'',t'] - I[q;t'',t'] + O\left((\xi'')^2\right)\right] \\
&= \frac{1}{N(\Omega)}\int_{q(t')=q'}^{q(t'')=q''}\mathcal{D}[q(t)]\exp\left[\frac{i}{\hbar}I[q;t'',t']\right] \\
&\quad\times\frac{\delta I[q;t'',t']}{\delta q(t'')} \\
&= \frac{1}{N(\Omega)}\int_{q(t')=q'}^{q(t'')=q''}\mathcal{D}[q(t)]\exp\left[\frac{i}{\hbar}I[q;t'',t']\right] \\
&\quad\times\frac{\partial L(q(t''),\dot{q}(t''))}{\partial\dot{q}(t'')} \\
&= < q'', t''|\,\frac{\partial L(\hat{q}(t''),\dot{\hat{q}}(t''))}{\partial\dot{\hat{q}}(t'')}\,|q', t' > .
\end{aligned} \tag{10.1.17}$$

From Eqs. (10.1.13) and (10.1.17), and the assumed completeness of the eigenket $|q', t' >$, we obtain the *operator identity*,

$$\hat{p}(t) = \frac{\partial L(\hat{q}(t),\dot{\hat{q}}(t))}{\partial\dot{\hat{q}}(t)}. \tag{10.1.18}$$

Equation (10.1.18) is the definition of the momentum $\hat{p}(t)$ canonically conjugate to $\hat{q}(t)$.

We now consider the *equal time canonical (anti-)commutator* of $\hat{p}(t)$ and $\hat{q}(t)$. For the *Bose system*, we have

$$< q'', t''|\,\hat{q}_B(t'')\,|q', t' > = q''_B < q'', t''|\, q', t' > . \tag{10.1.19}$$

From Eqs. (10.1.13) and (10.1.19), we have,

$$\begin{aligned}< q'', t''|\,\hat{p}_B(t'')\hat{q}_B(t'')\,|q', t' > &= \frac{\hbar}{i}\frac{\partial}{\partial q''_B}(q''_B < q'', t''|\,q', t' >) \\ &= \frac{\hbar}{i} < q'', t''|\,q', t' > \\ &\quad + q''_B < q'', t''|\,\hat{p}_B(t'')\,|q', t' > \\ &= \frac{\hbar}{i} < q'', t''|\,q', t' > \\ &\quad + < q'', t''|\,\hat{q}_B(t'')\hat{p}_B(t'')\,|q', t' > .\end{aligned} \tag{10.1.20}$$

Thus, from the assumed completeness of the eigenket $|q', t' >$, we obtain,

$$[\hat{p}_B(t), \hat{q}_B(t)] = \frac{\hbar}{i}, \tag{10.1.21}$$

where the commutator, $[A, B]$, is defined by

$$[A, B] \equiv AB - BA. \tag{10.1.22}$$

For the *Fermi system*, we have a minus sign in front of the second terms of the third line and the fifth line of Eq. (10.1.20) which originates from the anti-commuting Fermion number, so that we obtain,

$$\{\hat{p}_F(t), \hat{q}_F(t)\} = \frac{\hbar}{i}, \tag{10.1.23}$$

where the anti-commutator, $\{A, B\}$, is defined by

$$\{A, B\} \equiv AB + BA. \tag{10.1.24}$$

We define the *Hamiltonian* as the Legendre transform of the Lagrangian,

$$H(\hat{q}(t), \hat{p}(t)) = \hat{p}(t)\dot{\hat{q}}(t) - L(\hat{q}(t), \dot{\hat{q}}(t)), \tag{10.1.25}$$

where Eq. (10.1.18) is solved for $\dot{\hat{q}}(t)$ as a function of $\hat{q}(t)$ and $\hat{p}(t)$, and this $\dot{\hat{q}}(t)$ is substituted into the right-hand side of Eq. (10.1.25).

Proof that the *Heisenberg equation of motion* follows from Eqs. (10.1.11), (10.1.18), (10.1.21), (10.1.23) and (10.1.25) is left as an exercise for the reader. Another proof that the transformation function $< q'', t''|\,q', t' >$ satisfies the *Schrödinger equation* with respect to q'' and t'' with a Lagrangian of the form,

$$L(q(t), \dot{q}(t)) = \frac{1}{2}m\dot{q}(t)^2 - V(q(t)), \tag{10.1.26}$$

is discussed in the monograph by M. Masujima, cited at the beginning of this section.

We have thus deduced the canonical formalism of quantum mechanics from Feynman's action principle for the nonsingular Lagrangian. The extension of the present discussion to the case of quantum field theory with the nonsingular Lagrangian density $\mathcal{L}(\phi(x), \partial_\mu \phi(x))$ is straightforward, resulting in the normal dependence of the Hamiltonian density $\mathcal{H}(\hat{\phi}(x), \vec{\nabla}\hat{\phi}(x), \hat{\pi}(x))$ and the momentum $\hat{\pi}(x)$ canonically conjugate to $\hat{\phi}(x)$.

In view of the discussions in this section and Section 10.3 which follows, we eventually show the equivalence of path integral quantization and canonical quantization at least for the nonsingular Lagrangian (density); canonical quantization $\Leftrightarrow$ path integral quantization. It is in fact astonishingly difficult to establish this equivalence for the singular Lagrangian (density).

10.2 Feynman's Variational Principle in Quantum Statistical Mechanics

In this section, we shall briefly consider *Feynman's variational principle in quantum statistical mechanics* which is based on the analytic continuation in time from a real time to an imaginary time of Feynman's action principle in quantum mechanics.

We consider the canonical ensemble with the Hamiltonian $\hat{H}(\{\vec{q}_j, \vec{p}_j\}_{j=1}^N)$ at finite temperature. The density matrix $\hat{\rho}_{\mathrm{C}}(\beta)$ of this system satisfies the Bloch equation,

$$-\hbar \frac{\partial}{\partial \tau} \hat{\rho}_{\mathrm{C}}(\tau) = \hat{H}(\{\vec{q}_j, \vec{p}_j\}_{j=1}^N) \hat{\rho}_{\mathrm{C}}(\tau), \qquad 0 \le \tau \le \beta, \tag{10.2.1}$$

with its formal solution given by

$$\hat{\rho}_{\mathrm{C}}(\tau) = \exp\left[-\frac{\tau \hat{H}\left(\{\vec{q}_j, \vec{p}_j\}_{j=1}^N\right)}{\hbar}\right] \hat{\rho}_{\mathrm{C}}(0). \tag{10.2.2}$$

We compare the Bloch equation and the density matrix, Eqs. (10.2.1) and (10.2.2), with the Schrödinger equation for the state vector $|\psi, t>$,

$$i\hbar \frac{d}{dt} |\psi, t> = \hat{H}(\{\vec{q}_j, \vec{p}_j\}_{j=1}^N) |\psi, t> , \tag{10.2.3}$$

and its formal solution given by

$$|\psi, t> = \exp\left[-\frac{it\hat{H}(\{\vec{q}_j, \vec{p}_j\}_{j=1}^N)}{\hbar}\right] |\psi, 0> . \tag{10.2.4}$$

We find that by the analytic continuation,

$$t = -i\tau, \qquad 0 \le \tau \le \beta \equiv \frac{\hbar}{k_{\mathrm{B}} T}, \tag{10.2.5}$$

where $k_{\mathrm{B}} =$ Boltzmann constant, $T =$ absolute temperature,

the (real time) Schrödinger equation and its formal solution, Eqs. (10.2.3) and (10.2.4), are analytically continued into the Bloch equation and the density matrix, Eqs. (10.2.1) and (10.2.2), respectively. Under the analytic continuation, Eq. (10.2.5), we divide the interval $[0, \beta]$ into the n equal subintervals, and use the resolution of the identity in both the q-representation and the p-representation. In this way, we obtain the following list of correspondence. Here, we assume a Hamiltonian $\hat{H}(\{\vec{q}_j, \vec{p}_j\}_{j=1}^N)$ of the following form,

$$\hat{H}(\{\vec{q}_j, \vec{p}_j\}_{j=1}^N) = \sum_{j=1}^{N} \frac{1}{2m} \vec{p}_j^{\,2} + \sum_{j>k} V(\vec{q}_j, \vec{q}_k). \tag{10.2.6}$$

Table 10.1: List of Correspondence

Quantum Mechanics	Quantum Statistical Mechanics
Schrödinger equation $i\hbar \frac{\partial}{\partial t} \lvert\psi, t\rangle = H(\{\vec{q}_j, \vec{p}_j\}_{j=1}^N) \lvert\psi, t\rangle$	Bloch equation $-\hbar \frac{\partial}{\partial \tau} \hat{\rho}_{\mathrm{C}}(\tau) = H(\{\vec{q}_j, \vec{p}_j\}_{j=1}^N) \hat{\rho}_{\mathrm{C}}(\tau)$
Schrödinger state vector $\lvert\psi, t\rangle$ $= \exp\left[-itH(\{\vec{q}_j, \vec{p}_j\}_{j=1}^N)/\hbar\right] \lvert\psi, 0\rangle$	Density matrix $\hat{\rho}_{\mathrm{C}}(\tau)$ $= \exp\left[-\tau H(\{\vec{q}_j, \vec{p}_j\}_{j=1}^N)/\hbar\right] \hat{\rho}_{\mathrm{C}}(0)$
Minkowskian Lagrangian $L_{\mathrm{M}}(\{q_j(t), \dot{q}_j(t)\}_{j=1}^N)$ $= \sum_{j=1}^{N} \frac{1}{2} m \dot{q}_j^2(t) - \sum_{j \rangle k}^{N} V(\vec{q}_j, \vec{q}_k)$	Euclidean Lagrangian $L_{\mathrm{E}}(\{q_j(\tau), \dot{q}_j(\tau)\}_{j=1}^N)$ $= -\sum_{j=1}^{N} \frac{1}{2} m \dot{q}_j^2(\tau) - \sum_{j \rangle k}^{N} V(\vec{q}_j, \vec{q}_k)$
Minkowskian action functional $iI_{\mathrm{M}}[\{\vec{q}_j\}_{j=1}^N; \vec{q}_f, \vec{q}_i]$ $= i \int_{t_i}^{t_f} dt L_{\mathrm{M}}(\{q_j(t), \dot{q}_j(t)\}_{j=1}^N)$	Euclidean action functional $I_{\mathrm{E}}[\{\vec{q}_j\}_{j=1}^N; \vec{q}_f, \vec{q}_i]$ $= \int_0^{\beta} d\tau L_{\mathrm{E}}(\{q_j(\tau), \dot{q}_j(\tau)\}_{j=1}^N)$
Transformation function $\langle \vec{q}_f, t_f \vert \vec{q}_i, t_i \rangle = \int_{\vec{q}(t_i)=\vec{q}_i}^{\vec{q}(t_f)=\vec{q}_f} \mathcal{D}[\vec{q}] \times$ $\times \exp\left[iI_{\mathrm{M}}[\{\vec{q}_j\}_{j=1}^N; \vec{q}_f, \vec{q}_i]/\hbar\right]$	Transformation function $Z_{f,i} = \int_{\vec{q}(0)=\vec{q}_i}^{\vec{q}(\beta)=\vec{q}_f} \mathcal{D}[\vec{q}] \times$ $\times \exp\left[I_{\mathrm{E}}[\{\vec{q}_j\}_{j=1}^N; \vec{q}_f, \vec{q}_i]/\hbar\right]$
Vacuum to vacuum transition amplitude $\langle 0, \mathrm{out} \vert 0, \mathrm{in} \rangle$ $= \int \mathcal{D}[\vec{q}] \exp\left[iI_{\mathrm{M}}[\{\vec{q}_j\}_{j=1}^N]/\hbar\right]$	Partition function $Z_{\mathrm{C}}(\beta) = \mathrm{Tr}\hat{\rho}_{\mathrm{C}}(\beta)$ $= \text{“} \int d\vec{q}_f d\vec{q}_i \delta(\vec{q}_f - \vec{q}_i) Z_{f,i} \text{”}$
Vacuum expectation value $\langle O(\hat{q}) \rangle = \frac{\int \mathcal{D}[\vec{q}] O(\vec{q}) \exp\left[iI_{\mathrm{M}}[\{\vec{q}_j\}_{j=1}^N]/\hbar\right]}{\int \mathcal{D}[\vec{q}] \exp\{iI_{\mathrm{M}}[\{\vec{q}_j\}_{j=1}^N]/\hbar\}}$	Thermal expectation value $\langle O(\hat{q}) \rangle = \frac{\text{“} \mathrm{Tr}\hat{\rho}_{\mathrm{C}}(\beta) O(\hat{q}) \text{”}}{\mathrm{Tr}\hat{\rho}_{\mathrm{C}}(\beta)}$

In the list, we have entries enclosed in double quotes, whose precise expressions are given, respectively, by

Partition function:

$$\begin{aligned}Z_{\mathrm{C}}(\beta) &= \mathrm{Tr}\hat{\rho}_{\mathrm{C}}(\beta) = \text{``}\int d^3\vec{q}_f\, d^3\vec{q}_i \delta^3(\vec{q}_f - \vec{q}_i) Z_{f,i}\text{''} \\ &= \frac{1}{N!}\sum_P \delta_P \int d^3\vec{q}_f\, d^3\vec{q}_i \delta^3(\vec{q}_f - \vec{q}_{Pi}) Z_{f,Pi} \\ &= \frac{1}{N!}\sum_P \delta_P \int d^3\vec{q}_f\, d^3\vec{q}_{Pi} \delta^3(\vec{q}_f - \vec{q}_{Pi}) \\ &\quad\times \int_{\vec{q}(0)=\vec{q}_{Pi}}^{\vec{q}(\beta)=\vec{q}_f} \mathcal{D}[\vec{q}] \exp\left[\frac{I_{\mathrm{E}}[\{\vec{q}_j\}_{j=1}^N; \vec{q}_f, \vec{q}_{Pi}]}{\hbar}\right],\end{aligned} \tag{10.2.7}$$

and

Thermal expectation value:

$$\begin{aligned}\langle \hat{O}(\vec{q})\rangle &= \frac{\mathrm{Tr}\hat{\rho}_{\mathrm{C}}(\beta)\hat{O}(\hat{q})}{\mathrm{Tr}\hat{\rho}_{\mathrm{C}}(\beta)} \\ &= \frac{\text{``}\int d^3\vec{q}_f\, d^3\vec{q}_i \delta^3(\vec{q}_f - \vec{q}_i) Z_{f,i} \langle i|\, O(\vec{q})\,|f\rangle\text{''}}{\text{``}\int d^3\vec{q}_f\, d^3\vec{q}_i \delta^3(\vec{q}_f - \vec{q}_i) Z_{f,i}\text{''}} \\ &= \frac{1}{Z_{\mathrm{C}}(\beta)}\frac{1}{N!}\sum_P \delta_P \int d^3\vec{q}_f\, d^3\vec{q}_{Pi} \delta^3(\vec{q}_f - \vec{q}_{Pi}) Z_{f,Pi} \langle \vec{q}_{Pi}|\, \hat{O}(\vec{q})\,|\vec{q}_f\rangle\,.\end{aligned} \tag{10.2.8}$$

Here, $\vec{q}_i$ and $\vec{q}_f$ represent the initial position $\{\vec{q}_j(0)\}_{j=1}^N$ and the final position $\{\vec{q}_j(\beta)\}_{j=1}^N$ of N identical particles, P represents the permutation of $\{1, \ldots, N\}$, Pi represents the permutation of the initial position $\{\vec{q}(0)\}_{j=1}^N$ and δ_P represents the signature of the permutation P, respectively.

In this manner, we obtain the path integral representation of the partition function, $Z_{\mathrm{C}}(\beta)$, and the thermal expectation value, $\langle \hat{O}(\vec{q})\rangle$. This functional $I_{\mathrm{E}}[\{\vec{q}_j\}_{j=1}^N; \vec{q}_f, \vec{q}_i]$ of Eq. (10.2.7) can be obtained from $I_{\mathrm{M}}[\{\vec{q}_j\}_{j=1}^N; \vec{q}_f, \vec{q}_i]$ by replacing t with $-i\tau$. Since $\hat{\rho}_{\mathrm{C}}(\beta)$ is a solution of Eq. (10.2.1), the asymptotic form of $Z_{\mathrm{C}}(\beta)$ for a large τ interval from τ_i to τ_f is

$$Z_{\mathrm{C}}(\beta) \sim \exp\left[-\frac{E_0(\tau_f - \tau_i)}{\hbar}\right] \quad \text{as} \quad \tau_f - \tau_i \to \infty.$$

Therefore we must estimate $Z_{\mathrm{C}}(\beta)$ for large $\tau_f - \tau_i$.

We choose any real I_1 which approximates $I_{\mathrm{E}}[\{\vec{q}_j\}_{j=1}^N; \vec{q}_f, \vec{q}_i]$ and write $Z_{\mathrm{C}}(\beta)$ as

$$\begin{aligned}&\int \mathcal{D}[\vec{q}(\zeta)] \exp\left[\frac{I_{\mathrm{E}}[\{\vec{q}_j\}_{j=1}^N; \vec{q}_f, \vec{q}_i]}{\hbar}\right] \\ &\qquad = \int \mathcal{D}[\vec{q}(\zeta)] \exp\left[\frac{(I_{\mathrm{E}}[\{\vec{q}_j\}_{j=1}^N; \vec{q}_f, \vec{q}_i] - I_1)}{\hbar}\right] \exp\left[\frac{I_1}{\hbar}\right].\end{aligned} \tag{10.2.9}$$

The expression (10.2.9) can be regarded as the average of $\exp[(I_{\mathrm{E}} - I_1)/\hbar]$ with respect to the positive weight $\exp[I_1/\hbar]$. This observation motivates the variational principle based on Jensen's inequality. Since the exponential function is convex, for any real quantities f, the average of $\exp[f]$ exceeds the exponential of the average $< f >$,

$$< \exp[f] > \geq \exp[< f >]. \tag{10.2.10}$$

Hence, if in Eq. (10.2.9) we replace $I_{\mathrm{E}}[\{\vec{q}_j\}_{j=1}^{N}; \vec{q}_f, \vec{q}_i] - I_1$ by its average

$$< I_{\mathrm{E}}[\{\vec{q}_j\}_{j=1}^{N}; \vec{q}_f, \vec{q}_i] - I_1 > \\ = \frac{\int \mathcal{D}[\vec{q}(\zeta)](I_{\mathrm{E}}[\{\vec{q}_j\}_{j=1}^{N}; \vec{q}_f, \vec{q}_i] - I_1) \exp\left[\frac{I_1}{\hbar}\right]}{\int \mathcal{D}[\vec{q}(\zeta)] \exp\left[\frac{I_1}{\hbar}\right]}, \tag{10.2.11}$$

we will underestimate the value of Eq. (10.2.9). If E is computed from

$$\int \mathcal{D}[\vec{q}(\zeta)] \exp\left[\frac{< I_{\mathrm{E}}[\{\vec{q}_j\}_{j=1}^{N}; \vec{q}_f, \vec{q}_i] - I_1 >}{\hbar}\right] \exp\left[\frac{I_1}{\hbar}\right] \\ \sim \exp\left[-\frac{E(\tau_f - \tau_i)}{\hbar}\right], \tag{10.2.12}$$

we know that E exceeds the true E_0,

$$E \geq E_0. \tag{10.2.13}$$

If there are any free parameters in I_1, we choose as the best values those which minimize E.

Since $< I_{\mathrm{E}}[\{\vec{q}_j\}_{j=1}^{N}; \vec{q}_f, \vec{q}_i] - I_1 >$ defined in Eq. (10.2.11) is proportional to $\tau_f - \tau_i$, we write

$$< I_{\mathrm{E}}[\{\vec{q}_j\}_{j=1}^{N}; \vec{q}_f, \vec{q}_i] - I_1 > = s(\tau_f - \tau_i). \tag{10.2.14}$$

The factor $\exp[< I_{\mathrm{E}}[\{\vec{q}_j\}_{j=1}^{N}; \vec{q}_f, \vec{q}_i] - I_1 > /\hbar]$ in Eq. (10.2.12) is constant and can be taken outside the integral. We suppose the lowest energy E_1 for the action functional I_1 is known,

$$\int \mathcal{D}[\vec{q}(\zeta)] \exp\left[\frac{I_1}{\hbar}\right] \sim \exp\left[-\frac{E_1(\tau_f - \tau_i)}{\hbar}\right] \quad \text{as} \quad \tau_f - \tau_i \to \infty. \tag{10.2.15}$$

Then we have

$$E = E_1 - s$$

from Eq. (10.2.12), with s given by Eqs. (10.2.11) and (10.2.14).

If we choose the following *trial action functional*,

$$I_1 = -\frac{1}{2}\int \left(\frac{d\vec{q}}{d\tau}\right)^2 d\tau, \tag{10.2.16}$$

we have what corresponds to the *plane wave Born approximation* in standard perturbation theory. Another choice is

$$I_1 = -\frac{1}{2}\int \left(\frac{d\vec{q}}{d\tau}\right)^2 d\tau + \int V_{\text{trial}}(\vec{q}(\tau))\, d\tau, \tag{10.2.17}$$

where $V_{\text{trial}}(\vec{q}(\tau))$ is a *trial potential* to be chosen. This corresponds to the *distorted wave Born approximation* in standard perturbation theory.

If we choose a *Coulomb potential* as the trial potential,

$$V_{\text{trial}}(R) = \frac{Z}{R}, \tag{10.2.18}$$

we vary the parameter Z.

If we choose a *harmonic potential* as the trial potential,

$$V_{\text{trial}}(R) = \frac{1}{2}kR^2, \tag{10.2.19}$$

we vary the parameter k.

One problem with the trial potential used in Eq. (10.2.17) is that the particle with the coordinate $\vec{q}(\tau)$ is bound to a specific origin. A better choice would be the inter-particle potential of the form $V_{\text{trial}}(\vec{q}(\tau) - \vec{q}(\sigma))$ where an electron at $\vec{q}(\tau)$ is bound to another electron at $\vec{q}(\sigma)$.

10.3 Schwinger–Dyson Equation in Quantum Field Theory

In quantum field theory, we also have the notion of Green's functions, which is quite distinct from the ordinary Green's functions in mathematical physics in one important aspect: the governing equation of motion of the connected part of the two-point "full" Green's function in quantum field theory is the *closed system of the coupled nonlinear integro-differential equations*. We illustrate this point in some detail for the *Yukawa coupling* of the fermion field and the boson field.

We shall establish the relativistic notation. We employ the natural unit system in which

$$\hbar = c = 1. \tag{10.3.1}$$

We define the Minkowski space–time metric tensor $\eta_{\mu\nu}$ by

$$\eta_{\mu\nu} \equiv \text{diag}(1; -1, -1, -1) \equiv \eta^{\mu\nu}, \qquad \mu, \nu = 0, 1, 2, 3. \tag{10.3.2}$$

We define the contravariant components and the covariant components of the space–time coordinates x by

$$x^\mu \equiv (x^0, x^1, x^2, x^3), \tag{10.3.3a}$$

$$x_\mu \equiv \eta_{\mu\nu} x^\nu = (x^0, -x^1, -x^2, -x^3). \tag{10.3.3b}$$

We define the differential operators ∂_μ and ∂^μ by

$$\partial_\mu \equiv \frac{\partial}{\partial x^\mu} = \left(\frac{\partial}{\partial x^0}, \frac{\partial}{\partial x^1}, \frac{\partial}{\partial x^2}, \frac{\partial}{\partial x^3} \right), \qquad \partial^\mu \equiv \frac{\partial}{\partial x_\mu} = \eta^{\mu\nu} \partial_\nu. \tag{10.3.4}$$

We define the four-scalar product by

$$x \cdot y \equiv x^\mu y_\mu = \eta_{\mu\nu} x^\mu y^\nu = x^0 \cdot y^0 - \vec{x} \cdot \vec{y}. \tag{10.3.5}$$

We adopt the convention that the Greek indices μ, ν, ... run over 0, 1, 2 and 3, the Lattin indices i, j, ... run over 1, 2 and 3, and the repeated indices are summed over.

We consider the quantum field theory described by the total Lagrangian density $\mathcal{L}_{\text{tot}}$ of the form,

$$\begin{aligned} \mathcal{L}_{\text{tot}} = {} & \frac{1}{4} [\hat{\bar{\psi}}_\alpha(x), D_{\alpha\beta}(x) \hat{\psi}_\beta(x)] + \frac{1}{4} [D^{\mathrm{T}}_{\beta\alpha}(-x) \hat{\bar{\psi}}_\alpha(x), \hat{\psi}_\beta(x)] \\ & + \frac{1}{2} \hat{\phi}(x) K(x) \hat{\phi}(x) + \mathcal{L}_{\text{int}}(\hat{\phi}(x), \hat{\psi}(x), \hat{\bar{\psi}}(x)), \end{aligned} \tag{10.3.6}$$

where we have

$$D_{\alpha\beta}(x) = (i\gamma_\mu \partial^\mu - m + i\varepsilon)_{\alpha\beta}, \quad D^{\mathrm{T}}_{\beta\alpha}(-x) = (-i\gamma^{\mathrm{T}}_\mu \partial^\mu - m + i\varepsilon)_{\beta\alpha}, \tag{10.3.7a}$$

$$K(x) = -\partial^2 - \kappa^2 + i\varepsilon, \tag{10.3.7b}$$

$$\{\gamma^\mu, \gamma^\nu\} = 2\eta^{\mu\nu}, \quad (\gamma^\mu)^\dagger = \gamma^0 \gamma^\mu \gamma^0, \quad \hat{\bar{\psi}}_\alpha(x) = (\hat{\psi}^\dagger(x) \gamma^0)_\alpha, \tag{10.3.8}$$

$$I_{\text{tot}}[\hat{\phi}, \hat{\psi}, \hat{\bar{\psi}}] = \int d^4x \mathcal{L}_{\text{tot}}((10.3.6)), \quad I_{\text{int}}[\hat{\phi}, \hat{\psi}, \hat{\bar{\psi}}] = \int d^4x \mathcal{L}_{\text{int}}((10.3.6)). \tag{10.3.9}$$

We have Euler–Lagrange equations of motion for the field operators,

$$\hat{\psi}_\alpha(x): \quad \frac{\delta \hat{I}_{\text{tot}}}{\delta \hat{\bar{\psi}}_\alpha(x)} = 0, \quad \text{or} \quad D_{\alpha\beta}(x) \hat{\psi}_\beta(x) + \frac{\delta \hat{I}_{\text{int}}}{\delta \hat{\bar{\psi}}_\alpha(x)} = 0, \tag{10.3.10a}$$

$$\hat{\bar{\psi}}_\beta(x): \quad \frac{\delta \hat{I}_{\text{tot}}}{\delta \hat{\psi}_\beta(x)} = 0, \quad \text{or} \quad -D^{\mathrm{T}}_{\beta\alpha}(-x) \hat{\bar{\psi}}_\alpha(x) + \frac{\delta \hat{I}_{\text{int}}}{\delta \hat{\psi}_\beta(x)} = 0, \tag{10.3.10b}$$

$$\hat{\phi}(x): \quad \frac{\delta I_{\text{tot}}}{\delta \hat{\phi}(x)} = 0, \quad \text{or} \quad K(x) \hat{\phi}(x) + \frac{\delta \hat{I}_{\text{int}}}{\delta \hat{\phi}(x)} = 0. \tag{10.3.10c}$$

We have the equal time canonical (anti-)commutators,

$$\delta(x^0 - y^0)\{\hat{\psi}_\beta(x), \hat{\bar{\psi}}_\alpha(x)\} = \gamma^0_{\beta\alpha}\delta^4(x-y), \tag{10.3.11a}$$

$$\delta(x^0 - y^0)\{\hat{\psi}_\beta(x), \hat{\psi}_\alpha(y)\} = \delta(x^0 - y^0)\{\hat{\bar{\psi}}_\beta(x), \hat{\bar{\psi}}_\alpha(x)\} = 0, \tag{10.3.11b}$$

$$\delta(x^0 - y^0)[\hat{\phi}(x), \partial_0^y\hat{\phi}(y)] = i\delta^4(x-y), \tag{10.3.11c}$$

$$\delta(x^0 - y^0)[\hat{\phi}(x), \hat{\phi}(y)] = \delta(x^0 - y^0)[\partial_0^x\hat{\phi}(x), \partial_0^y\hat{\phi}(y)] = 0, \tag{10.3.11d}$$

and the rest of the equal time mixed canonical commutators are equal to 0. We define the *generating functional of the "full" Green's functions* by

$$\begin{aligned}
Z[J,\bar{\eta},\eta] &\equiv <0,\text{out}|\,\mathrm{T}(\exp\left[i\{(J\hat{\phi}) + (\bar{\eta}\hat{\psi}) + (\hat{\bar{\psi}}\eta)\}\right])\,|0,\text{in}> \\
&\equiv \sum_{l,m,n=0}^{\infty} \frac{i^{l+m+n}}{l!m!n!} <0,\text{out}|\,\mathrm{T}((\bar{\eta}\hat{\psi})^l (J\hat{\phi})^m (\hat{\bar{\psi}}\eta)^n)\,|0,\text{in}> \\
&\equiv \sum_{l,m,n=0}^{\infty} \frac{i^{l+m+n}}{l!m!n!} J(y_1)\cdot\ldots\cdot J(y_m)\bar{\eta}_{\alpha_l}(x_l)\cdot\ldots\cdot\bar{\eta}_{\alpha_1}(x_1) \\
&\quad \times <0,\text{out}|\,\mathrm{T}\{\hat{\psi}_{\alpha_1}(x_1)\cdot\ldots\cdot\hat{\psi}_{\alpha_l}(x_l)\hat{\phi}(y_1)\cdot\cdot\hat{\phi}(y_m) \\
&\quad \times \hat{\bar{\psi}}_{\beta_1}(z_1)\cdot\ldots\cdot\hat{\bar{\psi}}_{\beta_n}(z_n)\}\,|0,\text{in}>\eta_{\beta_n}(z_n)\cdot\ldots\cdot\eta_{\beta_1}(z_1) \\
&\equiv \sum_{n=0}^{\infty}\frac{i^n}{n!} <0,\text{out}|\,\mathrm{T}(\bar{\eta}\hat{\psi} + J\hat{\phi} + \hat{\bar{\psi}}\eta)^n\,|0,\text{in}>\,,
\end{aligned} \tag{10.3.12}$$

where the repeated continuous space–time indices x_1 through z_n are to be integrated over, and we introduced the abbreviations in Eq. (10.3.12),

$$J\hat{\phi} \equiv \int d^4y J(y)\hat{\phi}(y), \quad \bar{\eta}\hat{\psi} \equiv \int d^4x\bar{\eta}(x)\hat{\psi}(x), \quad \hat{\bar{\psi}}\eta \equiv \int d^4z\hat{\bar{\psi}}(z)\eta(z).$$

The *time-ordered product* is defined by

$$\begin{aligned}
&\mathrm{T}\{\hat{\Psi}(x_1)\cdot\ldots\cdot\hat{\Psi}(x_n)\} \\
&\quad \equiv \sum_{\substack{\text{all possible} \\ \text{permutations } P}} \delta_P\theta(x^0_{P1} - x^0_{P2})\cdot\ldots\cdot\theta(x^0_{P(n-1)} - x^0_{Pn})\hat{\Psi}(x_{P1})\cdot\ldots\cdot\hat{\Psi}(x_{Pn}),
\end{aligned}$$

with

$$\delta_P = \begin{cases} 1 & P \text{ even and odd} \quad \text{for a } \textbf{Boson}, \\ 1 & P \text{ even} \quad \text{for a } \textbf{Fermion}, \\ -1 & P \text{ odd} \quad \text{for a } \textbf{Fermion}. \end{cases}$$

We observe that

$$\frac{\delta}{\delta\bar{\eta}_\beta(x)}(\bar{\eta}\hat{\psi})^l = l\hat{\psi}_\beta(x)(\bar{\eta}\hat{\psi})^{l-1}, \tag{10.3.13a}$$

$$\frac{\delta}{\delta\eta_\alpha(x)}(\hat{\bar{\psi}}\eta)^n = -n\hat{\bar{\psi}}_\alpha(x)(\hat{\bar{\psi}}\eta)^{n-1}, \tag{10.3.13b}$$

$$\frac{\delta}{\delta J(x)}(J\hat{\phi})^m = m\hat{\phi}(x)(J\hat{\phi})^{m-1}. \tag{10.3.13c}$$

We also observe from the definition of $Z[J,\bar{\eta},\eta]$,

$$\frac{1}{i}\frac{\delta}{\delta\bar{\eta}_\beta(x)}Z[J,\bar{\eta},\eta] = \left\langle 0,\text{out}\middle|\text{T}\left(\hat{\psi}_\beta(x)\exp\left[i\int d^4z\{J(z)\hat{\phi}(z) + \bar{\eta}_\alpha(z)\hat{\psi}_\alpha(z) + \hat{\bar{\psi}}_\beta(z)\eta_\beta(z)\}\right]\right)\middle|0,\text{in}\right\rangle, \tag{10.3.14a}$$

$$i\frac{\delta}{\delta\eta_\alpha(x)}Z[J,\bar{\eta},\eta] = \left\langle 0,\text{out}\middle|\text{T}\left(\hat{\bar{\psi}}_\alpha(x)\exp\left[i\int d^4z\{J(z)\hat{\phi}(z) + \bar{\eta}_\alpha(z)\hat{\psi}_\alpha(z) + \hat{\bar{\psi}}_\beta(z)\eta_\beta(z)\}\right]\right)\middle|0,\text{in}\right\rangle, \tag{10.3.14b}$$

$$\frac{1}{i}\frac{\delta}{\delta J(x)}Z[J,\bar{\eta},\eta] = \left\langle 0,\text{out}\middle|\text{T}\left(\hat{\phi}(x)\exp\left[i\int d^4z\{J(z)\hat{\phi}(z) + \bar{\eta}_\alpha(z)\hat{\psi}_\alpha(z) + \hat{\bar{\psi}}_\beta(z)\eta_\beta(z)\}\right]\right)\middle|0,\text{in}\right\rangle. \tag{10.3.14c}$$

From the definition of the time-ordered product and the equal time canonical (anti-)commutators, Eqs. (10.3.11a through d), we have at the operator level,

Fermion:

$$\begin{aligned} &D_{\alpha\beta}(x)\text{T}\left(\hat{\psi}_\beta(x)\exp\left[i\{J\hat{\phi} + \bar{\eta}_\alpha\hat{\psi}_\alpha + \hat{\bar{\psi}}_\beta\eta_\beta\}\right]\right) \\ &= \text{T}\left(D_{\alpha\beta}(x)\hat{\psi}_\beta(x)\exp\left[i\{J\hat{\phi} + \bar{\eta}\hat{\psi} + \hat{\bar{\psi}}\eta\}\right]\right) \\ &\qquad - \eta_\alpha(x)\text{T}\left(\exp\left[i\{J\hat{\phi} + \bar{\eta}\hat{\psi} + \hat{\bar{\psi}}\eta\}\right]\right), \end{aligned} \tag{10.3.15}$$

Anti-Fermion:

$$\begin{aligned} &-D^{\text{T}}_{\beta\alpha}(-x)\text{T}\left(\hat{\bar{\psi}}_\alpha(x)\exp\left[i\{J\hat{\phi} + \bar{\eta}_\alpha\hat{\psi}_\alpha + \hat{\bar{\psi}}_\beta\eta_\beta\}\right]\right) \\ &= \text{T}\left(-D^{\text{T}}_{\beta\alpha}(-x)\hat{\bar{\psi}}_\alpha(x)\exp\left[i\{J\hat{\phi} + \bar{\eta}\hat{\psi} + \hat{\bar{\psi}}\eta\}\right]\right) \\ &\qquad + \bar{\eta}_\beta(x)\text{T}\left(\exp\left[i\{J\hat{\phi} + \bar{\eta}\hat{\psi} + \hat{\bar{\psi}}\eta\}\right]\right), \end{aligned} \tag{10.3.16}$$

Boson:

$$\begin{aligned}&K(x)\mathrm{T}\left(\hat{\phi}(x)\exp\left[i\{J\hat{\phi}+\bar{\eta}_\alpha\hat{\psi}_\alpha+\widehat{\bar{\psi}}_\beta\eta_\beta\}\right]\right)\\&=\mathrm{T}\left(K(x)\hat{\phi}(x)\exp\left[i\{J\hat{\phi}+\bar{\eta}\hat{\psi}+\widehat{\bar{\psi}}\eta\}\right]\right)\\&\qquad-J(x)\mathrm{T}\left(\exp\left[i\{J\hat{\phi}+\bar{\eta}\hat{\psi}+\widehat{\bar{\psi}}\eta\}\right]\right).\end{aligned}\quad(10.3.17)$$

Applying Euler–Lagrange equations of motion, Eqs. (10.3.10a through c), to the first terms on the right-hand sides of Eqs. (10.3.15), (10.3.16) and (10.3.17), and taking the vacuum expectation values, we obtain the equations of motion of the generating functional $Z[J,\bar{\eta},\eta]$ of the "full" Green's functions as

$$\left\{D_{\alpha\beta}(x)\frac{1}{i}\frac{\delta}{\delta\bar{\eta}_\beta(x)}+\frac{\delta I_{\text{int}}[\frac{1}{i}\frac{\delta}{\delta J},\frac{1}{i}\frac{\delta}{\delta\bar{\eta}},i\frac{\delta}{\delta\eta}]}{\delta\left(i\frac{\delta}{\delta\eta_\alpha(x)}\right)}+\eta_\alpha(x)\right\}Z[J,\bar{\eta},\eta]=0,\quad(10.3.18a)$$

$$\left\{-D^{\mathrm{T}}_{\beta\alpha}(-x)i\frac{\delta}{\delta\eta_\alpha(x)}+\frac{\delta I_{\text{int}}[\frac{1}{i}\frac{\delta}{\delta J},\frac{1}{i}\frac{\delta}{\delta\bar{\eta}},i\frac{\delta}{\delta\eta}]}{\delta\left(\frac{1}{i}\frac{\delta}{\delta\bar{\eta}_\beta(x)}\right)}-\bar{\eta}_\beta(x)\right\}Z[J,\bar{\eta},\eta]=0,\quad(10.3.18b)$$

$$\left\{K(x)\frac{1}{i}\frac{\delta}{\delta J(x)}+\frac{\delta I_{\text{int}}[\frac{1}{i}\frac{\delta}{\delta J},\frac{1}{i}\frac{\delta}{\delta\bar{\eta}},i\frac{\delta}{\delta\eta}]}{\delta\left(\frac{1}{i}\frac{\delta}{\delta J(x)}\right)}+J(x)\right\}Z[J,\bar{\eta},\eta]=0.\quad(10.3.18c)$$

Equivalently, from Eqs. (10.3.10a), (10.3.10b) and (10.3.10c), we can write Eqs. (10.3.18a), (10.3.18b) and (10.3.18c) as

$$\left\{\frac{\delta I_{\text{tot}}[\frac{1}{i}\frac{\delta}{\delta J},\frac{1}{i}\frac{\delta}{\delta\bar{\eta}},i\frac{\delta}{\delta\eta}]}{\delta\left(i\frac{\delta}{\delta\eta_\alpha(x)}\right)}+\eta_\alpha(x)\right\}Z[J,\bar{\eta},\eta]=0,\quad(10.3.19a)$$

$$\left\{\frac{\delta I_{\text{tot}}[\frac{1}{i}\frac{\delta}{\delta J},\frac{1}{i}\frac{\delta}{\delta\bar{\eta}},i\frac{\delta}{\delta\eta}]}{\delta\left(\frac{1}{i}\frac{\delta}{\delta\bar{\eta}_\beta(x)}\right)}-\bar{\eta}_\beta(x)\right\}Z[J,\bar{\eta},\eta]=0,\quad(10.3.19b)$$

$$\left\{\frac{\delta I_{\text{tot}}[\frac{1}{i}\frac{\delta}{\delta J},\frac{1}{i}\frac{\delta}{\delta\bar{\eta}},i\frac{\delta}{\delta\eta}]}{\delta\left(\frac{1}{i}\frac{\delta}{\delta J(x)}\right)}+J(x)\right\}Z[J,\bar{\eta},\eta]=0.\quad(10.3.19c)$$

We note that the coefficients of the external hook terms, $\eta_\alpha(x)$, $\bar{\eta}_\beta(x)$ and $J(x)$, in Eqs. (10.3.19a through c) are ± 1, which is a reflection of the fact that we are dealing with canonical quantum field theory and originates from the equal time canonical (anti-)commutators.

With this preparation, we shall discuss the Schwinger theory of Green's function with the interaction Lagrangian density $\mathcal{L}_{\text{int}}(\hat{\phi}(x),\hat{\psi}(x),\widehat{\bar{\psi}}(x))$ of the Yukawa coupling in mind,

$$\mathcal{L}_{\text{int}}(\hat{\phi}(x),\hat{\psi}(x),\widehat{\bar{\psi}}(x))=-G_0\widehat{\bar{\psi}}_\alpha(x)\gamma_{\alpha\beta}(x)\hat{\psi}_\beta(x)\hat{\phi}(x),\quad(10.3.20)$$

with

$$G_0 = \begin{cases} g_0 \\ f \\ e \end{cases}, \quad \gamma(x) = \begin{cases} \gamma_5 \\ \gamma_5 \tau_i \\ \gamma_\mu \end{cases}, \quad \hat{\psi}(x) = \begin{cases} \hat{\psi}_\alpha(x) \\ \hat{\psi}_{\mathrm{N},\alpha}(x) \\ \hat{\psi}_\alpha(x) \end{cases}, \quad \phi(x) = \begin{cases} \hat{\phi}(x) \\ \hat{\phi}_i(x) \\ \hat{A}_\mu(x) \end{cases}. \tag{10.3.21}$$

We define the *vacuum expectation values, $< F >^{J,\bar{\eta},\eta}$ and $< F >^{J}$, of the operator function $F(\hat{\phi}(x), \hat{\psi}(x), \hat{\bar{\psi}}(x))$ in the presence of the external hook terms $\{J, \bar{\eta}, \eta\}$* by

$$< F >^{J,\bar{\eta},\eta} \equiv \frac{1}{Z[J,\bar{\eta},\eta]} F\left(\frac{1}{i}\frac{\delta}{\delta J(x)}, \frac{1}{i}\frac{\delta}{\delta \bar{\eta}(x)}, i\frac{\delta}{\delta \eta(x)}\right) Z[J,\bar{\eta},\eta], \tag{10.3.22a}$$

$$< F >^{J} = \left. < F >^{J,\bar{\eta},\eta} \right|_{\bar{\eta}=\eta=0}. \tag{10.3.22b}$$

We define the connected parts of the two-point "full" Green's functions in the presence of the external hook $J(x)$ by

Fermion:

$$\begin{aligned} S'^{J}_{\mathrm{F},\alpha\beta}(x_1, x_2) &\equiv \left. \frac{1}{i}\frac{\delta}{\delta \bar{\eta}_\alpha(x_1)} i \frac{\delta}{\delta \eta_\beta(x_2)} \frac{1}{i} \ln Z[J,\bar{\eta},\eta] \right|_{\bar{\eta}=\eta=0} \\ &= \frac{1}{i}\left(\frac{1}{i}\frac{\delta}{\delta \bar{\eta}_\alpha(x_1)}\right) \left. < \hat{\bar{\psi}}_\beta(x_2) >^{J,\bar{\eta},\eta} \right|_{\bar{\eta}=\eta=0} \\ &= \frac{1}{i}\Big\{ \left. < \hat{\psi}_\alpha(x_1)\hat{\bar{\psi}}_\beta(x_2) >^{J,\bar{\eta},\eta} \right|_{\bar{\eta}=\eta=0} \\ &\quad - \left. < \hat{\psi}_\alpha(x_1) >^{J,\bar{\eta},\eta} < \hat{\bar{\psi}}_\beta(x_2) >^{J,\bar{\eta},\eta} \right|_{\bar{\eta}=\eta=0} \Big\} \\ &= \frac{1}{i} < \hat{\psi}_\alpha(x_1)\hat{\bar{\psi}}_\beta(x_2) >^{J} \\ &\equiv \frac{1}{i} < 0, \mathrm{out}|\, \mathrm{T}(\hat{\psi}_\alpha(x_1)\hat{\bar{\psi}}_\beta(x_2))\, |0, \mathrm{in} >^{J}_{\mathrm{C}}, \end{aligned} \tag{10.3.23}$$

$$\left. < \hat{\psi}_\alpha(x_1) >^{J,\bar{\eta},\eta} \right|_{\bar{\eta}=\eta=0} = \left. < \hat{\bar{\psi}}_\beta(x_2) >^{J,\bar{\eta},\eta} \right|_{\bar{\eta}=\eta=0} = 0, \tag{10.3.24}$$

and

Boson:

$$\begin{aligned} D'^{J}_{\mathrm{F}}(x_1, x_2) &\equiv \left. \frac{1}{i}\frac{\delta}{\delta J(x_1)} \frac{1}{i}\frac{\delta}{\delta J(x_2)} \frac{1}{i} \ln Z[J,\bar{\eta},\eta] \right|_{\bar{\eta}=\eta=0} \\ &= \frac{1}{i}\left(\frac{1}{i}\frac{\delta}{\delta J(x_1)} < \hat{\phi}(x_2) >^{J}\right) \\ &= \frac{1}{i}\left\{ < \hat{\phi}(x_1)\hat{\phi}(x_2) >^{J} - < \hat{\phi}(x_1) >^{J} < \hat{\phi}(x_2) >^{J} \right\} \\ &\equiv \frac{1}{i} < 0, \mathrm{out}|\, \mathrm{T}(\hat{\phi}(x_1)\hat{\phi}(x_2))\, |0, \mathrm{in} >^{J}_{\mathrm{C}}, \end{aligned} \tag{10.3.25}$$

$$\left. < \hat{\phi}(x) >^{J} \right|_{J=0} = 0. \tag{10.3.26}$$

We have the equations of motion of $Z[J,\bar{\eta},\eta]$, Eqs. (10.3.18a through c), when the interaction Lagrangian density $\mathcal{L}_{\text{int}}(\hat{\phi}(x),\hat{\psi}(x),\hat{\bar{\psi}}(x))$ is given by Eq. (10.3.20) as

$$\left\{D_{\alpha\beta}(x)\left(\frac{1}{i}\frac{\delta}{\delta\bar{\eta}_{\beta}(x)}\right)-G_{0}\gamma_{\alpha\beta}(x)\left(\frac{1}{i}\frac{\delta}{\delta\bar{\eta}_{\beta}(x)}\right)\left(\frac{1}{i}\frac{\delta}{\delta J(x)}\right)\right\}Z[J,\bar{\eta},\eta] \\ =-\eta_{\alpha}(x)Z[J,\bar{\eta},\eta], \quad (10.3.27a)$$

$$\left\{-D^{\mathrm{T}}_{\beta\alpha}(-x)\left(i\frac{\delta}{\delta\eta_{\alpha}(x)}\right)+G_{0}\left(i\frac{\delta}{\delta\eta_{\alpha}(x)}\right)\gamma_{\alpha\beta}(x)\left(\frac{1}{i}\frac{\delta}{\delta J(x)}\right)\right\}Z[J,\bar{\eta},\eta] \\ =+\bar{\eta}_{\beta}(x)Z[J,\bar{\eta},\eta], \quad (10.3.27b)$$

$$\left\{K(x)\left(\frac{1}{i}\frac{\delta}{\delta J(x)}\right)-G_{0}\left(i\frac{\delta}{\delta\eta_{\alpha}(x)}\right)\gamma_{\alpha\beta}(x)\left(\frac{1}{i}\frac{\delta}{\delta\bar{\eta}_{\beta}(x)}\right)\right\}Z[J,\bar{\eta},\eta] \\ =-J(x)Z[J,\bar{\eta},\eta]. \quad (10.3.27c)$$

Dividing Eqs. (10.3.27a through c) by $Z[J,\bar{\eta},\eta]$, and referring to Eqs. (10.3.22a) and (10.3.22b), we obtain the equations of motion for

$$<\hat{\psi}_{\beta}(x)>^{J,\bar{\eta},\eta},\quad <\hat{\bar{\psi}}_{\alpha}(x)>^{J,\bar{\eta},\eta}\ \text{and}\ <\hat{\phi}(x)>^{J,\bar{\eta},\eta},$$

as

$$D_{\alpha\beta}(x)<\hat{\psi}_{\beta}(x)>^{J,\bar{\eta},\eta}-G_{0}\gamma_{\alpha\beta}(x)<\hat{\psi}_{\beta}(x)\hat{\phi}(x)>^{J,\bar{\eta},\eta}=-\eta_{\alpha}(x), \quad (10.3.28)$$

$$-D^{\mathrm{T}}_{\beta\alpha}(-x)<\hat{\bar{\psi}}_{\alpha}(x)>+G_{0}\gamma_{\alpha\beta}(x)<\hat{\bar{\psi}}_{\alpha}(x)\hat{\phi}(x)>^{J,\bar{\eta},\eta}=+\bar{\eta}_{\beta}(x), \quad (10.3.29)$$

$$K(x)<\hat{\phi}(x)>^{J,\bar{\eta},\eta}-G_{0}\gamma_{\alpha\beta}(x)<\hat{\bar{\psi}}_{\alpha}(x)\hat{\psi}_{\beta}(x)>^{J,\bar{\eta},\eta}=-J(x). \quad (10.3.30)$$

We take the following functional derivatives,

$$i\frac{\delta}{\delta\eta_{\varepsilon}(y)}\text{Eq. (10.3.28)}\Big|_{\bar{\eta}=\eta=0}:$$

$$D_{\alpha\beta}(x)<\hat{\bar{\psi}}_{\varepsilon}(y)\hat{\psi}_{\beta}(x)>^{J} \\ -G_{0}\gamma_{\alpha\beta}(x)<\hat{\bar{\psi}}_{\varepsilon}(y)\hat{\psi}_{\beta}(x)\hat{\phi}(x)>^{J}=-i\delta_{\alpha\varepsilon}\delta^{4}(x-y),$$

$$\frac{1}{i}\frac{\delta}{\delta\bar{\eta}_{\varepsilon}(y)}\text{Eq. (10.3.29)}\Big|_{\bar{\eta}=\eta=0}:$$

$$-D^{\mathrm{T}}_{\beta\alpha}(-x)<\hat{\psi}_{\varepsilon}(y)\hat{\bar{\psi}}_{\alpha}(x)>^{J} \\ +G_{0}\gamma_{\alpha\beta}(x)<\hat{\psi}_{\varepsilon}(y)\hat{\bar{\psi}}_{\alpha}(x)\hat{\phi}(x)>^{J}=-i\delta_{\beta\varepsilon}\delta^{4}(x-y),$$

$$\frac{1}{i}\frac{\delta}{\delta J(y)}\text{Eq. (10.3.30)}\bigg|_{\bar{\eta}=\eta=0}:$$

$$K(x)\{<\hat{\phi}(y)\hat{\phi}(x)>^J - <\phi(y)>^J<\phi(x)>^J\} \\ - G_0\gamma_{\alpha\beta}(x)\frac{1}{i}\frac{\delta}{\delta J(y)} <\hat{\bar{\psi}}_\alpha(x)\hat{\psi}_\beta(x)>^J = i\delta^4(x-y).$$

These equations are a part of the infinite system of coupled equations. We observe the following identities,

$$<\hat{\bar{\psi}}_\varepsilon(y)\hat{\psi}_\beta(x)>^J = -iS'^J_{\mathrm{F},\beta\varepsilon}(x,y),$$

$$<\hat{\psi}_\varepsilon(y)\hat{\bar{\psi}}_\alpha(x)>^J = iS'^J_{\mathrm{F},\varepsilon\alpha}(y,x),$$

$$<\hat{\bar{\psi}}_\varepsilon(y)\hat{\psi}_\beta(x)\hat{\phi}(x)>^J = -i\left(<\hat{\phi}(x)>^J + \frac{1}{i}\frac{\delta}{\delta J(x)}\right)S'^J_{\mathrm{F},\beta\varepsilon}(x,y),$$

$$<\hat{\psi}_\varepsilon(y)\hat{\bar{\psi}}_\alpha(x)\hat{\phi}(x)>^J = i\left(<\hat{\phi}(x)>^J + \frac{1}{i}\frac{\delta}{\delta J(x)}\right)S'^J_{\mathrm{F},\varepsilon\alpha}(y,x).$$

With these identities, we obtain the equations of motion of the connected parts of the two-point "full" Green's functions in the presence of the external hook $J(x)$,

$$\left\{D_{\alpha\beta}(x) - G_0\gamma_{\alpha\beta}(x)\left(<\hat{\phi}(x)>^J + \frac{1}{i}\frac{\delta}{\delta J(x)}\right)\right\}S'^J_{\mathrm{F},\beta\varepsilon}(x,y) \\ = \delta_{\alpha\varepsilon}\delta^4(x-y), \quad (10.3.31\text{a})$$

$$\left\{D^{\mathrm{T}}_{\beta\alpha}(-x) - G_0\gamma_{\alpha\beta}(x)\left(<\hat{\phi}(x)>^J + \frac{1}{i}\frac{\delta}{\delta J(x)}\right)\right\}S'^J_{\mathrm{F},\varepsilon\alpha}(y,x) \\ = \delta_{\beta\varepsilon}\delta^4(x-y), \quad (10.3.32\text{a})$$

$$K(x)D'^J_{\mathrm{F}}(x,y) + G_0\gamma_{\alpha\beta}(x)\frac{1}{i}\frac{\delta}{\delta J(y)}S'^J_{\mathrm{F},\beta\alpha}(x,x_\pm) = \delta^4(x-y). \quad (10.3.33\text{a})$$

Since the transpose of Eq. (10.3.32a) is Eq. (10.3.31a), we have to consider only Eqs. (10.3.31a) and (10.3.33a). We may get the impression that we have the equations of motion of the two-point "full" Green's functions, $S'^J_{\mathrm{F},\alpha\beta}(x,y)$ and $D'^J_{\mathrm{F}}(x,y)$, in closed form, at first sight. Because of the presence of the functional derivatives $\delta/i\delta J(x)$ and $\delta/i\delta J(y)$, however, Eqs. (10.3.31a), (10.3.32a) and (10.3.33a) involve the three-point "full" Green's functions and are merely a part of the infinite system of the coupled nonlinear equations of motion of the "full" Green's functions.

From this point onward, we use the variables, "1", "2", "3", ... to represent the continuous space–time indices, x, y, z, ..., the spinor indices, α, β, γ, ..., as well as other internal indices, i, j, k, ...

Using the "free" Green's functions, $S_0^{\mathrm{F}}(1-2)$ and $D_0^{\mathrm{F}}(1-2)$, defined by

$$D(1)S_0^{\mathrm{F}}(1-2) = 1, \tag{10.3.34a}$$

$$K(1)D_0^{\mathrm{F}}(1-2) = 1, \tag{10.3.34b}$$

we rewrite the functional differential equations satisfied by the "full" Green's functions, $S_{\mathrm{F}}^{'J}(1,2)$ and $D_{\mathrm{F}}^{'J}(1,2)$, Eqs. (10.3.31a) and (10.3.33a), into the integral equations,

$$S_{\mathrm{F}}^{'J}(1,2) = S_0^{\mathrm{F}}(1-2) + S_0^{\mathrm{F}}(1-3)(G_0\gamma(3))\left(<\hat{\phi}(3)>^J + \frac{1}{i}\frac{\delta}{\delta J(3)}\right)S_{\mathrm{F}}^{'J}(3,2), \tag{10.3.31b}$$

$$D_{\mathrm{F}}^{'J}(1,2) = D_0^{\mathrm{F}}(1-2) + D_0^{\mathrm{F}}(1-3)\left(-G_0\mathrm{tr}\gamma(3)\frac{1}{i}\frac{\delta}{\delta J(2)}S_{\mathrm{F}}^{'J}(3,3_\pm)\right). \tag{10.3.33b}$$

We compare Eqs. (10.3.31b) and (10.3.33b) with the defining integral equations of the *proper self-energy parts*, $\boldsymbol{\Sigma}^*$ and $\boldsymbol{\Pi}^*$, due to Dyson, in the presence of the external hook $J(x)$,

$$S_{\mathrm{F}}^{'J}(1,2) = S_0^{\mathrm{F}}(1-2) + S_0^{\mathrm{F}}(1-3)(G_0\gamma(3) < \phi(3) >^J)S_{\mathrm{F}}^{'J}(3,2) + S_0^{\mathrm{F}}(1-3)\boldsymbol{\Sigma}^*(3,4)S_{\mathrm{F}}^{'J}(4,2), \tag{10.3.35}$$

$$D_{\mathrm{F}}^{'J}(1,2) = D_0^{\mathrm{F}}(1-2) + D_0^{\mathrm{F}}(1-3)\boldsymbol{\Pi}^*(3,4)D_{\mathrm{F}}^{'J}(4,2), \tag{10.3.36}$$

obtaining

$$G_0\gamma(1)\frac{1}{i}\frac{\delta}{\delta J(1)}S_{\mathrm{F}}^{'J}(1,2) = \boldsymbol{\Sigma}^*(1,3)S_{\mathrm{F}}^{'J}(3,2) \equiv \boldsymbol{\Sigma}^*(1)S_{\mathrm{F}}^{'J}(1,2), \tag{10.3.37}$$

$$-G_0\mathrm{tr}\gamma(1)\frac{1}{i}\frac{\delta}{\delta J(2)}S_{\mathrm{F}}^{'J}(1,1_\pm) = \boldsymbol{\Pi}^*(1,3)D_{\mathrm{F}}^{'J}(3,2) \equiv \boldsymbol{\Pi}^*(1)D_{\mathrm{F}}^{'J}(1,2). \tag{10.3.38}$$

Thus we can write the functional differential equations, Eqs. (10.3.31a) and (10.3.33a), compactly as

$$\{D(1) - G_0\gamma(1) < \hat{\phi}(1) >^J - \boldsymbol{\Sigma}^*(1)\}S_{\mathrm{F}}^{'J}(1,2) = \delta(1-2), \tag{10.3.39}$$

$$\{K(1) - \boldsymbol{\Pi}^*(1)\}D_{\mathrm{F}}^{'J}(1,2) = \delta(1-2). \tag{10.3.40}$$

Defining the *"Nucleon" differential operator* and *"Meson" differential operator* by

$$D_{\mathrm{N}}(1,2) \equiv \{D(1) - G_0\gamma(1) < \hat{\phi}(1) >^J\}\delta(1-2) - \boldsymbol{\Sigma}^*(1,2), \tag{10.3.41}$$

and

$$D_{\mathrm{M}}(1,2) \equiv K(1)\delta(1-2) - \boldsymbol{\Pi}^*(1,2), \tag{10.3.42}$$

we can write the differential equations, Eqs. (10.3.39) and (10.3.40), as

$$D_{\mathrm{N}}(1,3)S_{\mathrm{F}}^{'J}(3,2)=\delta(1-2), \quad \text{or} \quad D_{\mathrm{N}}(1,2)=(S_{\mathrm{F}}^{'J}(1,2))^{-1}, \tag{10.3.43}$$

and

$$D_{\mathrm{M}}(1,3)D_{\mathrm{F}}^{'J}(3,2)=\delta(1-2), \quad \text{or} \quad D_{\mathrm{M}}(1,2)=(D_{\mathrm{F}}^{'J}(1,2))^{-1}. \tag{10.3.44}$$

Next, we take the functional derivative of Eq. (10.3.39),
$\frac{1}{i}\frac{\delta}{\delta J(3)}$ Eq. (10.3.39):

$$\begin{aligned}\{D(1)-G_0\gamma(1)<\hat{\phi}(1)>^J-\mathbf{\Sigma}^*(1)\}\frac{1}{i}\frac{\delta}{\delta J(3)}S_{\mathrm{F}}^{'J}(1,2)\\=\left\{iG_0\gamma(1)D_{\mathrm{F}}^{'J}(1,3)+\frac{1}{i}\frac{\delta}{\delta J(3)}\mathbf{\Sigma}^*(1)\right\}S_{\mathrm{F}}^{'J}(1,2).\end{aligned} \tag{10.3.45}$$

Solving Eq. (10.3.45) for $\delta S_{\mathrm{F}}^{'J}(1,2)/i\delta J(3)$ and using Eqs. (10.3.39), (10.3.41) and (10.3.43), we obtain

$$\begin{aligned}&\frac{1}{i}\frac{\delta}{\delta J(3)}S_{\mathrm{F}}^{'J}(1,2)=S_{\mathrm{F}}^{'J}(1,4)\left\{iG_0\gamma(4)D_{\mathrm{F}}^{'J}(4,3)+\frac{1}{i}\frac{\delta}{\delta J(3)}\mathbf{\Sigma}^*(4)\right\}S_{\mathrm{F}}^{'J}(4,2)\\&=iG_0S_{\mathrm{F}}^{'J}(1,4)\{\gamma(4)\delta(4-5)\delta(4-6)\\&\qquad+\frac{1}{G_0}\frac{\delta}{\delta<\hat{\phi}(6)>^J}\mathbf{\Sigma}^*(4,5)\}S_{\mathrm{F}}^{'J}(5,2)D_{\mathrm{F}}^{'J}(6,3).\end{aligned} \tag{10.3.46}$$

Comparing Eq. (10.3.46) with the definition of the *vertex operator* $\mathbf{\Gamma}(4,5;6)$ of Dyson,

$$\frac{1}{i}\frac{\delta}{\delta J(3)}S_{\mathrm{F}}^{'J}(1,2)\equiv iG_0S_{\mathrm{F}}^{'J}(1,4)\mathbf{\Gamma}(4,5;6)S_{\mathrm{F}}^{'J}(5,2)D_{\mathrm{F}}^{'J}(6,3), \tag{10.3.47}$$

we obtain

$$\mathbf{\Gamma}(1,2;3)=\gamma(1)\delta(1-2)\delta(1-3)+\frac{1}{G_0}\frac{\delta}{\delta<\hat{\phi}(3)>^J}\mathbf{\Sigma}^*(1,2), \tag{10.3.48}$$

while we can write the left-hand side of Eq. (10.3.47) as

$$\frac{1}{i}\frac{\delta}{\delta J(3)}S_{\mathrm{F}}^{'J}(1,2)=iD_{\mathrm{F}}^{'J}(6,3)\frac{\delta}{\delta<\hat{\phi}(6)>^J}S_{\mathrm{F}}^{'J}(1,2). \tag{10.3.49}$$

From this, we have

$$-\frac{1}{G_0}\frac{\delta}{\delta<\hat{\phi}(6)>^J}S_{\mathrm{F}}^{'J}(1,2)=-S_{\mathrm{F}}^{'J}(1,4)\mathbf{\Gamma}(4,5;6)S_{\mathrm{F}}^{'J}(5,2),$$

and we obtain the compact representation of $\mathbf{\Gamma}(1,2;3)$,

$$\mathbf{\Gamma}(1,2;3) = -\frac{1}{G_0}\frac{\delta}{\delta < \hat{\phi}(3) >^J}(S_{\mathrm{F}}^{\prime J}(1,2))^{-1} = -\frac{1}{G_0}\frac{\delta}{\delta < \hat{\phi}(3) >^J}D_{\mathrm{N}}(1,2) = (10.3.48). \quad (10.3.50)$$

Lastly, from Eqs. (10.3.38) and (10.3.39), which define $\mathbf{\Sigma}^*(1,2)$ and $\mathbf{\Pi}^*(1,2)$ indirectly, and the defining equation of $\mathbf{\Gamma}(1,2;3)$, Eq. (10.3.47), we have

$$\mathbf{\Sigma}^*(1,3)S_{\mathrm{F}}^{\prime J}(3,2) = -iG_0^2\gamma(1)S_{\mathrm{F}}^{\prime J}(1,4)\mathbf{\Gamma}(4,5;6)S_{\mathrm{F}}^{\prime J}(5,2)D_{\mathrm{F}}^{\prime J}(6,1), \quad (10.3.51)$$

$$\mathbf{\Pi}^*(1,3)D_{\mathrm{F}}^{\prime J}(3,2) = iG_0^2\mathrm{tr}\gamma(1)S_{\mathrm{F}}^{\prime J}(1,4)\mathbf{\Gamma}(4,5;6)S_{\mathrm{F}}^{\prime J}(5,1)D_{\mathrm{F}}^{\prime J}(6,2). \quad (10.3.52)$$

Namely, we obtain

$$\mathbf{\Sigma}^*(1,2) = -iG_0^2\gamma(1)S_{\mathrm{F}}^{\prime J}(1,3)\mathbf{\Gamma}(3,2;4)D_{\mathrm{F}}^{\prime J}(4,1), \quad (10.3.53)$$

$$\mathbf{\Pi}^*(1,2) = iG_0^2\mathrm{tr}\gamma(1)S_{\mathrm{F}}^{\prime J}(1,3)\mathbf{\Gamma}(3,4;2)S_{\mathrm{F}}^{\prime J}(4,1). \quad (10.3.54)$$

Equation (10.3.30) can be expressed after setting $\eta = \bar{\eta} = 0$ as

$$K(1) < \hat{\phi}(1) >^J + iG_0\mathrm{tr}(\gamma(1)S_{\mathrm{F}}^{\prime J}(1,1)) = -J(1). \quad (10.3.55)$$

The system of equations, (10.3.41), (10.3.42), (10.3.43), (10.3.44), (10.3.48), (10.3.53), (10.3.54) and (10.3.55), is called the Schwinger–Dyson equation. This system of nonlinear coupled integro-differential equations is exact and closed. Starting from the zeroth-order term of $\mathbf{\Gamma}(1,2;3)$, we can develop the covariant perturbation theory by iteration. In the first-order approximation, after setting $J = 0$, we have the following expressions,

$$\mathbf{\Sigma}^*(1-2) \cong -iG_0^2\gamma(1)S_0^{\mathrm{F}}(1-2)\gamma(2)D_0^{\mathrm{F}}(2-1), \quad (10.3.56)$$

$$\mathbf{\Pi}^*(1-2) \cong iG_0^2\mathrm{tr}\{\gamma(1)S_0^{\mathrm{F}}(1-2)\gamma(2)S_0^{\mathrm{F}}(2-1)\}, \quad (10.3.57)$$

and

$$\mathbf{\Gamma}(1,2;3) \cong \gamma(1)\delta(1-2)\delta(1-3) - iG_0^2\gamma(1)S_0^{\mathrm{F}}(1-3)\gamma(3)S_0^{\mathrm{F}}(3-2)\gamma(2)D_0^{\mathrm{F}}(2-1). \quad (10.3.58)$$

We point out that the covariant perturbation theory based on the Schwinger–Dyson equation is somewhat different in spirit from the standard covariant perturbation theory due to Feynman and Dyson. The former is capable of dealing with the bound state problem in general as will be shown shortly. Its power is demonstrated in the positronium problem.

Summary of Schwinger–Dyson equation

$$D_{\rm N}(1,3)S_{\rm F}^{\prime J}(3,2) = \delta(1-2), \quad D_{\rm M}(1,3)D_{\rm F}^{\prime J}(3,2) = \delta(1-2),$$

$$D_{\rm N}(1,2) \equiv \{D(1) - G_0\gamma(1) < \hat{\phi}(1) >^J\}\delta(1-2) - \mathbf{\Sigma}^*(1,2),$$

$$D_{\rm M}(1,2) \equiv K(1)\delta(1-2) - \mathbf{\Pi}^*(1,2),$$

$$K(1) < \hat{\phi}(1) >^J + iG_0{\rm tr}(\gamma(1)S_{\rm F}^{\prime J}(1,1)) = -J(1),$$

$$\mathbf{\Sigma}^*(1,2) \equiv -iG_0^2\gamma(1)S_{\rm F}^{\prime J}(1,3)\mathbf{\Gamma}(3,2;4)D_{\rm F}^{\prime J}(4,1),$$

$$\mathbf{\Pi}^*(1,2) \equiv iG_0^2{\rm tr}\{\gamma(1)S_{\rm F}^{\prime J}(1,3)\mathbf{\Gamma}(3,4;2)S_{\rm F}^{\prime J}(4,1)\},$$

$$\begin{aligned}\mathbf{\Gamma}(1,2;3) &\equiv -\frac{1}{G_0}\frac{\delta}{\delta < \hat{\phi}(3) >^J}(S_{\rm F}^{\prime J}(1,2))^{-1} \\ &= \gamma(1)\delta(1-2)\delta(1-3) + \frac{1}{G_0}\frac{\delta}{\delta < \hat{\phi}(3) >^J}\mathbf{\Sigma}^*(1,2).\end{aligned}$$

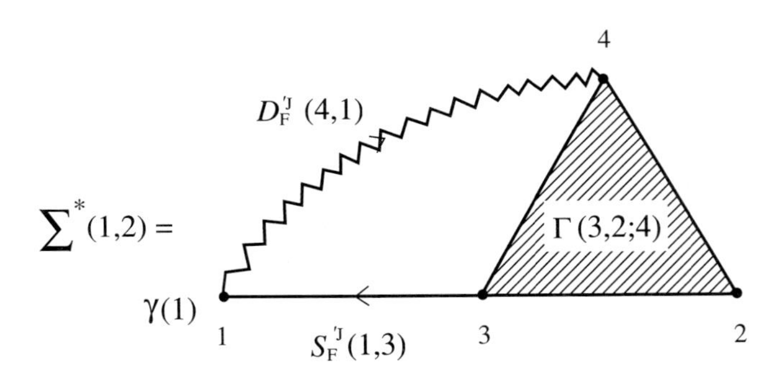

Fig. 10.1: Graphical representation of the proper self-energy part, $\mathbf{\Sigma}^*(1,2)$.

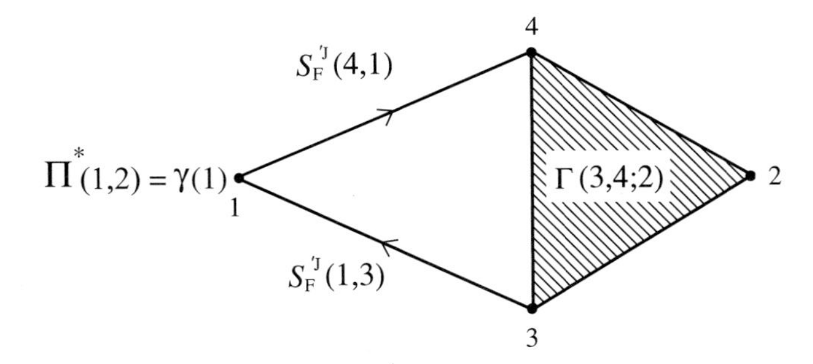

Fig. 10.2: Graphical representation of the proper self-energy part, $\mathbf{\Pi}^*(1,2)$.

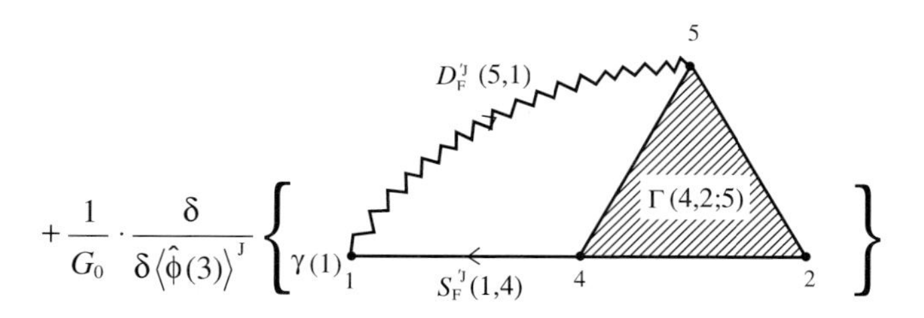

Fig. 10.3: Graphical representation of the vertex operator, $\mathbf{\Gamma}(1,2;3)$.

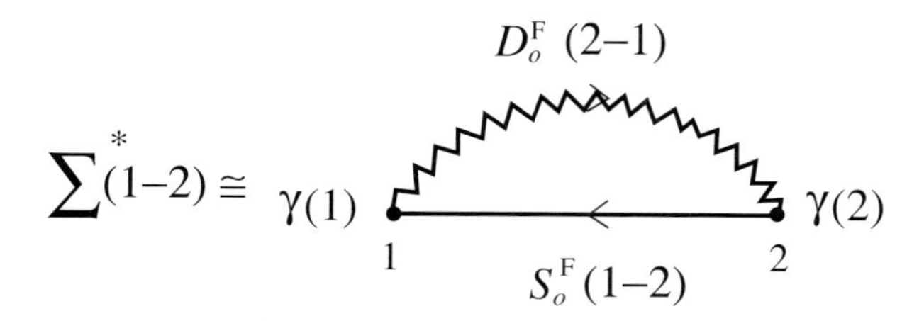

Fig. 10.4: The first-order approximation for the proper self-energy part, $\mathbf{\Sigma}^*(1,2)$.

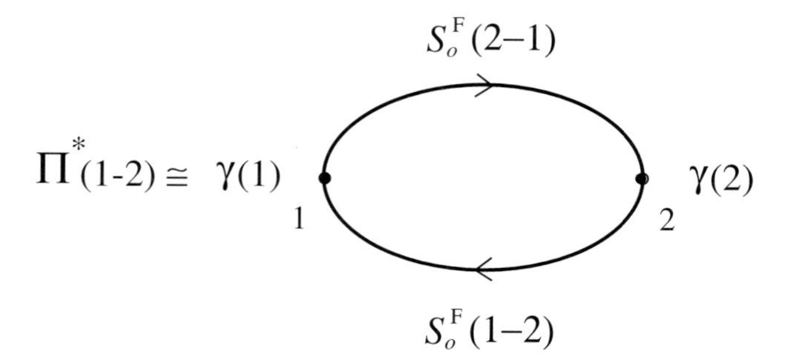

Fig. 10.5: The first-order approximation of the proper self-energy part, $\mathbf{\Pi}^*(1,2)$.

$$\Gamma(1,2;3) \cong \gamma(1)\delta(1-2)\delta(1-3) \quad {}^{1}_{2}\bullet\, 3$$

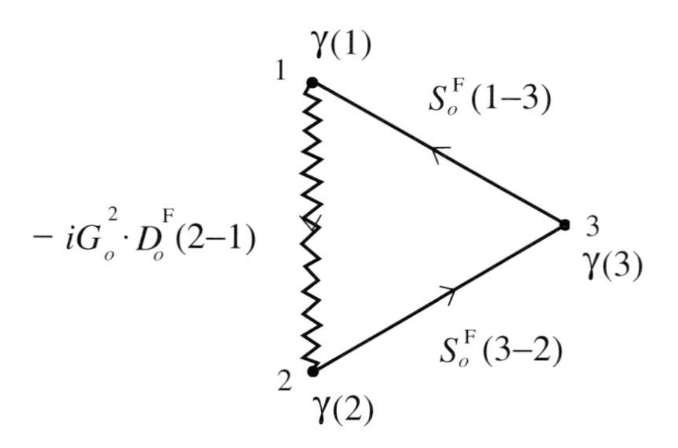

Fig. 10.6: The first-order approximation of the vertex operator, $\mathbf{\Gamma}(1,2;3)$.

* * *

We consider the two-body (four-point) nucleon "full" Green's function $S_{\mathrm{F}}^{\prime J}(1,2;3,4)$ with the Yukawa coupling, Eqs. (10.3.20) and (10.3.21), in mind, defined by

$$\begin{aligned}
S_{\mathrm{F}}^{\prime J}(1,2;3,4) \equiv & \frac{1}{i}\frac{\delta}{\delta\bar{\eta}(1)}\frac{1}{i}\frac{\delta}{\delta\bar{\eta}(2)}i\frac{\delta}{\delta\eta(4)}i\frac{\delta}{\delta\eta(3)}\frac{1}{i}\ln Z[J,\bar{\eta},\eta]\Big|_{\bar{\eta}=\eta=0} \\
= & \left(\frac{1}{i}\right)^2 \Big\{ <\hat{\psi}(1)\hat{\psi}(2)\hat{\bar{\psi}}(3)\hat{\bar{\psi}}(4)>^J \\
& - <\hat{\psi}(1)\hat{\psi}(2)>^J <\hat{\bar{\psi}}(3)\hat{\bar{\psi}}(4)>^J \\
& + <\hat{\psi}(1)\hat{\bar{\psi}}(3)>^J <\hat{\psi}(2)\hat{\bar{\psi}}(4)>^J \\
& - <\hat{\psi}(1)\hat{\bar{\psi}}(4)>^J <\hat{\psi}(2)\hat{\bar{\psi}}(3)>^J \Big\} \\
\equiv & \left(\frac{1}{i}\right)^2 <0,\mathrm{out}|\, \mathrm{T}(\hat{\psi}(1)\hat{\psi}(2)\hat{\bar{\psi}}(3)\hat{\bar{\psi}}(4))\, |0,\mathrm{in}>_{\mathrm{C}}^J .
\end{aligned} \tag{10.3.59}$$

Operating the *"Nucleon differential operators"*, $D_{\mathrm{N}}(1,5)$ and $D_{\mathrm{N}}(2,6)$, on $S_{\mathrm{F}}^{\prime J}(5,6;3,4)$ and using the Schwinger–Dyson equation derived above, we obtain

$$\begin{aligned}
&\{D_{\mathrm{N}}(1,5)D_{\mathrm{N}}(2,6) - I(1,2;5,6)\}S_{\mathrm{F}}^{\prime J}(5,6;3,4) \\
&\qquad = \delta(1-3)\delta(2-4) - \delta(1-4)\delta(2-3),
\end{aligned} \tag{10.3.60}$$

where the operator $I(1,2;3,4)$ is called the proper interaction kernel and satisfies the following integral equations,

$$
\begin{aligned}
I(1,2;5,6)S_{\mathrm{F}}^{\prime J}(5,6;3,4) \\
&= g_0^2 \mathrm{tr}^{(\mathrm{M})}[\gamma(1)\Gamma(2)D_{\mathrm{F}}^{\prime J}(5,6)]S_{\mathrm{F}}^{\prime J}(5,6;3,4) \\
&+ g_0^2 \mathrm{tr}^{(\mathrm{M})}[\gamma(1)S_{\mathrm{F}}^{\prime J}(1,5)\frac{1}{i}\frac{\delta}{\delta J(5)}]I(5,2;6,7)S_{\mathrm{F}}^{\prime J}(6,7;3,4)
\end{aligned} \tag{10.3.61}
$$

$$
\begin{aligned}
&= g_0^2 \mathrm{tr}^{(\mathrm{M})}[\gamma(2)\Gamma(1)D_{\mathrm{F}}^{\prime J}(5,6)]S_{\mathrm{F}}^{\prime J}(5,6;3,4) \\
&+ g_0^2 \mathrm{tr}^{(\mathrm{M})}[\gamma(2)S_{\mathrm{F}}^{\prime J}(2,5)\frac{1}{i}\frac{\delta}{\delta J(5)}]I(1,5;6,7)S_{\mathrm{F}}^{\prime J}(6,7;3,4).
\end{aligned} \tag{10.3.62}
$$

Here $\mathrm{tr}^{(\mathrm{M})}$ indicates that the trace should be taken only over the meson coordinate. Equation (10.3.60) can be cast into the integral equation after a little algebra as

$$
\begin{aligned}
S_{\mathrm{F}}^{\prime J}(1,2;3,4) &= S_{\mathrm{F}}^{\prime J}(1,3)S_{\mathrm{F}}^{\prime J}(2,4) - S_{\mathrm{F}}^{\prime J}(1,4)S_{\mathrm{F}}^{\prime J}(2,3) \\
&+ S_{\mathrm{F}}^{\prime J}(1,5)S_{\mathrm{F}}^{\prime J}(2,6)I(5,6;7,8)S_{\mathrm{F}}^{\prime J}(7,8;3,4).
\end{aligned} \tag{10.3.63}
$$

The system of equations, (10.3.60) (or (10.3.63)) and (10.3.61) (or (10.3.62)), is called the Bethe–Salpeter equation. If we set $t_1 = t_2 = t > t_3 = t_4 = t'$ in Eq. (10.3.59), $S_{\mathrm{F}}^{\prime J}(1,2;3,4)$ represents the transition probability amplitude whereby the nucleons originally located at $\vec{x}_3$ and $\vec{x}_4$ at time t' are to be found at $\vec{x}_1$ and $\vec{x}_2$ at the later time t. In the integral equations for $I(1,2;3,4)$, Eqs. (10.3.61) and (10.3.62), the first terms on the right-hand sides represent the scattering state and the second terms represent the bound state. The bound state problem is formulated by dropping the first terms on the right-hand sides of Eqs. (10.3.61) and (10.3.62). The proper interaction kernel $I(1,2;3,4)$ assumes the following form in the first-order approximation,

$$
I(1,2;3,4) \cong g_0^2\gamma(1)\gamma(2)D_0^{\mathrm{F}}(1-2)(\delta(1-3)\delta(2-4) - \delta(1-4)\delta(2-3)). \tag{10.3.64}
$$

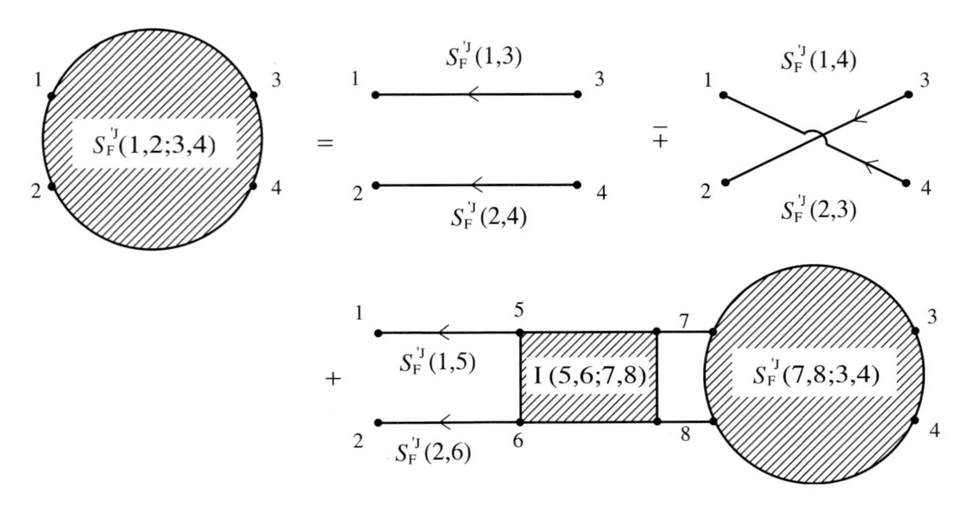

Fig. 10.7: Graphical representation of the Bethe–Salpeter equation.

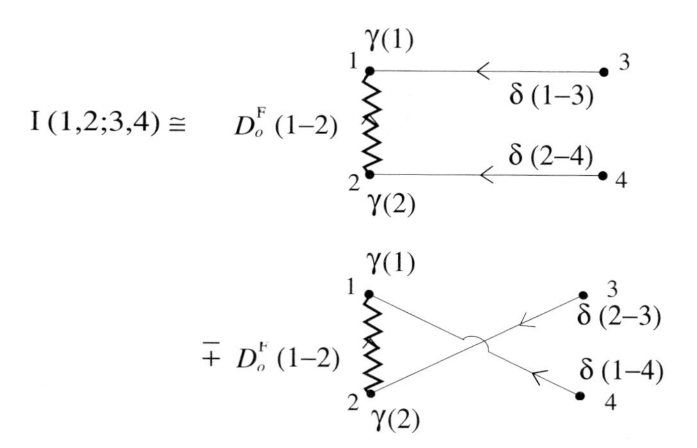

Fig. 10.8: The first-order approximation of the proper interaction kernel,$I(1,2;3,4)$.

As a byproduct of the derivation of Schwinger–Dyson equations, we can derive the functional integral representation of the generating functional $Z[J,\bar{\eta},\eta]$ from Eqs. (10.3.19a), (10.3.19b) and (10.3.19c) in the following way.

We define the functional Fourier transform $\tilde{Z}[\phi,\psi,\bar{\psi}]$ of the generating functional $Z[J,\bar{\eta},\eta]$ by

$$Z[J,\bar{\eta},\eta] \equiv \int \mathcal{D}[\phi]\mathcal{D}[\psi]\mathcal{D}[\bar{\psi}]\tilde{Z}[\phi,\psi,\bar{\psi}] \exp\left[i(J\phi+\bar{\eta}\psi+\bar{\psi}\eta)\right]. \tag{10.3.65}$$

By functional integral by parts, we obtain the identity,

$$\begin{aligned}
&\begin{bmatrix} \eta_\alpha(x) \\ -\bar{\eta}_\beta(x) \\ J(x) \end{bmatrix} Z[J,\bar{\eta},\eta] \\
&= \int \mathcal{D}[\phi]\mathcal{D}[\psi]\mathcal{D}[\bar{\psi}]\tilde{Z}[\phi,\psi,\bar{\psi}] \begin{bmatrix} \frac{1}{i}\frac{\delta}{\delta\bar{\psi}_\alpha(x)} \\ \frac{1}{i}\frac{\delta}{\delta\psi_\beta(x)} \\ \frac{1}{i}\frac{\delta}{\delta\phi(x)} \end{bmatrix} \times \exp\left[i(J\phi+\bar{\eta}\psi+\bar{\psi}\eta)\right] \qquad (10.3.66) \\
&= \int \mathcal{D}[\phi]\mathcal{D}[\psi]\mathcal{D}[\bar{\psi}] \left\{ \begin{bmatrix} -\frac{1}{i}\frac{\delta}{\delta\bar{\psi}_\alpha(x)} \\ -\frac{1}{i}\frac{\delta}{\delta\psi_\beta(x)} \\ -\frac{1}{i}\frac{\delta}{\delta\phi(x)} \end{bmatrix} \tilde{Z}[\phi,\psi,\bar{\psi}] \right\} \exp\left[i(J\phi+\bar{\eta}\psi+\bar{\psi}\eta)\right].
\end{aligned}$$

With the identity derived above, we have the equations of motion of the functional Fourier transform $\tilde{Z}[\phi, \psi, \bar{\psi}]$ of the generating functional $Z[J, \bar{\eta}, \eta]$ from Eqs. (10.3.19a through c) as

$$\left\{ \frac{\delta I_{\text{tot}}[\phi, \psi, \bar{\psi}]}{\delta \bar{\psi}_{\alpha}(x)} - \frac{1}{i} \frac{\delta}{\delta \bar{\psi}_{\alpha}(x)} \right\} \tilde{Z}[\phi, \psi, \bar{\psi}] = 0, \tag{10.3.67a}$$

$$\left\{ \frac{\delta I_{\text{tot}}[\phi, \psi, \bar{\psi}]}{\delta \psi_{\beta}(x)} - \frac{1}{i} \frac{\delta}{\delta \psi_{\beta}(x)} \right\} \tilde{Z}[\phi, \psi, \bar{\psi}] = 0, \tag{10.3.67b}$$

$$\left\{ \frac{\delta I_{\text{tot}}[\phi, \psi, \bar{\psi}]}{\delta \phi(x)} - \frac{1}{i} \frac{\delta}{\delta \phi(x)} \right\} \tilde{Z}[\phi, \psi, \bar{\psi}] = 0. \tag{10.3.67c}$$

We divide Eqs. (10.3.67a through c) by $\tilde{Z}[\phi, \psi, \bar{\psi}]$, and obtain

$$\frac{\delta}{\delta \bar{\psi}_{\alpha}(x)} \ln \tilde{Z}[\phi, \psi, \bar{\psi}] = i \frac{\delta}{\delta \bar{\psi}_{\alpha}(x)} I_{\text{tot}}[\phi, \psi, \bar{\psi}], \tag{10.3.68a}$$

$$\frac{\delta}{\delta \psi_{\beta}(x)} \ln \tilde{Z}[\phi, \psi, \bar{\psi}] = i \frac{\delta}{\delta \psi_{\beta}(x)} I_{\text{tot}}[\phi, \psi, \bar{\psi}], \tag{10.3.68b}$$

$$\frac{\delta}{\delta \phi(x)} \ln \tilde{Z}[\phi, \psi, \bar{\psi}] = i \frac{\delta}{\delta \phi(x)} I_{\text{tot}}[\phi, \psi, \bar{\psi}]. \tag{10.3.68c}$$

We can immediately integrate Eqs. (10.3.68a through c) with the result,

$$\begin{aligned} \tilde{Z}[\phi, \psi, \bar{\psi}] &= \frac{1}{C_V} \exp\left[i I_{\text{tot}}[\phi, \psi, \bar{\psi}] \right] \\ &= \frac{1}{C_V} \exp\left[i \int d^4 z \mathcal{L}_{\text{tot}}(\phi(z), \psi(z), \bar{\psi}(z)) \right], \end{aligned} \tag{10.3.69}$$

where C_V is the integration constant. From Eq. (10.3.69), we shall obtain the generating functional $Z[J, \bar{\eta}, \eta]$ in the functional integral representation,

$$\begin{aligned} Z[J, \bar{\eta}, \eta] = \frac{1}{C_V} \int \mathcal{D}[\phi] \mathcal{D}[\psi] \mathcal{D}[\bar{\psi}] \exp\Big[i \int d^4 z \Big\{ & \mathcal{L}_{\text{tot}}(\phi(z), \psi(z), \bar{\psi}(z)) \\ & + J(z)\phi(z) + \bar{\eta}(z)\psi(z) + \bar{\psi}(z)\eta(z) \Big\} \Big]. \end{aligned} \tag{10.3.70}$$

We employ the following normalization,

$$Z[J = \bar{\eta} = \eta = 0] = \langle 0, \text{out} | 0, \text{in} \rangle \quad \text{with} \quad C_V = 1. \tag{10.3.71}$$

In this manner, as a byproduct of the derivation of the Schwinger–Dyson equations, we have succeeded in deriving the functional integral representation of the generating functional $Z[J, \bar{\eta}, \eta]$ of Green's functions from the canonical formalism of quantum theory. In Section 10.1, we have derived the canonical formalism of quantum theory from the functional (path) integral formalism of quantum theory, adopting the transformation function in the functional (path) integral representation as Feynman's action principle, therefore establishing the equivalence of canonical formalism and functional (path) integral formalism at least for the nonsingular Lagrangian (density) system. It is astonishingly hard to establish this equivalence for the singular Lagrangian (density).

10.4 Schwinger–Dyson Equation in Quantum Statistical Mechanics

We consider the grand canonical ensemble of the Fermion (mass m) and the Boson (mass κ) with Euclidean Lagrangian density in contact with the particle source μ,

$$\begin{aligned}
&\mathcal{L}'_{\mathrm{E}}(\psi_{\mathrm{E}\alpha}(\tau,\vec{x}),\partial_\mu\psi_{\mathrm{E}\alpha}(\tau,\vec{x}),\phi(\tau,\vec{x}),\partial_\mu\phi(\tau,\vec{x}))\\
&\quad=\bar{\psi}_{\mathrm{E}\,\alpha}(\tau,\vec{x})\left\{i\gamma^k\partial_k+i\gamma^4\left(\frac{\partial}{\partial\tau}-\mu\right)-m\right\}_{\alpha,\beta}\psi_{\mathrm{E}\,\beta}(\tau,\vec{x})\\
&\qquad+\frac{1}{2}\phi(\tau,\vec{x})\left(\frac{\partial^2}{\partial x_\nu^2}-\kappa^2\right)\phi(\tau,\vec{x})+\frac{1}{2}g\mathrm{Tr}\{\gamma[\bar{\psi}_{\mathrm{E}}(\tau,\vec{x}),\psi_{\mathrm{E}}(\tau,\vec{x})]\}\phi(\tau,\vec{x}).
\end{aligned}\tag{10.4.1}$$

The density matrix $\hat{\rho}_{\mathrm{GC}}(\beta)$ of the grand canonical ensemble in the Schrödinger Picture is given by

$$\hat{\rho}_{\mathrm{GC}}(\beta)=\exp\left[-\beta(\hat{H}-\mu\hat{N})\right],\qquad\beta=\frac{1}{k_{\mathrm{B}}T},\tag{10.4.2}$$

where the total Hamiltonian $\hat{H}$ is split into two parts,

$$\begin{aligned}
\hat{H}_0 &= \text{free Hamiltonian for Fermion (mass } m\text{) and Boson (mass } \kappa\text{)},\\
\hat{H}_1 &= -\int d^3\vec{x}\,\hat{\jmath}(\vec{x})\hat{\phi}(\vec{x}),
\end{aligned}\tag{10.4.3}$$

with the "current" given by

$$\hat{\jmath}(\vec{x})=\frac{1}{2}g\mathrm{Tr}\{\gamma[\hat{\bar{\psi}}_{\mathrm{E}}(\vec{x}),\hat{\psi}_{\mathrm{E}}(\vec{x})]\},\tag{10.4.4}$$

and

$$\hat{N}=\frac{1}{2}\int d^3\vec{x}\mathrm{Tr}\{-\gamma^4[\hat{\bar{\psi}}_{\mathrm{E}}(\vec{x}),\hat{\psi}_{\mathrm{E}}(\vec{x})]\}.\tag{10.4.5}$$

By the standard method of quantum field theory, we use the Interaction Picture with $\hat{N}$ included in the free part, and obtain

$$\hat{\rho}_{\mathrm{GC}}(\beta)=\hat{\rho}_0(\beta)\hat{S}(\beta),\tag{10.4.6a}$$

$$\hat{\rho}_0(\beta) = \exp\left[-\beta(\hat{H}_0 - \mu\hat{N})\right], \tag{10.4.7}$$

$$\hat{S}(\beta) = \mathrm{T}_\tau\left\{\exp\left[-\int_0^\beta d\tau \int d^3\vec{x}\hat{\mathcal{H}}_1(\tau,\vec{x})\right]\right\}, \tag{10.4.8a}$$

and

$$\hat{\mathcal{H}}_1(\tau,\vec{x}) = -\hat{\jmath}^{(\mathrm{I})}(\tau,\vec{x})\hat{\phi}^{(\mathrm{I})}(\tau,\vec{x}). \tag{10.4.9}$$

We know that the Interaction Picture operator $\hat{f}^{(\mathrm{I})}(\tau,\vec{x})$ is related to the Schrödinger Picture operator $\hat{f}(\vec{x})$ by

$$\hat{f}^{(\mathrm{I})}(\tau,\vec{x}) = \hat{\rho}_0^{-1}(\tau)\cdot\hat{f}(\vec{x})\cdot\hat{\rho}_0(\tau). \tag{10.4.10}$$

We introduce the external hook $\{J(\tau,\vec{x}),\bar{\eta}_\alpha(\tau,\vec{x}),\eta_\beta(\tau,\vec{x})\}$ in the Interaction Picture, and obtain

$$\begin{aligned}\hat{\mathcal{H}}_1^{\mathrm{int}}(\tau,\vec{x}) = -\{[\hat{\jmath}^{(\mathrm{I})}(\tau,\vec{x}) + J(\tau,\vec{x})]\hat{\phi}^{(\mathrm{I})}(\tau,\vec{x})\\ + \bar{\eta}_\alpha(\tau,\vec{x})\hat{\psi}^{(\mathrm{I})}_{\mathrm{E}\,\alpha}(\tau,\vec{x}) + \widehat{\bar{\psi}}_{\mathrm{E}\,\beta}(\tau,\vec{x})\eta_\beta(\tau,\vec{x}))\}.\end{aligned} \tag{10.4.11}$$

We replace Eqs. (10.4.6a), (10.4.7) and (10.4.8a) with

$$\hat{\rho}_{\mathrm{GC}}(\beta;[J,\bar{\eta},\eta]) = \hat{\rho}_0(\beta)\hat{S}(\beta;[J,\bar{\eta},\eta]), \tag{10.4.6b}$$

$$\hat{\rho}_0(\beta) = \exp\left[-\beta(\hat{H}_0 - \mu\hat{N})\right], \tag{10.4.7}$$

$$\hat{S}(\beta;[J,\bar{\eta},\eta]) = \mathrm{T}_\tau\{\exp\left[-\int_0^\beta d\tau \int d^3\vec{x}\hat{\mathcal{H}}_1^{\mathrm{int}}(\tau,\vec{x})\right]\}. \tag{10.4.8b}$$

Here we have
(a) $0 \le \tau \le \beta$.

$$\begin{aligned}
&\frac{\delta}{\delta J(\tau,\vec{x})}\hat{\rho}_{\mathrm{GC}}(\beta;[J,\bar{\eta},\eta])\Big|_{J=\bar{\eta}=\eta=0}\\
&\quad= \hat{\rho}_0(\beta)\ \mathrm{T}_\tau\{\hat{\phi}^{(\mathrm{I})}(\tau,\vec{x})\exp\left[-\int_0^\beta d\tau\int d^3\vec{x}\right]\hat{\mathcal{H}}_1^{\mathrm{int}}(\tau,\vec{x})]\}\Big|_{J=\bar{\eta}=\eta=0}\\
&\quad= \hat{\rho}_0(\beta)\mathrm{T}_\tau\{\exp\left[-\int_\tau^\beta d\tau\int d^3\vec{x}\hat{\mathcal{H}}_1^{\mathrm{int}}\right]\}\\
&\qquad\times\hat{\phi}^{(\mathrm{I})}(\tau,\vec{x})\ \mathrm{T}_\tau\{\exp[-\int_0^\tau d\tau\int d^3\vec{x}\hat{\mathcal{H}}_1^{\mathrm{int}}]\}\Big|_{J=\bar{\eta}=\eta=0}\\
&\quad= \hat{\rho}_0(\beta)\ \hat{S}(\beta;[J,\bar{\eta},\eta])\hat{S}(-\tau;[J,\bar{\eta},\eta])\hat{\phi}^{(\mathrm{I})}(\tau,\vec{x})\hat{S}(\tau;[J,\bar{\eta},\eta])\Big|_{J=\bar{\eta}=\eta=0}\\
&\quad= \hat{\rho}_{\mathrm{GC}}(\beta;[J,\bar{\eta},\eta])\{\hat{\rho}_0(\tau)\hat{S}(\tau;[J,\bar{\eta},\eta])\}^{-1}\hat{\phi}(\vec{x})\ \{\hat{\rho}_0(\tau)\hat{S}(\tau;[J,\bar{\eta},\eta])\}\Big|_{J=\bar{\eta}=\eta=0}\\
&\quad= \hat{\rho}_{\mathrm{GC}}(\beta)\hat{\phi}(\tau,\vec{x}).
\end{aligned}$$

Thus we obtain

$$\frac{\delta}{\delta J(\tau,\vec{x})}\hat{\rho}_{\mathrm{GC}}(\beta;[J,\bar{\eta},\eta])\bigg|_{J=\bar{\eta}=\eta=0} = \hat{\rho}_{\mathrm{GC}}(\beta)\hat{\phi}(\tau,\vec{x}). \tag{10.4.12}$$

Likewise we obtain

$$\frac{\delta}{\delta\bar{\eta}_\alpha(\tau,\vec{x})}\hat{\rho}_{\mathrm{GC}}(\beta;[J,\bar{\eta},\eta])\bigg|_{J=\bar{\eta}=\eta=0} = \hat{\rho}_{\mathrm{GC}}(\beta)\hat{\psi}_{\mathrm{E}\,\alpha}(\tau,\vec{x}), \tag{10.4.13}$$

$$\frac{\delta}{\delta\eta_\beta(\tau,\vec{x})}\hat{\rho}_{\mathrm{GC}}(\beta;[J,\bar{\eta},\eta])\bigg|_{J=\bar{\eta}=\eta=0} = -\hat{\rho}_{\mathrm{GC}}(\beta)\hat{\bar{\psi}}_{\mathrm{E}\,\beta}(\tau,\vec{x}), \tag{10.4.14}$$

and

$$\begin{aligned}\frac{\delta^2}{\delta\bar{\eta}_\alpha(\tau,\vec{x})\delta\eta_\beta(\tau',\vec{x})}\hat{\rho}_{\mathrm{GC}}(\beta;[J,\bar{\eta},\eta])\bigg|_{J=\bar{\eta}=\eta=0}\\ = -\hat{\rho}_{\mathrm{GC}}(\beta)\mathrm{T}_\tau\{\hat{\psi}_{\mathrm{E}\,\alpha}(\tau,\vec{x})\hat{\bar{\psi}}_{\mathrm{E}\,\beta}(\tau,\vec{x})\}.\end{aligned} \tag{10.4.15}$$

The Heisenberg Picture operator $\hat{f}(\tau,\vec{x})$ is related to the Schrödinger Picture operator $\hat{f}(\vec{x})$ by

$$\hat{f}(\tau,\vec{x}) = \hat{\rho}_{\mathrm{GC}}^{-1}(\tau)\cdot\hat{f}(\vec{x})\cdot\hat{\rho}_{\mathrm{GC}}(\tau). \tag{10.4.16}$$

(b) $\tau \notin [0,\beta]$.

As for $\tau \notin [0,\beta]$, the functional derivative of $\hat{\rho}_{\mathrm{GC}}(\beta;[J,\bar{\eta},\eta])$ with respect to $\{J,\bar{\eta},\eta\}$ vanishes.

In order to derive the equation of motion for the partition function, we use the "equation of motion" of $\hat{\psi}_{\mathrm{E}\,\alpha}(\tau,\vec{x})$ and $\hat{\phi}(\tau,\vec{x})$,

$$\left\{i\gamma^k\partial_k + i\gamma^4\left(\frac{\partial}{\partial\tau}-\mu\right) - m + g\gamma\hat{\phi}(\tau,\vec{x})\right\}_{\beta,\alpha}\hat{\psi}_{\mathrm{E}\,\alpha}(\tau,\vec{x}) = 0, \tag{10.4.17}$$

$$\hat{\bar{\psi}}_{\mathrm{E}\,\beta}(\tau,\vec{x})\left\{i\gamma^k\partial_k + i\gamma^4\left(\frac{\partial}{\partial\tau}-\mu\right) - m + g\gamma\hat{\phi}(\tau,\vec{x})\right\}^{\mathrm{T}}_{\beta,\alpha} = 0, \tag{10.4.18}$$

$$\left(\frac{\partial^2}{\partial x_\nu^2} - \kappa^2\right)\hat{\phi}(\tau,\vec{x}) + g\mathrm{Tr}\left\{\gamma\hat{\bar{\psi}}_{\mathrm{E}}(\tau,\vec{x})\hat{\psi}_{\mathrm{E}}(\tau,\vec{x})\right\} = 0, \tag{10.4.19}$$

and the equal "time" canonical (anti-)commuters,

$$\delta(\tau-\tau')\left\{\hat{\psi}_{\mathrm{E}\,\alpha}(\tau,\vec{x}),\hat{\bar{\psi}}_{\mathrm{E}\,\beta}(\tau,\vec{x})\right\} = \delta_{\alpha\beta}\delta(\tau-\tau')\delta(\vec{x}-\vec{x}'), \tag{10.4.20a}$$

$$\delta(\tau-\tau')\left[\hat{\phi}(\tau,\vec{x}),\frac{\partial}{\partial\tau'}\hat{\phi}(\tau',\vec{x}')\right] = \delta(\tau-\tau')\delta^3(\vec{x}-\vec{y}), \tag{10.4.20b}$$

with all the rest of equal "time" (anti-)commutators equal to 0. We obtain the equations of motion of the partition function of the grand canonical ensemble

$$Z_{\mathrm{GC}}(\beta;[J,\bar{\eta},\eta]) = \mathrm{Tr}\hat{\rho}_{\mathrm{GC}}(\beta;[J,\bar{\eta},\eta]) \tag{10.4.21}$$

in the presence of the external hook $\{J,\bar{\eta},\eta\}$ from Eqs. (10.4.12), (10.4.13) and (10.4.14) as

$$\left\{ i\gamma^k \partial_k + i\gamma^4 \left(\frac{\partial}{\partial \tau} - \mu \right) - m + g\gamma \frac{1}{i} \frac{\delta}{\delta J(\tau, \vec{x})} \right\}_{\beta,\alpha} \frac{1}{i} \frac{\delta}{\delta \bar{\eta}_\alpha(\tau, \vec{x})} Z_{\mathrm{GC}}(\beta; [J, \bar{\eta}, \eta])$$

$$= -\eta_\beta(\tau, \vec{x}) Z_{\mathrm{GC}}(\beta; [J, \bar{\eta}, \eta]), \tag{10.4.22}$$

$$\left\{ i\gamma^k \partial_k + i\gamma^4 \left(\frac{\partial}{\partial \tau} + \mu \right) + m - g\gamma \frac{1}{i} \frac{\delta}{\delta J(\tau, \vec{x})} \right\}^{\mathrm{T}}_{\beta,\alpha} i \frac{\delta}{\delta \eta_\beta(\tau, \vec{x})} Z_{\mathrm{GC}}(\beta; [J, \bar{\eta}, \eta])$$

$$= \bar{\eta}_\alpha(\tau, \vec{x}) Z_{\mathrm{GC}}(\beta; [J, \bar{\eta}, \eta]), \tag{10.4.23}$$

$$\left\{ \left(\frac{\partial^2}{\partial x_\nu^2} - \kappa^2 \right) \frac{1}{i} \frac{\delta}{\delta J(\tau, \vec{x})} - g\gamma_{\beta\alpha} \frac{\delta^2}{\delta \bar{\eta}_\alpha(\tau, \vec{x}) \delta \eta_\beta(\tau, \vec{x})} \right\} Z_{\mathrm{GC}}(\beta; [J, \bar{\eta}, \eta])$$

$$= J(\tau, \vec{x}) Z_{\mathrm{GC}}(\beta; [J, \bar{\eta}, \eta]) \tag{10.4.24}$$

We can solve the functional differential equations, (10.4.22), (10.4.23) and (10.4.24), by the method of Section 10.3. As in Section 10.3, we define the functional Fourier transform $\tilde{Z}_{\mathrm{GC}}(\beta; [\phi, \psi_{\mathrm{E}}, \bar{\psi}_{\mathrm{E}}])$ of $Z_{\mathrm{GC}}(\beta; [J, \bar{\eta}, \eta])$ by

$$Z_{\mathrm{GC}}(\beta; [J, \bar{\eta}, \eta]) \equiv \int \mathcal{D}[\bar{\psi}_{\mathrm{E}}] \mathcal{D}[\psi_{\mathrm{E}}] \mathcal{D}[\phi] \tilde{Z}_{\mathrm{GC}}(\beta; [\phi, \psi_{\mathrm{E}}, \bar{\psi}_{\mathrm{E}}])$$

$$\times \exp \left[i \int_0^\beta d\tau \int d^3\vec{x} \left\{ J(\tau, \vec{x}) \phi(\tau, \vec{x}) + \bar{\eta}_\alpha(\tau, \vec{x}) \psi_{\mathrm{E}\,\alpha}(\tau, \vec{x}) + \bar{\psi}_{\mathrm{E}\,\beta}(\tau, \vec{x}) \eta_\beta(\tau, \vec{x}) \right\} \right].$$

We obtain the equations of motion satisfied by the functional Fourier transform $\tilde{Z}_{\mathrm{GC}}(\beta; [\phi, \psi_{\mathrm{E}}, \bar{\psi}_{\mathrm{E}}])$ from Eqs. (10.4.22), (10.4.23) and (10.4.24), after the functional integral by parts on the right-hand sides involving $\bar{\eta}_\alpha$, η_β and J as

$$\frac{\delta}{\delta \psi_{\mathrm{E}\,\alpha}(\tau, \vec{x})} \ln \tilde{Z}_{\mathrm{GC}}(\beta; [\phi, \psi_{\mathrm{E}}, \bar{\psi}_{\mathrm{E}}]) = \frac{\delta}{\delta \psi_{\mathrm{E}\,\alpha}(\tau, \vec{x})} \int_0^\beta d\tau \int d^3\vec{x} \mathcal{L}'_{\mathrm{E}}((10.4.1)),$$

$$\frac{\delta}{\delta \bar{\psi}_{\mathrm{E}\,\beta}(\tau, \vec{x})} \ln \tilde{Z}_{\mathrm{GC}}(\beta; [\phi, \psi_{\mathrm{E}}, \bar{\psi}_{\mathrm{E}}]) = \frac{\delta}{\delta \bar{\psi}_{\mathrm{E}\,\beta}(\tau, \vec{x})} \int_0^\beta d\tau \int d^3\vec{x} \mathcal{L}'_{\mathrm{E}}((10.4.1)),$$

$$\frac{\delta}{\delta \phi(\tau, \vec{x})} \ln \tilde{Z}_{\mathrm{GC}}(\beta; [\phi, \psi_{\mathrm{E}}, \bar{\psi}_{\mathrm{E}}]) = \frac{\delta}{\delta \phi(\tau, \vec{x})} \int_0^\beta d\tau \int d^3\vec{x} \mathcal{L}'_{\mathrm{E}}((10.4.1)),$$

which we can immediately integrate to obtain

$$\tilde{Z}_{\mathrm{GC}}(\beta; [\phi, \psi_{\mathrm{E}}, \bar{\psi}_{\mathrm{E}}]) = C \exp \left[\int_0^\beta d\tau \int d^3\vec{x} \mathcal{L}'_{\mathrm{E}}((10.4.1)) \right]. \tag{10.4.25a}$$

Thus we have the path integral representation of $Z_{\text{GC}}(\beta;[J,\bar{\eta},\eta])$ as

$$
\begin{aligned}
Z_{\text{GC}}(\beta;[J,\bar{\eta},\eta]) &= C\int \mathcal{D}[\bar{\psi}_{\text{E}}]\mathcal{D}[\psi_{\text{E}}]\mathcal{D}[\phi]\exp\left[\int_0^\beta d\tau\int d^3\vec{x}\Big\{\mathcal{L}'_{\text{E}}((10.4.1))\right.\\
&\left.+iJ(\tau,\vec{x})\phi(\tau,\vec{x})+i\bar{\eta}_\alpha(\tau,\vec{x})\psi_{\text{E}\,\alpha}(\tau,\vec{x})+i\bar{\psi}_{\text{E}\,\beta}(\tau,\vec{x})\eta_\beta(\tau,\vec{x})\Big\}\right]\\
&= Z_0\exp\left[-g\gamma_{\beta\alpha}\int_0^\beta d\tau\int d^3\vec{x}\, i\frac{\delta}{\delta\eta_\beta(\tau,\vec{x})}\frac{1}{i}\frac{\delta}{\delta\bar{\eta}_\alpha(\tau,\vec{x})}\frac{1}{i}\frac{\delta}{\delta J(\tau,\vec{x})}\right]\\
&\times\exp\left[\int_0^\beta d\tau\int d^3\vec{x}\int_0^\beta d\tau'\int d^3\vec{x}'\Big\{-\frac{1}{2}J(\tau,\vec{x})D_0(\tau-\tau',\vec{x}-\vec{x}')J(\tau',\vec{x}')\right.\\
&\left.+\bar{\eta}_\alpha(\tau,\vec{x})\underset{\alpha\beta}{S_0}(\tau-\tau',\vec{x}-\vec{x}')\eta_\beta(\tau',\vec{x}')\Big\}\right].
\end{aligned}
\tag{10.4.25b}
$$

The normalization constant Z_0 is so chosen that

$$
\begin{aligned}
Z_0 &= Z_{\text{GC}}(\beta;J=\bar{\eta}=\eta=0,g=0) &(10.4.26)\\
&= \prod_{|\vec{p}|,|\vec{k}|}\{1+\exp[-\beta(\varepsilon_{\vec{p}}-\mu)]\}\{1+\exp[-\beta(\varepsilon_{\vec{p}}+\mu)]\}\{1-\exp[-\beta\omega_{\vec{k}}]\}^{-1},
\end{aligned}
$$

with

$$
\varepsilon_{\vec{p}}=(\vec{p}^2+m^2)^{\frac{1}{2}},\qquad \omega_{\vec{k}}=(\vec{k}^2+\kappa^2)^{\frac{1}{2}}. \tag{10.4.27}
$$

$D_0(\tau-\tau',\vec{x}-\vec{x}')$ and $\underset{\alpha\beta}{S_0}(\tau-\tau',\vec{x}-\vec{x}')$ are the "free" temperature Green's functions of the Bose field and the Fermi field, respectively, and are given by

$$
\begin{aligned}
D_0(\tau-\tau',\vec{x}-\vec{x}') &= \int\frac{d^3\vec{k}}{(2\pi)^3 2\omega_{\vec{k}}}\Big\{(f_{\vec{k}}+1)\exp\left[i\vec{k}(\vec{x}-\vec{x}')-\omega_{\vec{k}}(\tau-\tau')\right]\\
&\quad+f_{\vec{k}}\exp\left[-i\vec{k}(\vec{x}-\vec{x}')+\omega_{\vec{k}}(\tau-\tau')\right]\Big\},
\end{aligned}
$$

$$
\underset{\alpha\beta}{S_0}(\tau-\tau',\vec{x}-\vec{x}') = (i\gamma^\nu\partial_\nu+m)_{\alpha,\beta}\int\frac{d^3\vec{k}}{(2\pi)^3 2\varepsilon_{\vec{k}}}
\times\begin{cases}
\Big\{(N^+_{\vec{k}}-1)\exp\left[i\vec{k}(\vec{x}-\vec{x}')-(\varepsilon_{\vec{k}}-\mu)(\tau-\tau')\right]\\
\quad+N^-_{\vec{k}}\exp\left[-i\vec{k}(\vec{x}-\vec{x}')+(\varepsilon_{\vec{k}}+\mu)(\tau-\tau')\right]\Big\}, & \text{for } \tau>\tau',\\
\Big\{N^+_{\vec{k}}\exp\left[-i\vec{k}(\vec{x}-\vec{x}')-(\varepsilon_{\vec{k}}-\mu)(\tau-\tau')\right]\\
\quad+(N^-_{\vec{k}}-1)\exp\left[i\vec{k}(\vec{x}-\vec{x}')+(\varepsilon_{\vec{k}}+\mu)(\tau-\tau')\right]\Big\}, & \text{for } \tau<\tau',
\end{cases}
$$

$$\partial_4 \equiv \frac{\partial}{\partial \tau} - \mu, \qquad f_{\vec{k}} = \frac{1}{\exp\left[\beta\omega_{\vec{k}}\right] - 1}, \qquad N_{\vec{k}}^{\pm} = \frac{1}{\exp[\beta(\varepsilon_{\vec{k}} \mp \mu)] + 1}.$$

The $f_{\vec{k}}$ is the density of the state of the Bose particles at energy $\omega_{\vec{k}}$, and the $N_{\vec{k}}^{\pm}$ is the density of the state of the (anti-)Fermi particles at energy $\varepsilon_{\vec{k}}$.

We have two ways of expressing $Z_{\text{GC}}(\beta; [J, \bar{\eta}, \eta])$, Eq. (10.4.25b):

$$\begin{aligned}
&Z_{\text{GC}}(\beta; [J, \bar{\eta}, \eta]) \\
&= Z_0 \exp\left[-\frac{1}{2}\int_0^\beta d\tau \int d^3\vec{x} \int_0^\beta d\tau' \int d^3\vec{x}' \left\{J(\tau, \vec{x}) - g\gamma_{\beta\alpha} i \frac{\delta}{\delta\eta_\beta(\tau, \vec{x})} \frac{1}{i} \frac{\delta}{\delta\bar{\eta}_\alpha(\tau, \vec{x})}\right\}\right. \\
&\left.\times D_0(\tau - \tau', \vec{x} - \vec{x}') \left\{J(\tau', \vec{x}') - g\gamma_{\beta\alpha} i \frac{\delta}{\delta\eta_\beta(\tau', \vec{x}')} \frac{1}{i} \frac{\delta}{\delta\bar{\eta}_\alpha(\tau', \vec{x}')}\right\}\right] \\
&\times \exp\left[\int_0^\beta d\tau \int d^3\vec{x} \int_0^\beta d\tau' \int d^3\vec{x}' \bar{\eta}_\alpha(\tau, \vec{x}) S_{\underset{\alpha\beta}{0}}(\tau - \tau', \vec{x} - \vec{x}') \eta_\beta(\tau', \vec{x}')\right]
\end{aligned} \tag{10.4.28}$$

$$\begin{aligned}
&= Z_0 \left\{\text{Det}\left(1 + gS_0(\tau, \vec{x})\gamma \frac{1}{i} \frac{\delta}{\delta J(\tau, \vec{x})}\right)\right\}^{-1} \exp\left[\int_0^\beta d\tau \int d^3\vec{x} \int_0^\beta d\tau' \int d^3\vec{x}'\right. \\
&\left.\times\, \bar{\eta}_\alpha(\tau, \vec{x}) \left(1 + gS_0(\tau, \vec{x})\gamma \frac{1}{i} \frac{\delta}{\delta J(\tau, \vec{x})}\right)^{-1}_{\alpha\varepsilon} S_{\underset{\varepsilon\beta}{0}}(\tau - \tau', \vec{x} - \vec{x}') \eta_\beta(\tau', \vec{x}')\right] \\
&\times \exp\left[-\frac{1}{2}\int_0^\beta d\tau \int d^3\vec{x} \int_0^\beta d\tau' \int d^3\vec{x}' J(\tau, \vec{x}) D_0(\tau - \tau', \vec{x} - \vec{x}') J(\tau', \vec{x}')\right].
\end{aligned} \tag{10.4.29}$$

The thermal expectation value of the τ-ordered function

$$f_{\tau\text{-ordered}}(\hat{\psi}, \hat{\bar{\psi}}, \hat{\phi})$$

in the grand canonical ensemble is given by

$$\begin{aligned}
&< f_{\tau\text{-ordered}}(\hat{\psi}, \hat{\bar{\psi}}, \hat{\phi}) > \equiv \frac{\text{Tr}\left\{\hat{\rho}_{\text{GC}}(\beta) f_{\tau\text{-ordered}}(\hat{\psi}, \hat{\bar{\psi}}, \hat{\phi})\right\}}{\text{Tr}\hat{\rho}_{\text{GC}}(\beta)} \\
&\equiv \frac{1}{Z_{\text{GC}}(\beta; [J, \bar{\eta}, \eta])} f\left(\frac{1}{i}\frac{\delta}{\delta\bar{\eta}}, i\frac{\delta}{\delta\eta}, \frac{1}{i}\frac{\delta}{\delta J}\right) Z_{\text{GC}}(\beta; [J, \bar{\eta}, \eta])\bigg|_{J=\bar{\eta}=\eta=0}.
\end{aligned} \tag{10.4.30}$$

According to this formula, the one-body "full" temperature Green's functions of the Bose field and the Fermi field, $D(\tau-\tau',\vec{x}-\vec{x}')$ and $S_{\alpha\beta}(\tau-\tau',\vec{x}-\vec{x}')$, are given, respectively, by

$$D(\tau-\tau',\vec{x}-\vec{x}') = \frac{\mathrm{Tr}\left\{\hat{\rho}_{\mathrm{GC}}(\beta)\mathrm{T}_\tau\left(\hat{\phi}(\tau,\vec{x})\hat{\phi}(\tau',\vec{x}')\right)\right\}}{\mathrm{Tr}\hat{\rho}_{\mathrm{GC}}(\beta)}$$
$$= -\frac{1}{Z_{\mathrm{GC}}(\beta;[J,\bar{\eta},\eta])}\frac{1}{i}\frac{\delta}{\delta J(\tau,\vec{x})}\frac{1}{i}\frac{\delta}{\delta J(\tau',\vec{x}')}Z_{\mathrm{GC}}(\beta;[J,\bar{\eta},\eta])\bigg|_{J=\bar{\eta}=\eta=0}, \quad (10.4.31)$$

$$S_{\alpha\beta}(\tau-\tau',\vec{x}-\vec{x}') = \frac{\mathrm{Tr}\left\{\hat{\rho}_{\mathrm{GC}}(\beta)\hat{\psi}_\alpha(\tau,\vec{x})\hat{\bar{\psi}}_\beta(\tau',\vec{x}')\right\}}{\mathrm{Tr}\hat{\rho}_{\mathrm{GC}}(\beta)}$$
$$= -\frac{1}{Z_{\mathrm{GC}}(\beta;[J,\bar{\eta},\eta])}\frac{1}{i}\frac{\delta}{\delta\bar{\eta}_\alpha(\tau,\vec{x})}i\frac{\delta}{\delta\eta_\beta(\tau',\vec{x}')}Z_{\mathrm{GC}}(\beta;[J,\bar{\eta},\eta])\bigg|_{J=\bar{\eta}=\eta=0}. \quad (10.4.32)$$

From the cyclicity of Tr and the (anti-)commutativity of $\hat{\phi}(\tau,\vec{x})$ $(\hat{\psi}_\alpha(\tau,\vec{x}))$ under the T_τ-ordering symbol, we have

$$D(\tau-\tau'<0,\vec{x}-\vec{x}') = +D(\tau-\tau'+\beta,\vec{x}-\vec{x}'), \quad (10.4.33)$$

and

$$S_{\alpha\beta}(\tau-\tau'<0,\vec{x}-\vec{x}') = -S_{\alpha\beta}(\tau-\tau'+\beta,\vec{x}-\vec{x}'), \quad (10.4.34)$$

where

$$0\le\tau,\tau'\le\beta,$$

i.e., the Boson (Fermion) "full" temperature Green's function is (anti-)periodic with period β. From this, we have the Fourier decompositions as

$$\hat{\phi}(\tau,\vec{x}) = \frac{1}{\beta}\sum_n\int\frac{d^3\vec{k}}{(2\pi)^3 2\omega_{\vec{k}}}\left\{\exp\left[i(\vec{k}\vec{x}-\omega_n\tau)\right]a(\omega_n,\vec{k})\right.$$
$$\left.+\exp\left[-i(\vec{k}\vec{x}-\omega_n\tau)\right]a^\dagger(\omega_n,\vec{k})\right\}, \quad (10.4.35)$$
$$\omega_n = \frac{2n\pi}{\beta}, \qquad n = \text{integer},$$

$$\hat{\psi}_\alpha(\tau,\vec{x}) = \frac{1}{\beta}\sum_n\int\frac{d^3\vec{k}}{(2\pi)^3 2\varepsilon_{\vec{k}}}\{\exp\left[i(\vec{k}\vec{x}-\omega_n\tau)\right]u_{n\alpha}(\vec{k})b(\omega_n,\vec{k})$$
$$+\exp\left[-i(\vec{k}\vec{x}-\omega_n\tau)\right]\bar{v}_{n\alpha}(\vec{k})d^\dagger(\omega_n,\vec{k})\}, \quad (10.4.36)$$
$$\omega_n = \frac{(2n+1)\pi}{\beta}, \qquad n = \text{integer},$$

where

$$[a(\omega_n,\vec{k}),a^\dagger(\omega_{n'},\vec{k}')] = 2\omega_{\vec{k}}(2\pi)^3\delta^3(\vec{k}-\vec{k}')\delta_{n,n'}, \tag{10.4.37}$$

$$[a(\omega_n,\vec{k}),a(\omega_{n'},\vec{k}')] = [a^\dagger(\omega_n,\vec{k}),a^\dagger(\omega_{n'},\vec{k}')] = 0, \tag{10.4.38}$$

$$\begin{aligned}\left\{b(\omega_n,\vec{k}),b^\dagger(\omega_{n'},\vec{k}')\right\} &= \left\{d(\omega_n,\vec{k}),d^\dagger(\omega_{n'},\vec{k}')\right\}\\ &= 2\varepsilon_{\vec{k}}(2\pi)^3\delta^3(\vec{k}-\vec{k}')\delta_{n,n'},\end{aligned} \tag{10.4.39}$$

$$\text{the rest of the anti-commutators} = 0. \tag{10.4.40}$$

We shall now address ourselves to the problem of finding the equation of motion of the one-body Boson and Fermion Green's functions. We define the one-body Boson and Fermion "full" temperature Green's functions, $D^J(x,y)$ and $S^J_{\alpha,\beta}(x,y)$, by:
for *Boson field Green's function*

$$\begin{aligned} D^J(x,y) &\equiv -<\mathrm{T}_\tau(\hat{\phi}(x)\hat{\phi}(y))>^J\Big|_{\bar{\eta}=\eta=0}\\ &\equiv -\frac{\delta^2}{\delta J(x)\delta J(y)}\ln Z_{\mathrm{GC}}(\beta;[J,\bar{\eta},\eta])\Big|_{\bar{\eta}=\eta=0}\\ &\equiv -\frac{\delta}{\delta J(x)}<\hat{\phi}(y)>^J\Big|_{\bar{\eta}=\eta=0}, \end{aligned} \tag{10.4.41}$$

and for *Fermion field Green's function*

$$\begin{aligned} S^J_{\alpha,\beta}(x,y) &\equiv +<\mathrm{T}_\tau(\hat{\psi}_\alpha(x)\hat{\bar{\psi}}_\beta(y))>^J\Big|_{\bar{\eta}=\eta=0}\\ &\equiv -\frac{1}{Z_{\mathrm{GC}}(\beta;[J,\bar{\eta},\eta])}\frac{\delta}{\delta\bar{\eta}_\alpha(x)}\frac{\delta}{\delta\eta_\beta(y)}Z_{\mathrm{GC}}(\beta;[J,\bar{\eta},\eta])\Big|_{\bar{\eta}=\eta=0}. \end{aligned} \tag{10.4.42}$$

From Eqs. (10.4.22), (10.4.23) and (10.4.24), we obtain a summary of the Schwinger–Dyson equation satisfied by $D^J(x,y)$ and $S^J_{\alpha,\beta}(x,y)$:

Summary of Schwinger–Dyson equation in configuration space

$$\left(i\gamma^\nu\partial_\nu - m + g\gamma<\hat{\phi}(x)>^J\right)_{\alpha\varepsilon}S^J_{\varepsilon\beta}(x,y) - \int d^4z\boldsymbol{\Sigma}^*_{\alpha\varepsilon}(x,z)S^J_{\varepsilon\beta}(z,y)$$
$$= \delta_{\alpha\beta}\delta^4(x-y),$$
$$\left(\frac{\partial^2}{\partial x_\nu^2}-\kappa^2\right)<\hat{\phi}(x)>^J$$
$$= \frac{1}{2}g\gamma_{\beta\alpha}\left\{S^J_{\alpha\beta}(\tau,\vec{x};\tau-\varepsilon,\vec{x}) + S^J_{\alpha\beta}(\tau,\vec{x};\tau+\varepsilon,\vec{x})\right\}\Big|_{\varepsilon\to0+},$$
$$\left(\frac{\partial^2}{\partial x_\nu^2}-\kappa^2\right)D^J(x,y) - \int d^4z\boldsymbol{\Pi}^*(x,z)D^J(z,y) = \delta^4(x-y),$$

$$\Sigma^*_{\alpha\beta}(x,y) = g^2 \int d^4u d^4v \gamma_{\alpha\delta} S^J_{\delta\nu}(x,u) \Gamma_{\nu\beta}(u,y;v) D^J(v,x),$$

$$\Pi^*(x,y) = g^2 \int d^4u d^4v \gamma_{\alpha\beta} S^J_{\beta\delta}(x,u) \Gamma_{\delta\nu}(u,v;y) S^J_{\nu\alpha}(v,x),$$

$$\Gamma_{\alpha\beta}(x,y;z) = \gamma_{\alpha\beta}(z)\delta^4(x-y)\delta^4(x-z) + \frac{1}{g}\frac{\delta\Sigma^*_{\alpha\beta}(x,y)}{\delta < \hat{\phi}(z) >^J}.$$

This system of nonlinear coupled integro-differential equations is exact and closed. Starting from the zeroth-order term of $\Gamma_{\alpha\beta}(x,y;z)$, we can develop Feynman–Dyson type graphical perturbation theory for quantum statistical mechanics in the configuration space, by iteration. We here employed the following abbreviation,

$$x \equiv (\tau_x, \vec{x}), \qquad \int d^4x \equiv \int_0^\beta d\tau_x \int d^3\vec{x}.$$

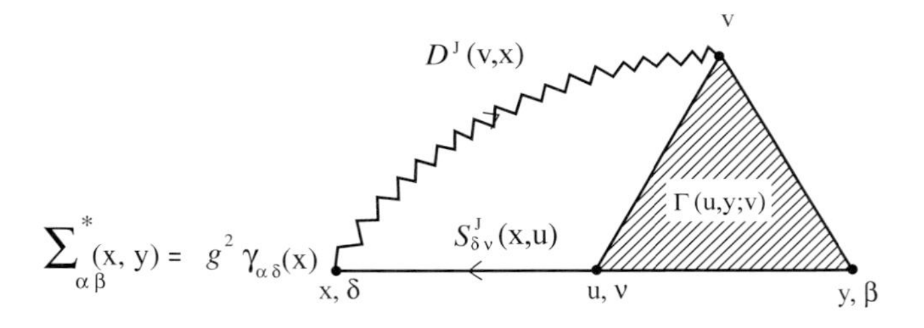

Fig. 10.9: Graphical representation of the proper self-energy part, $\Sigma^*_{\alpha\beta}(x,y)$.

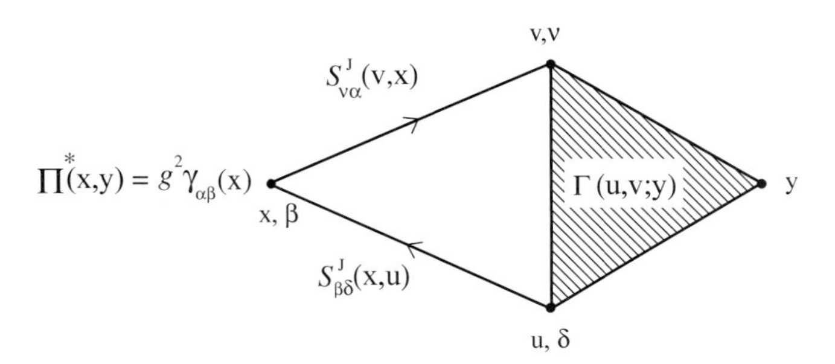

Fig. 10.10: Graphical representation of the proper self-energy part, $\Pi^*(x,y)$.

We note that $S^J_{\alpha\beta}(x,y)$ and $D^J(x,y)$ are determined by Eqs. (10.4.3a through f) only for

$$\tau_x - \tau_y \in [-\beta, \beta],$$

and we assume that they are defined by the periodic boundary condition with the period 2β for other

$$\tau_x - \tau_y \notin [-\beta, \beta].$$

Next we set

$$J \equiv 0,$$

and hence we have

$$< \hat{\phi}(x) >^{J\equiv 0} \equiv 0,$$

restoring the translational invariance of the system. We Fourier transform $S_{\alpha\beta}(x)$ and $D(x)$,

$$S_{\alpha\beta}(x) = \frac{1}{\beta}\sum_{p_4}\int \frac{d^3\vec{p}}{(2\pi)^3} S_{\alpha\beta}(p_4,\vec{p}) \exp\left[i(\vec{p}\vec{x} - p_4\tau_x)\right], \tag{10.4.43a}$$

$$D(x) = \frac{1}{\beta}\sum_{p_4}\int \frac{d^3\vec{p}}{(2\pi)^3} D(p_4,\vec{p}) \exp\left[i(\vec{p}\vec{x} - p_4\tau_x)\right], \tag{10.4.43b}$$

$$\begin{aligned} \mathbf{\Gamma}_{\alpha,\beta}(x,y;z) = \mathbf{\Gamma}_{\alpha,\beta}(x-y,x-z) = \frac{1}{\beta^2}\sum_{p_4,k_4}\int \frac{d^3\vec{p}d^3\vec{k}}{(2\pi)^6}\mathbf{\Gamma}_{\alpha,\beta}(p,k) \\ \times \exp\Big[i\left\{\vec{p}(\vec{x}-\vec{y}) - p_4(\tau_x-\tau_y)\right\} \\ - i\left\{\vec{k}(\vec{x}-\vec{z}) - k_4(\tau_x-\tau_z)\right\}\Big], \end{aligned} \tag{10.4.43c}$$

$$p_4 = \begin{cases} \dfrac{(2n+1)\pi}{\beta}, & \textit{Fermion}, \quad n = \text{integer}, \\ \dfrac{2n\pi}{\beta}, & \textit{Boson}, \quad n = \text{integer}. \end{cases} \tag{10.4.43d}$$

We have the Schwinger–Dyson equation in momentum space:

Summary of Schwinger–Dyson equation in momentum space

$$\left\{-\vec{\gamma}\vec{p} + \gamma^4(p_4 - i\mu) - (m + \mathbf{\Sigma}^*(p))\right\}_{\alpha\varepsilon} S_{\varepsilon\beta}(p) = \delta_{\alpha\beta}\sum_n \delta\left(p_4 - \frac{(2n+1)\pi}{\beta}\right),$$

$$\left\{-k_\nu^2 - (\kappa^2 + \mathbf{\Pi}^*(k))\right\} D(k) = \sum_n \delta\left(k_4 - \frac{2n\pi}{\beta}\right),$$

$$\mathbf{\Sigma}^*_{\alpha,\beta}(p) = g^2\frac{1}{\beta}\sum_{k_4}\int \frac{d^3\vec{k}}{(2\pi)^3}\gamma_{\alpha\delta}S_{\delta\varepsilon}(p+k)\mathbf{\Gamma}_{\varepsilon\beta}(p+k,k)D(k),$$

$$\mathbf{\Pi}^*(k) = g^2\frac{1}{\beta}\sum_{p_4}\int \frac{d^3\vec{p}}{(2\pi)^3}\gamma_{\mu\nu}S_{\nu\lambda}(p+k)\mathbf{\Gamma}_{\lambda\rho}(p+k,k)S_{\rho\mu}(p),$$

$$\mathbf{\Gamma}_{\alpha\beta}(p,k) = \sum_{n,,m}\gamma_{\alpha\beta}\delta\left(p_4 - \frac{(2n+1)\pi}{\beta}\right)\delta\left(k_4 - \frac{(2m+1)\pi}{\beta}\right) + \mathbf{\Lambda}_{\alpha\beta}(p,k),$$

where $\mathbf{\Lambda}_{\alpha\beta}(p,k)$ represents the sum of the vertex diagram, except for the first term. This system of nonlinear coupled integral equations is exact and closed. Starting from the zeroth-order term of $\mathbf{\Gamma}_{\alpha\beta}(p,k)$, we can develop a Feynman–Dyson type graphical perturbation theory for quantum statistical mechanics in the momentum space, by iteration.

From the Schwinger–Dyson equation, we can derive the Bethe–Goldstone diagram rule of the many-body problems at finite temperature in quantum statistical mechanics, nuclear physics and condensed matter physics. For details of this diagram rule, we refer the reader to A.L. Fetter and J.D. Walecka.

10.5 Weyl's Gauge Principle

In *electrodynamics*, we have a property known as the *gauge invariance*. In the *theory of the gravitational field*, we have a property known as the *scale invariance*. Before the birth of quantum mechanics, H. Weyl attempted to construct the unified theory of classical electrodynamics and the gravitational field. But he failed to accomplish his goal. After the birth of quantum mechanics, he realized that the gauge invariance of electrodynamics is not related to the scale invariance of the gravitational field, but is related to the invariance of the matter field $\phi(x)$ under the local phase transformation. The matter field in interaction with the electromagnetic field has a property known as the *charge conservation law* or the *current conservation law*. In this section, we discuss Weyl's gauge principle for the $U(1)$ gauge field and the non-Abelian gauge field, and Kibble's gauge principle for the gravitational field.

Weyl's gauge principle: Electrodynamics is described by the total Lagrangian density with the use of the four-vector potential $A_\mu(x)$ by

$$\mathcal{L}_{\text{tot}} = \mathcal{L}_{\text{matter}}\big(\phi(x), \partial_\mu\phi(x)\big) + \mathcal{L}_{\text{int}}\big(\phi(x), A_\mu(x)\big) + \mathcal{L}_{\text{gauge}}\big(A_\mu(x), \partial_\nu A_\mu(x)\big). \quad (10.5.1)$$

This system is invariant under the local $U(1)$ transformations,

$$A_\mu(x) \to A'_\mu(x) \equiv A_\mu(x) - \partial_\mu\varepsilon(x), \quad (10.5.2)$$

$$\phi(x) \to \phi'(x) \equiv \exp\big[iq\varepsilon(x)\big]\phi(x). \quad (10.5.3)$$

The interaction Lagrangian density $\mathcal{L}_{\text{int}}\big(\phi(x), A_\mu(x)\big)$ is generated by the substitution,

$$\partial_\mu\phi(x) \to D_\mu\phi(x) \equiv \big(\partial_\mu + iqA_\mu(x)\big)\phi(x), \quad (10.5.4)$$

in the original matter field Lagrangian density $\mathcal{L}_{\text{matter}}\big(\phi(x), \partial_\mu\phi(x)\big)$. The derivative $D_\mu\phi(x)$ is called the *covariant derivative* of $\phi(x)$ and transforms exactly as $\phi(x)$,

$$D_\mu\phi(x) \to \big(D_\mu\phi(x)\big)' = \exp\big[iq\varepsilon(x)\big]D_\mu\phi(x), \quad (10.5.5)$$

under the local $U(1)$ transformations, Eqs. (10.5.2) and (10.5.3).

The physical meaning of this local $U(1)$ invariance lies in its weaker version, the global $U(1)$ invariance, namely,

$$\varepsilon(x) = \varepsilon, \quad \text{space–time independent constant.}$$

The global $U(1)$ invariance of the matter field Lagrangian density,

$$\mathcal{L}_{\text{matter}}\big(\phi(x), \partial_\mu \phi(x)\big),$$

under the global $U(1)$ transformation of $\phi(x)$,

$$\phi(x) \to \phi''(x) = \exp\left[iq\varepsilon\right]\phi(x), \quad \varepsilon = \text{constant}, \tag{10.5.6}$$

in its infinitesimal version,

$$\delta\phi(x) = iq\varepsilon\phi(x), \quad \varepsilon = \text{infinitesimal constant}, \tag{10.5.7}$$

results in

$$\frac{\partial \mathcal{L}_{\text{matter}}\big(\phi(x), \partial_\mu \phi(x)\big)}{\partial \phi(x)}\delta\phi(x) + \frac{\partial \mathcal{L}_{\text{matter}}\big(\phi(x), \partial_\mu \phi(x)\big)}{\partial\big(\partial_\mu \phi(x)\big)}\delta\big(\partial_\mu\phi(x)\big) = 0. \tag{10.5.8}$$

With the use of the Euler–Lagrange equation of motion for $\phi(x)$,

$$\frac{\partial \mathcal{L}_{\text{matter}}\big(\phi(x), \partial_\mu \phi(x)\big)}{\partial \phi(x)} - \partial_\mu \frac{\partial \mathcal{L}_{\text{matter}}\big(\phi(x), \partial_\mu \phi(x)\big)}{\partial\big(\partial_\mu \phi(x)\big)} = 0,$$

we obtain the current conservation law,

$$\partial_\mu J^\mu_{\text{matter}}(x) = 0, \tag{10.5.9}$$

$$\varepsilon J^\mu_{\text{matter}}(x) = \frac{\partial \mathcal{L}_{\text{matter}}\big(\phi(x), \partial_\mu \phi(x)\big)}{\partial\big(\partial_\mu \phi(x)\big)}\delta\phi(x). \tag{10.5.10}$$

This in its integrated form becomes the charge conservation law,

$$\frac{d}{dt}Q_{\text{matter}}(t) = 0, \tag{10.5.11}$$

$$Q_{\text{matter}}(t) = \int d^3\vec{x} J^0_{\text{matter}}(t, \vec{x}). \tag{10.5.12}$$

Weyl's gauge principle considers the analysis backwards. The extension of the "current conserving" global $U(1)$ invariance, Eqs. (10.5.7) and (10.5.8), of the matter field Lagrangian density $\mathcal{L}_{\text{matter}}\big(\phi(x), \partial_\mu \phi(x)\big)$ to the local $U(1)$ invariance necessitates:

1) the introduction of the $U(1)$ gauge field $A_\mu(x)$, and the replacement of the derivative $\partial_\mu\phi(x)$ in the matter field Lagrangian density with the covariant derivative $D_\mu\phi(x)$,

$$\partial_\mu\phi(x) \to D_\mu\phi(x) \equiv \big(\partial_\mu + iqA_\mu(x)\big)\phi(x), \tag{10.5.13}$$

and

2) the requirement that the covariant derivative $D_\mu\phi(x)$ transforms exactly like the matter field $\phi(x)$ under the local $U(1)$ phase transformation of $\phi(x)$,

$$D_\mu\phi(x) \to \big(D_\mu\phi(x)\big)' = \exp\big[iq\varepsilon(x)\big]D_\mu\phi(x), \tag{10.5.14}$$

under

$$\phi(x) \to \phi'(x) \equiv \exp\big[iq\varepsilon(x)\big]\phi(x). \tag{10.5.15}$$

From requirement 2), we obtain the transformation law of the $U(1)$ gauge field $A_\mu(x)$ immediately,

$$A_\mu(x) \to A'_\mu(x) \equiv A_\mu(x) - \partial_\mu \varepsilon(x). \tag{10.5.16}$$

From requirement 2), the local $U(1)$ invariance of $\mathcal{L}_{\text{matter}}\big(\phi(x), D_\mu\phi(x)\big)$ is also self-evident. In order to give dynamical content to the $U(1)$ gauge field, we introduce the field strength tensor $F_{\mu\nu}(x)$ to the gauge field Lagrangian density $\mathcal{L}_{\text{gauge}}\big(A_\mu(x), \partial_\nu A_\mu(x)\big)$ by the trick,

$$[D_\mu, D_\nu]\phi(x) = iq\big(\partial_\mu A_\nu(x) - \partial_\nu A_\mu(x)\big)\phi(x) \equiv iqF_{\mu\nu}(x)\phi(x). \tag{10.5.17}$$

From the transformation law of the $U(1)$ gauge field $A_\mu(x)$, Eq. (10.5.2), we observe that the field strength tensor $F_{\mu\nu}(x)$ is the locally invariant quantity,

$$\begin{aligned} F_{\mu\nu}(x) \to F'_{\mu\nu}(x) &= \partial_\mu A'_\nu(x) - \partial_\nu A'_\mu(x) \\ &= \partial_\mu A_\nu(x) - \partial_\nu A_\mu(x) = F_{\mu\nu}(x). \end{aligned} \tag{10.5.18}$$

As the gauge field Lagrangian density $\mathcal{L}_{\text{gauge}}\big(A_\mu(x), \partial_\nu A_\mu(x)\big)$, we choose

$$\mathcal{L}_{\text{gauge}}\big(A_\mu(x), \partial_\nu A_\mu(x)\big) = -\frac{1}{4}F_{\mu\nu}(x)F^{\mu\nu}(x). \tag{10.5.19}$$

In this manner, we obtain the total Lagrangian density of the matter-gauge system which is locally $U(1)$ invariant as

$$\mathcal{L}_{\text{tot}} = \mathcal{L}_{\text{matter}}\big(\phi(x), D_\mu\phi(x)\big) + \mathcal{L}_{\text{gauge}}\big(F_{\mu\nu}(x)\big). \tag{10.5.20}$$

We obtain the interaction Lagrangian density $\mathcal{L}_{\text{int}}\big(\phi(x), A_\mu(x)\big)$ as

$$\mathcal{L}_{\text{int}}\big(\phi(x), A_\mu(x)\big) = \mathcal{L}_{\text{matter}}\big(\phi(x), D_\mu\phi(x)\big) - \mathcal{L}_{\text{matter}}\big(\phi(x), \partial_\mu\phi(x)\big), \tag{10.5.21}$$

which is the universal coupling generated by Weyl's gauge principle. As a result of the local extension of the global $U(1)$ invariance, we have derived the electrodynamics from the current conservation law, Eq. (10.5.9), or the charge conservation law, Eq. (10.5.11).

We shall now consider the extension of the present discussion to the non-Abelian gauge field. We let the semi-simple Lie group G be the gauge group. We let the representation of G in the Hilbert space be $U(g)$, and its matrix representation on the field operator $\hat{\psi}_n(x)$ in the internal space be $D(g)$,

$$U(g)\hat{\psi}_n(x)U^{-1}(g) = D_{n,m}(g)\hat{\psi}_m(x), \quad g \in G. \tag{10.5.22}$$

For the element $g_\varepsilon \in G$ continuously connected to the identity of G by the parameter $\{\varepsilon_\alpha\}_{\alpha=1}^{N}$, we have

$$U(g_\varepsilon) = \exp\left[i\varepsilon_\alpha T_\alpha\right] = 1 + i\varepsilon_\alpha T_\alpha + \cdots, \quad T_\alpha : \text{generator of Lie group } G, \tag{10.5.23}$$

$$D(g_\varepsilon) = \exp\left[i\varepsilon_\alpha t_\alpha\right] = 1 + i\varepsilon_\alpha t_\alpha + \cdots, \quad t_\alpha : \text{realization of } T_\alpha \text{ on } \hat{\psi}_n(x), \tag{10.5.24}$$

$$[T_\alpha, T_\beta] = iC_{\alpha\beta\gamma}T_\gamma, \tag{10.5.25}$$

$$[t_\alpha, t_\beta] = iC_{\alpha\beta\gamma}t_\gamma. \tag{10.5.26}$$

We shall assume that the action functional $I_{\text{matter}}[\psi_n]$ of the matter field Lagrangian density $\mathcal{L}_{\text{matter}}(\psi_n(x), \partial_\mu\psi_n(x))$, given by

$$I_{\text{matter}}[\psi_n] \equiv \int d^4x \mathcal{L}_{\text{matter}}(\psi_n(x), \partial_\mu\psi_n(x)), \tag{10.5.27}$$

is invariant under the global G transformation,

$$\delta\psi_n(x) = i\varepsilon_\alpha(t_\alpha)_{n,m}\psi_m(x), \quad \varepsilon_\alpha = \text{infinitesimal constant.} \tag{10.5.28}$$

Namely, we have

$$\frac{\partial\mathcal{L}_{\text{matter}}(\psi_n(x), \partial_\mu\psi_n(x))}{\partial\psi_n(x)}\delta\psi_n(x) + \frac{\partial\mathcal{L}_{\text{matter}}(\psi_n(x), \partial_\mu\psi_n(x))}{\partial(\partial_\mu\psi_n(x))}\delta(\partial_\mu\psi_n(x)) = 0. \tag{10.5.29}$$

By the use of the Euler–Lagrange equation of motion,

$$\frac{\partial\mathcal{L}_{\text{matter}}(\psi(x), \partial_\mu\psi(x))}{\partial\psi_n(x)} - \partial_\mu\left(\frac{\partial\mathcal{L}_{\text{matter}}(\psi(x), \partial_\mu\psi(x))}{\partial(\partial_\mu\psi_n(x))}\right) = 0, \tag{10.5.30}$$

we have the current conservation law and the charge conservation law,

$$\partial_\mu J^\mu_{\alpha,\text{matter}}(x) = 0, \qquad \alpha = 1, \ldots, N, \tag{10.5.31a}$$

where the conserved matter current $J^\mu_{\alpha,\text{matter}}(x)$ is given by

$$\varepsilon_\alpha J^\mu_{\alpha,\text{matter}}(x) = \frac{\partial\mathcal{L}_{\text{matter}}(\psi(x), \partial_\mu\psi(x))}{\partial(\partial_\mu\psi_n(x))}\delta\psi_n(x), \tag{10.5.31b}$$

and

$$\frac{d}{dt}Q_\alpha^{\text{matter}}(t) = 0, \qquad \alpha = 1, \ldots, N, \tag{10.5.32a}$$

where the conserved matter charge $Q_\alpha^{\text{matter}}(t)$ is given by

$$Q_\alpha^{\text{matter}}(t) = \int d^3\vec{x} J^0_{\alpha,\text{matter}}(t, \vec{x}), \qquad \alpha = 1, \ldots, N. \tag{10.5.32b}$$

Invoking Weyl's gauge principle, we extend the global G invariance of the matter system to the local G invariance of the matter-gauge system under the local G phase transformation,

$$\delta\psi_n(x) = i\varepsilon_\alpha(x)(t_\alpha)_{n,m}\psi_m(x). \tag{10.5.33}$$

Weyl's gauge principle requires the following:

1) the introduction of the non-Abelian gauge field $A_{\alpha\mu}(x)$ and the replacement of the derivative $\partial_\mu\psi_n(x)$ in the matter field Lagrangian density with the covariant derivative $\left(D_\mu\psi(x)\right)_n$,

$$\partial_\mu\psi_n(x) \to \left(D_\mu\psi(x)\right)_n \equiv \left(\partial_\mu\delta_{n,m} + i(t_\gamma)_{n,m}A_{\gamma\mu}(x)\right)\psi_m(x), \tag{10.5.34}$$

and

2) the requirement that the covariant derivative $\left(D_\mu\psi(x)\right)_n$ transforms exactly as the matter field $\psi_n(x)$ under the local G phase transformation of $\psi_n(x)$, Eq. (10.5.33),

$$\delta\left(D_\mu\psi(x)\right)_n = i\varepsilon_\alpha(x)(t_\alpha)_{n,m}\left(D_\mu\psi(x)\right)_m, \tag{10.5.35}$$

where t_γ is the realization of the generator T_γ upon the multiplet $\psi_n(x)$.

From Eqs. (10.5.33) and (10.5.35), the infinitesimal transformation law of the non-Abelian gauge field $A_{\alpha\mu}(x)$ follows,

$$\delta A_{\alpha\mu}(x) = -\partial_\mu\varepsilon_\alpha(x) + i\varepsilon_\beta(x)(t_\beta^{\mathrm{adj}})_{\alpha\gamma}A_{\gamma\mu}(x) \tag{10.5.36a}$$

$$= -\partial_\mu\varepsilon_\alpha(x) + \varepsilon_\beta(x)C_{\beta\alpha\gamma}A_{\gamma\mu}(x). \tag{10.5.36b}$$

Then the local G invariance of the gauged matter field Lagrangian density $\mathcal{L}_{\text{matter}}\left(\psi(x), D_\mu\psi(x)\right)$ becomes self-evident as long as the ungauged matter field Lagrangian density $\mathcal{L}_{\text{matter}}\left(\psi(x), \partial_\mu\psi(x)\right)$ is globally G invariant.

In order to provide dynamical content to the non-Abelian gauge field $A_{\alpha\mu}(x)$, we introduce the field strength tensor $F_{\gamma\mu\nu}(x)$ by the following trick,

$$[D_\mu, D_\nu]\psi(x) \equiv i(t_\gamma)F_{\gamma\mu\nu}(x)\psi(x), \tag{10.5.37}$$

$$F_{\gamma\mu\nu}(x) = \partial_\mu A_{\gamma\nu}(x) - \partial_\nu A_{\gamma\mu}(x) - C_{\alpha\beta\gamma}A_{\alpha\mu}(x)A_{\beta\nu}(x). \tag{10.5.38}$$

We can easily show that the field strength tensor $F_{\gamma\mu\nu}(x)$ undergoes local G rotation under local G transformations, Eqs. (10.5.33) and (10.5.36a), under the adjoint representation,

$$\delta F_{\gamma\mu\nu}(x) = i\varepsilon_\alpha(x)(t_\alpha^{\mathrm{adj}})_{\gamma\beta}F_{\beta\mu\nu}(x) \tag{10.5.39a}$$

$$= \varepsilon_\alpha(x)C_{\alpha\gamma\beta}F_{\beta\mu\nu}(x). \tag{10.5.39b}$$

As the Lagrangian density of the non-Abelian gauge field $A_{\alpha\mu}(x)$, we choose

$$\mathcal{L}_{\text{gauge}}\big(A_{\gamma\mu}(x), \partial_\nu A_{\gamma\mu}(x)\big) \equiv -\frac{1}{4} F_{\gamma\mu\nu}(x) F_{\gamma}^{\mu\nu}(x). \tag{10.5.40}$$

The total Lagrangian density $\mathcal{L}_{\text{total}}$ of the matter-gauge system is given by

$$\mathcal{L}_{\text{total}} = \mathcal{L}_{\text{matter}}\big(\psi(x), D_\mu\psi(x)\big) + \mathcal{L}_{\text{gauge}}\big(F_{\gamma\mu\nu}(x)\big). \tag{10.5.41}$$

The interaction Lagrangian density $\mathcal{L}_{\text{int}}$ consists of two parts due to the nonlinearity of the field strength tensor $F_{\gamma\mu\nu}(x)$ with respect to $A_{\gamma\mu}(x)$,

$$\begin{aligned}\mathcal{L}_{\text{int}} = {} & \mathcal{L}_{\text{matter}}\big(\psi(x), D_\mu\psi(x)\big) - \mathcal{L}_{\text{matter}}\big(\psi(x), \partial_\mu\psi(x)\big) \\ & + \mathcal{L}_{\text{gauge}}\big(F_{\gamma\mu\nu}(x)\big) - \mathcal{L}_{\text{gauge}}^{\text{quad}}\big(F_{\gamma\mu\nu}(x)\big),\end{aligned} \tag{10.5.42}$$

which provides the universal coupling just as the $U(1)$ gauge field theory. The conserved current $J^{\mu}_{\alpha,\text{total}}(x)$ and the conserved charge $\left\{Q_{\alpha}^{\text{total}}(t)\right\}_{\alpha=1}^{N}$ after the extension to the local G invariance also consist of two parts,

$$J^{\mu}_{\alpha,\text{total}}(x) \equiv J^{\mu\ \text{gauged}}_{\alpha,\text{matter}}(x) + J^{\mu}_{\alpha,\text{gauge}}(x) \equiv \frac{\delta I_{\text{total}}[\psi, A_{\alpha\mu}]}{\delta A_{\alpha\mu}(x)}, \tag{10.5.43a}$$

$$I_{\text{total}}[\psi, A_{\alpha\mu}] = \int d^4x \Big\{\mathcal{L}_{\text{matter}}\big(\psi(x), D_\mu\psi(x)\big) + \mathcal{L}_{\text{gauge}}\big(F_{\gamma\mu\nu}(x)\big)\Big\}, \tag{10.5.43b}$$

$$Q_{\alpha}^{\text{total}}(t) = Q_{\alpha}^{\text{matter}}(t) + Q_{\alpha}^{\text{gauge}}(t) = \int d^3\vec{x}\, \left\{J^{0\ \text{gauged}}_{\alpha,\text{matter}}(t,\vec{x}) + J^{0}_{\alpha,\text{gauge}}(t,\vec{x})\right\}. \tag{10.5.44}$$

We note that the gauged matter current $J^{\mu\ \text{gauged}}_{\alpha,\text{matter}}(x)$ of Eq. (10.5.43a) is not identical to the ungauged matter current $J^{\mu}_{\alpha,\text{matter}}(x)$ of Eq. (10.5.31b):

$$\varepsilon_\alpha J^{\mu}_{\alpha,\text{matter}}(x) \text{ of (10.5.31b)} = \frac{\partial \mathcal{L}_{\text{matter}}\big(\psi(x), \partial_\mu\psi(x)\big)}{\partial\big(\partial_\mu\psi_n(x)\big)} \delta\psi_n(x),$$

whereas after the local G extension,

$$\begin{aligned}\varepsilon_\alpha J^{\mu\ \text{gauged}}_{\alpha,\text{matter}}(x) \text{ of (10.5.43a)} &= \frac{\partial \mathcal{L}_{\text{matter}}\big(\psi(x), D_\mu\psi(x)\big)}{\partial\big(D_\mu\psi_n(x)\big)_n} \delta\psi_n(x) \\ &= \frac{\partial \mathcal{L}_{\text{matter}}\big(\psi(x), D_\mu\psi(x)\big)}{\partial\big(D_\mu\psi(x)\big)_n} i\varepsilon_\alpha (t_\alpha)_{n,m}\psi_m(x) \\ &= \varepsilon_\alpha \frac{\partial \mathcal{L}_{\text{matter}}\big(\psi(x), D_\mu\psi(x)\big)}{\partial(D_\nu\psi(x))_n} \frac{\partial\big(D_\nu\psi(x)\big)_n}{\partial A_{\alpha\mu}(x)} \\ &= \varepsilon_\alpha \frac{\partial \mathcal{L}_{\text{matter}}\big(\psi(x), D_\mu\psi(x)\big)}{\partial A_{\alpha\mu}(x)} \\ &= \varepsilon_\alpha \frac{\delta}{\delta A_{\alpha\mu}(x)} I_{\text{matter}}^{\text{gauged}}[\psi, D_\mu\psi].\end{aligned} \tag{10.5.45}$$

Here we note that $I_{\text{matter}}^{\text{gauged}}[\psi, D_\mu\psi]$ is not identical to the ungauged matter action functional $I_{\text{matter}}[\psi_n]$ given by Eq. (10.5.27), but is the gauged matter action functional defined by

$$I_{\text{matter}}^{\text{gauged}}[\psi, D_\mu\psi] \equiv \int d^4x \mathcal{L}_{\text{matter}}(\psi(x), D_\mu\psi(x)). \tag{10.5.46}$$

We emphasize here that the conserved Noether current after the extension of the global G invariance to the local G invariance, is not the gauged matter current $J_{\alpha,\text{matter}}^{\mu\ \text{gauged}}(x)$ but the total current $J_{\alpha,\text{total}}^{\mu}(x)$, Eq. (10.5.43a). At the same time, we note that the strict conservation law of the total current $J_{\alpha,\text{total}}^{\mu}(x)$ is enforced at the expense of loss of covariance as we shall see. The origin of this problem is the self-interaction of the non-Abelian gauge field $A_{\alpha\mu}(x)$ and the nonlinearity of the Euler–Lagrange equation of motion for the non-Abelian gauge field $A_{\alpha\mu}(x)$.

We shall make a table of the global $U(1)$ transformation law and the global G transformation law.

Global $U(1)$ transformation law.		Global G transformation law.	
$\delta\psi_n(x) = i\varepsilon q_n\psi_n(x),$	charged.	$\delta\psi_n(x) = i\varepsilon_\alpha(t_\alpha)_{n,m}\psi_m(x),$	charged.
$\delta A_\mu(x) = 0,$	neutral.	$\delta A_{\alpha\mu}(x) = i\varepsilon_\beta(t_\beta^{\text{adj}})_{\alpha\gamma}A_{\gamma\mu}(x),$	charged.

(10.5.47)

In the global transformation law of internal symmetry, Eq. (10.5.47), the matter fields $\psi_n(x)$ which have the group charge undergo global ($U(1)$ or G) rotation. As for the gauge fields, $A_\mu(x)$ and $A_{\alpha\mu}(x)$, the Abelian gauge field $A_\mu(x)$ remains unchanged under global $U(1)$ transformation while the non-Abelian gauge field $A_{\alpha\mu}(x)$ undergoes global G rotation under global G transformation. Hence the Abelian gauge field $A_\mu(x)$ is $U(1)$-neutral while the non-Abelian gauge field $A_{\alpha\mu}(x)$ is G-charged. The field strength tensors, $F_{\mu\nu}(x)$ and $F_{\alpha\mu\nu}(x)$, behave as $A_\mu(x)$ and $A_{\alpha\mu}(x)$, under global $U(1)$ and G transformations. The field strength tensor $F_{\mu\nu}(x)$ is $U(1)$-neutral, while the field strength tensor $F_{\alpha\mu\nu}(x)$ is G-charged, which originates from their linearity and nonlinearity in $A_\mu(x)$ and $A_{\alpha\mu}(x)$, respectively.

Global $U(1)$ transformation law.		Global G transformation law.	
$\delta F_{\mu\nu}(x) = 0,$	neutral.	$\delta F_{\alpha\mu\nu}(x) = i\varepsilon_\beta(t_\beta^{\text{adj}})_{\alpha\gamma}F_{\gamma\mu\nu}(x),$	charged.

(10.5.48)

When we write the Euler–Lagrange equation of motion for each case, the linearity and the nonlinearity with respect to the gauge fields become clear.

Abelian $U(1)$ gauge field.		non-Abelian G gauge field.	
$\partial_\nu F^{\nu\mu}(x) = j_{\text{matter}}^{\mu}(x),$	linear.	$D_\nu^{\text{adj}}F_\alpha^{\nu\mu}(x) = j_{\alpha,\text{matter}}^{\mu}(x),$	nonlinear.

(10.5.49)

From the anti-symmetry of the field strength tensor with respect to the Lorentz indices, μ and ν, we have the following current conservation as an identity.

Abelian $U(1)$ gauge field.	non-Abelian G gauge field.
$\partial_\mu j_{\text{matter}}^{\mu}(x) = 0.$	$D_\mu^{\text{adj}} j_{\alpha,\text{matter}}^{\mu}(x) = 0.$

(10.5.50)

Here we have

$$D_\mu^{\text{adj}} = \partial_\mu + it_\gamma^{\text{adj}} A_{\gamma\mu}(x). \tag{10.5.51}$$

As the result of the extension to the local ($U(1)$ or G) invariance, in the case of the Abelian $U(1)$ gauge field, due to the neutrality of $A_\mu(x)$, the matter current $j^\mu_{\text{matter}}(x)$ alone which originates from the global $U(1)$ invariance is conserved, while in the case of the non-Abelian G gauge field, due to the G-charge of $A_{\alpha\mu}(x)$, the gauged matter current $j^\mu_{\alpha,\text{matter}}(x)$ alone which originates from the local G invariance is not conserved, but the sum with the gauge current $j^\mu_{\alpha,\text{gauge}}(x)$ which originates from the self-interaction of the non-Abelian gauge field $A_{\alpha\mu}(x)$ is conserved at the expense of the loss of covariance. A similar situation exists for the charge conservation law.

$$\begin{array}{ll} \text{Abelian } U(1) \text{ gauge field.} & \text{non-Abelian } G \text{ gauge field.} \\ \frac{d}{dt} Q^{\text{matter}}(t) = 0. & \frac{d}{dt} Q^{\text{tot}}_\alpha(t) = 0. \\ Q^{\text{matter}}(t) = \int d^3\vec{x}\; j^0_{\text{matter}}(t,\vec{x}). & Q^{\text{tot}}_\alpha(t) = \int d^3\vec{x}\; j^0_{\alpha,\text{tot}}(t,\vec{x}). \end{array} \tag{10.5.52}$$

Before plunging into the gravitational field, we discuss the finite gauge transformation property of the non-Abelian gauge field $A_{\alpha\mu}(x)$. Under the finite local G phase transformation of $\psi_n(x)$,

$$\psi_n(x) \to \psi'_n(x) = (\exp[i\varepsilon_\alpha(x)t_\alpha])_{n,m}\psi_m(x), \tag{10.5.53}$$

we demand that the covariant derivative $D_\mu\psi(x)$ defined by Eq. (10.5.34) transforms exactly as $\psi(x)$,

$$\begin{aligned} D_\mu\psi(x) \to (D_\mu\psi(x))' &= (\partial_\mu + it_\gamma A'_{\gamma\mu}(x))\psi'(x) \\ &= \exp[i\varepsilon_\alpha(x)t_\alpha]\, D_\mu\psi(x). \end{aligned} \tag{10.5.54}$$

From Eq. (10.5.54), we obtain the following equation,

$$\exp[-i\varepsilon_\alpha(x)t_\alpha]\,(\partial_\mu + it_\gamma A'_{\gamma\mu}(x))\exp[i\varepsilon_\alpha(x)t_\alpha]\,\psi(x) = (\partial_\mu + it_\gamma A_{\gamma\mu}(x))\psi(x).$$

Cancelling the $\partial_\mu\psi(x)$ term from both sides of the equation above, we obtain

$$\begin{aligned} &\exp[-i\varepsilon_\alpha(x)t_\alpha]\,(\partial_\mu \exp[i\varepsilon_\alpha(x)t_\alpha]) + \exp[-i\varepsilon_\alpha(x)t_\alpha]\,(it_\gamma A'_{\gamma\mu}(x))\exp[i\varepsilon_\alpha(x)t_\alpha] \\ &= it_\gamma A_{\gamma\mu}(x). \end{aligned}$$

Solving the above equation for $t_\gamma A'_{\gamma\mu}(x)$, we finally obtain the finite gauge transformation law of $A'_{\gamma\mu}(x)$,

$$\begin{aligned} t_\gamma A'_{\gamma\mu}(x) = \exp[i\varepsilon_\alpha(x)t_\alpha]\,\{t_\gamma A_{\gamma\mu}(x) \\ + \exp[-i\varepsilon_\beta(x)t_\beta]\,(i\partial_\mu \exp[i\varepsilon_\beta(x)t_\beta])\}\exp[-i\varepsilon_\alpha(x)t_\alpha]. \end{aligned} \tag{10.5.55}$$

At first sight, we get the impression that the finite gauge transformation law of $A'_{\gamma\mu}(x)$, (10.5.55), may depend on the specific realization $\{t_\gamma\}_{\gamma=1}^{N}$ of the generator $\{T_\gamma\}_{\gamma=1}^{N}$ upon the multiplet $\psi(x)$. Actually $A'_{\gamma\mu}(x)$ transforms under the adjoint representation $\{t_\gamma^{\text{adj}}\}_{\gamma=1}^{N}$. The infinitesimal version of the finite gauge transformation, (10.5.55), does reduce to the infinitesimal gauge transformation, (10.5.36a) and (10.5.36b), under the adjoint representation.

The first step of an extension of Weyl's gauge principle to the non-Abelian gauge group G was carried out by C.N. Yang and R.L. Mills for the $SU(2)$ isospin gauge group and the said gauge field is commonly called the Yang–Mills gauge field.

Furthermore, we can generalize Weyl's gauge principle to Utiyama's gauge principle and Kibble's gauge principle to obtain the Lagrangian density for the gravitational field. We note that Weyl's gauge principle, Utiyama's gauge principle and Kibble's gauge principle belong to the category of the invariant variational principle.

R. Utiyama derived the theory of the gravitational field from his version of the gauge principle, based on the requirement of the invariance of the action functional $I[\phi]$ under the *local six-parameter Lorentz transformation*. T.W.B. Kibble derived the theory of the gravitational field from his version of the gauge principle, based on the requirement of the invariance of the action functional $I[\phi]$ under the *local ten-parameter Poincaré transformation*, extending the treatment of Utiyama.

Gravitational field: We shall now discuss Kibble's gauge principle for the gravitational field. We let ϕ represent the set of generic matter field variables $\phi_a(x)$, which we regard as the elements of a column vector $\phi(x)$, and define the matter action functional $I_{\text{matter}}[\phi]$ in terms of the matter Lagrangian density $\mathcal{L}_{\text{matter}}(\phi, \partial_\mu\phi)$ as

$$I_{\text{matter}}[\phi] = \int d^4x \mathcal{L}_{\text{matter}}(\phi, \partial_\mu\phi). \tag{10.5.56}$$

We first discuss the infinitesimal transformation of both the coordinates x^μ and the matter field variables $\phi(x)$,

$$x^\mu \to x'^\mu = x^\mu + \delta x^\mu, \quad \phi(x) \to \phi'(x') = \phi(x) + \delta\phi(x), \tag{10.5.57}$$

where the invariance group G is not specified. It is convenient to allow the possibility that the matter Lagrangian density $\mathcal{L}_{\text{matter}}$ explicitly depends on the coordinates x^μ. Then, under the infinitesimal transformation, (10.5.57), we have

$$\delta\mathcal{L}_{\text{matter}} \equiv \frac{\partial\mathcal{L}_{\text{matter}}}{\partial\phi}\delta\phi + \frac{\partial\mathcal{L}_{\text{matter}}}{\partial(\partial_\mu\phi)}\delta(\partial_\mu\phi) + \left.\frac{\partial\mathcal{L}_{\text{matter}}}{\partial x^\mu}\right|_{\phi \text{ fixed}} \delta x^\mu.$$

It is also useful to consider the variation of $\phi(x)$ at a fixed value of x^μ,

$$\delta_0\phi \equiv \phi'(x) - \phi(x) = \delta\phi - \delta x^\mu \partial_\mu\phi. \tag{10.5.58}$$

It is obvious that δ_0 commutes with ∂_μ, so we have

$$\delta(\partial_\mu\phi) = \partial_\mu(\delta\phi) - (\partial_\mu\delta x^\nu)\partial_\nu\phi. \tag{10.5.59}$$

The matter action functional, (10.5.56), over a space–time region Ω is transformed under the transformations, (10.5.57), into

$$I'_{\text{matter}}[\Omega] \equiv \int_{\Omega} \mathcal{L}'_{\text{matter}}(x') \det(\partial_\nu {x'}^{\mu}) d^4x.$$

Thus the matter action functional $I_{\text{matter}}[\Omega]$ over an arbitrary region Ω is invariant if

$$\delta\mathcal{L}_{\text{matter}} + (\partial_\mu \delta x^\mu)\mathcal{L}_{\text{matter}} \equiv \delta_0\mathcal{L}_{\text{matter}} + \partial_\mu(\delta x^\mu \mathcal{L}_{\text{matter}}) \equiv 0. \tag{10.5.60}$$

We now consider the specific case of the Poincaré transformation,

$$\delta x^\mu = i\varepsilon^\mu_\nu x^\nu + \varepsilon^\mu, \quad \delta\phi = \frac{1}{2} i\varepsilon^{\mu\nu} S_{\mu\nu}\phi, \tag{10.5.61}$$

where $\{\varepsilon^\mu\}$ and $\{\varepsilon^{\mu\nu}\}$ with $\varepsilon^{\mu\nu} = -\varepsilon^{\nu\mu}$, are the 10 infinitesimal constant parameters of the Poincaré group, and $S_{\mu\nu}$ with $S_{\mu\nu} = -S_{\nu\mu}$, are the mixing matrices of the components of a column vector $\phi(x)$ satisfying

$$[S_{\mu\nu}, S_{\rho\sigma}] = i(\eta_{\nu\rho}S_{\mu\sigma} + \eta_{\mu\sigma}S_{\nu\rho} - \eta_{\nu\sigma}S_{\mu\rho} - \eta_{\mu\rho}S_{\nu\sigma}).$$

$\{S_{\mu\nu}\}$ will be identified as the spin matrices of the matter field ϕ later. From Eq. (10.5.59), we have

$$\delta(\partial_\mu\phi) = \frac{1}{2} i\varepsilon^{\rho\sigma} S_{\rho\sigma}\partial_\mu\phi - i\varepsilon^\rho_\mu \partial_\rho\phi. \tag{10.5.62}$$

Since we have $\partial_\mu(\delta x^\mu) = \varepsilon^\mu_\mu = 0$, the condition, (10.5.60), for the invariance of the matter action functional $I_{\text{matter}}[\phi]$ under the infinitesimal Poincaré transformations, (10.5.61), reduces to

$$\delta\mathcal{L}_{\text{matter}} \equiv 0,$$

and results in the 10 identities,

$$\frac{\partial\mathcal{L}_{\text{matter}}}{\partial x^\rho} \equiv \partial_\rho\mathcal{L}_{\text{matter}} - \frac{\partial\mathcal{L}_{\text{matter}}}{\partial\phi}\partial_\rho\phi - \frac{\partial\mathcal{L}_{\text{matter}}}{\partial(\partial_\mu\phi)}\partial_\rho\partial_\mu\phi \equiv 0, \tag{10.5.63}$$

$$\frac{\partial\mathcal{L}_{\text{matter}}}{\partial\phi} iS_{\rho\sigma}\phi + \frac{\partial\mathcal{L}_{\text{matter}}}{\partial(\partial_\mu\phi)}(iS_{\rho\sigma}\partial_\mu\phi + \eta_{\mu\rho}\partial_\sigma\phi - \eta_{\mu\sigma}\partial_\rho\phi) \equiv 0. \tag{10.5.64}$$

The conditions (10.5.63) express the translational invariance of the system and are equivalent to the requirement that $\mathcal{L}_{\text{matter}}$ is explicitly independent of x^μ. We use the Euler–Lagrange equations of motion in Eqs. (10.5.63) and (10.5.64), obtaining the ten conservation laws, which we write as

$$\partial_\mu T^\mu_\rho = 0, \quad \partial_\mu(S^\mu_{\rho\sigma} - x_\rho T^\mu_\sigma + x_\sigma T^\mu_\rho) = 0, \tag{10.5.65}$$

$$T^\mu_\rho \equiv \frac{\partial\mathcal{L}_{\text{matter}}}{\partial(\partial_\mu\phi)}\partial_\rho\phi - \delta^\mu_\rho\mathcal{L}_{\text{matter}}, \quad S^\mu_{\rho\sigma} \equiv -i\frac{\partial\mathcal{L}_{\text{matter}}}{\partial(\partial_\mu\phi)}S_{\rho\sigma}\phi. \tag{10.5.66}$$

The ten conservation laws, Eqs. (10.5.63) and (10.5.64), are the conservation laws of energy, momentum and angular momentum. Thus $\{S_{\mu\nu}\}$ are the spin matrices of the matter field $\phi(x)$.

We shall also examine the transformations in terms of the variation $\delta_0\phi$, which in this case is

$$\delta_0\phi = -i\varepsilon^\rho\partial_\rho\phi + \frac{1}{2}i\varepsilon^{\rho\sigma}\left(S_{\rho\sigma} + x_\rho\frac{1}{i}\partial_\sigma - x_\sigma\frac{1}{i}\partial_\rho\right)\phi. \tag{10.5.67}$$

On comparing with Weyl's gauge principle, the role of the realizations $\{t_\alpha\}$ of the generators $\{T_\alpha\}$ upon the multiplet ϕ is played by the differential operators,

$$\frac{1}{i}\partial_\rho, \quad \text{and} \quad S_{\rho\sigma} + x_\rho\frac{1}{i}\partial_\sigma - x_\sigma\frac{1}{i}\partial_\rho.$$

Then, by the definition of the currents, we expect the currents corresponding to ε^ρ and $\varepsilon^{\rho\sigma}$ to be given, respectively, by

$$J^\mu_\rho \equiv \frac{\partial\mathcal{L}_{\text{matter}}}{\partial(\partial_\mu\phi)}\partial_\rho\phi, \quad \text{and} \quad J^\mu_{\rho\sigma} \equiv S^\mu_{\rho\sigma} - x_\rho\frac{1}{i}J^\mu_\sigma + x_\sigma\frac{1}{i}J^\mu_\rho. \tag{10.5.68}$$

In terms of δ_0, however, the invariance condition (10.5.60) is not simply $\delta_0\mathcal{L}_{\text{matter}} \equiv 0$, and the additional term $\delta x^\rho\partial_\rho\mathcal{L}_{\text{matter}}$ results in the appearance of the term $\partial_\rho\mathcal{L}_{\text{matter}}$ in the identities (10.5.63) and thus for the term $\delta^\mu_\rho\mathcal{L}_{\text{matter}}$ in T^μ_ρ.

We shall now consider the local ten-parameter Poincaré transformation in which the ten arbitrary infinitesimal constants, $\{\varepsilon^\mu\}$ and $\{\varepsilon^{\mu\nu}\}$, in Eq. (10.5.61) become the 10 arbitrary infinitesimal functions, $\{\varepsilon^\mu(x)\}$ and $\{\varepsilon^{\mu\nu}(x)\}$. It is convenient to regard

$$\varepsilon^{\mu\nu}(x) \quad \text{and} \quad \xi^\mu(x) \equiv i\varepsilon^\mu_\nu(x)x^\nu + \varepsilon^\mu(x),$$

as the 10 independent infinitesimal functions. Such choice avoids the explicit appearance of x^μ. Furthermore, we can always choose $\varepsilon^\mu(x)$ such that

$$\xi^\mu(x) = 0 \quad \text{and} \quad \varepsilon^{\mu\nu}(x) \neq 0,$$

so that the coordinate and field transformations are completely separated.

Based on this fact, we use Latin indices for $\varepsilon^{ij}(x)$ and Greek indices for ξ^μ and x^μ. The Latin indices, $i, j, k, \cdots$, also assume the values 0, 1, 2 and 3. Then the transformations under consideration are

$$\delta x^\mu = \xi^\mu(x), \quad \text{and} \quad \delta\phi(x) = \frac{1}{2}i\varepsilon^{ij}(x)S_{ij}\phi(x), \tag{10.5.69}$$

or

$$\delta_0\phi(x) = -\xi^\mu(x)\partial_\mu\phi(x) + \frac{1}{2}i\varepsilon^{ij}(x)S_{ij}\phi(x). \tag{10.5.70}$$

This notation emphasizes the similarity of the $\varepsilon^{ij}(x)$ transformations to the linear transformations of Weyl's gauge principle. Actually, in Utiyama's gauge principle, the $\varepsilon^{ij}(x)$ transformations alone are considered in the local six-parameter Lorentz transformation. The $\xi^\mu(x)$ transformations correspond to the general coordinate transformation.

According to the convention we have just employed, the differential operator ∂_μ must have a Greek index. In the matter Lagrangian density $\mathcal{L}_{\text{matter}}$, we then have the two kinds of indices, and we shall regard $\mathcal{L}_{\text{matter}}$ as a given function of $\phi(x)$ and $\tilde{\partial}_k\phi(x)$, satisfying the identities, (10.5.63) and (10.5.64). The original matter Lagrangian density $\mathcal{L}_{\text{matter}}$ is obtained by setting

$$\tilde{\partial}_k\phi(x) = \delta_k^\mu \partial_\mu\phi(x).$$

The matter Lagrangian density $\mathcal{L}_{\text{matter}}$ is not invariant under the local ten-parameter transformations, (10.5.69) or (10.5.70), but we will later obtain an invariant expression by replacing $\tilde{\partial}_k\phi(x)$ with a suitable covariant derivative $D_k\phi(x)$ in the matter Lagrangian density $\mathcal{L}_{\text{matter}}$.

The transformation of $\partial_\mu\phi(x)$ is given by

$$\delta\partial_\mu\phi = \frac{1}{2}i\varepsilon^{ij}S_{ij}\partial_\mu\phi + \frac{1}{2}i(\partial_\mu\varepsilon^{ij})S_{ij}\phi - (\partial_\mu\xi^\nu)(\partial_\nu\phi), \tag{10.5.71}$$

and the original matter Lagrangian density $\mathcal{L}_{\text{matter}}$ transforms according to

$$\delta\mathcal{L}_{\text{matter}} \equiv -(\partial_\mu\xi^\rho)J_\rho^\mu - \frac{1}{2}i(\partial_\mu\varepsilon^{ij})S_{ij}^\mu.$$

We note that it is J_ρ^μ instead of T_ρ^μ which appears here. The reason for this is that we have not included the extra term $(\partial_\mu\delta x^\mu)\mathcal{L}_{\text{matter}}$ in Eq. (10.5.60). The left-hand side of Eq. (10.5.60) actually has the value

$$\delta\mathcal{L}_{\text{matter}} + (\partial_\mu\delta x^\mu)\mathcal{L}_{\text{matter}} \equiv -(\partial_\mu\xi^\rho)T_\rho^\mu - \frac{1}{2}i(\partial_\mu\varepsilon^{ij})S_{ij}^\mu.$$

We shall now look for the modified matter Lagrangian density $\mathcal{L}'_{\text{matter}}$ which makes the matter action functional $I_{\text{matter}}[\phi]$ invariant under (10.5.69) or (10.5.70). The extra term just mentioned is of a different kind in that it involves $\mathcal{L}_{\text{matter}}$ and not $\partial\mathcal{L}_{\text{matter}}/\partial(\tilde{\partial}_k\phi)$. In particular, the extra term includes the contributions from terms in $\mathcal{L}_{\text{matter}}$ which do not contain the derivatives. Thus it is clear that we cannot remove the extra term by replacing the derivative $\tilde{\partial}_\mu$ with a suitable covariant derivative D_μ. For this reason, we shall consider the problem in two stages. First we eliminate the noninvariance arising from the fact that $\partial_\mu\phi(x)$ is not a covariant quantity, and second, we obtain an expression $\mathcal{L}'_{\text{matter}}$ satisfying

$$\delta\mathcal{L}'_{\text{matter}} \equiv 0. \tag{10.5.72}$$

Because the invariance condition (10.5.60) for the matter action functional I_{matter} requires the matter Lagrangian density $\mathcal{L}'_{\text{matter}}$ to be an invariant scalar density rather than an invariant scalar, we shall make a further modification, replacing $\mathcal{L}'_{\text{matter}}$ with $\mathcal{L}''_{\text{matter}}$, which satisfies

$$\delta\mathcal{L}''_{\text{matter}} + (\partial_\mu\xi^\mu)\mathcal{L}''_{\text{matter}} \equiv 0. \tag{10.5.73}$$

The first part of this program can be accomplished by replacing $\tilde{\partial}_k\phi$ in $\mathcal{L}_{\text{matter}}$ with a covariant derivative $D_k\phi$ which transforms according to

$$\delta(D_k\phi) = \frac{1}{2}i\varepsilon^{ij}S_{ij}(D_k\phi) - i\varepsilon_k^i(D_i\phi). \tag{10.5.74}$$

The condition (10.5.72) follows from the identities, (10.5.63) and (10.5.64). To do this, it is necessary to introduce 40 new field variables towards the end. We first consider the ε^{ij} transformations, and eliminate the $\partial_\mu \varepsilon^{ij}$ term in (10.5.71) by setting

$$D_{|\mu}\phi \equiv \partial_\mu \phi + \frac{1}{2} A_\mu^{ij} S_{ij} \phi, \tag{10.5.75}$$

where A_μ^{ij} with

$$A_\mu^{ij} = -A_\mu^{ji}$$

are 24 new field variables.

We can then impose the condition

$$\delta(D_{|\mu}\phi) = \frac{1}{2} i\varepsilon^{ij} S_{ij}(D_{|\mu}\phi) - (\partial_\mu \xi^\nu)(D_{|\nu}\phi), \tag{10.5.76}$$

which determines the transformation properties of A_μ^{ij} uniquely. They are

$$\delta A_\mu^{ij} = -\partial_\mu \varepsilon^{ij} + \varepsilon_k^i A_\mu^{kj} + \varepsilon_k^j A_\mu^{ik} - (\partial_\mu \xi^\nu) A_\nu^{ij}. \tag{10.5.77}$$

The position of the last term in Eq. (10.5.71) is rather different. The term involving $\partial_\mu \varepsilon^{ij}$ is inhomogeneous in the sense that it contains ϕ rather than $\partial_\mu \phi$, but this is not true of the last term. Correspondingly, the transformation law for $D_{|\mu}\phi$, (10.5.76), is already homogeneous. This means that to force the covariant derivative $D_k\phi$ to transform according to Eq. (10.5.74), we must add to $D_{|\mu}\phi$ not a term in ϕ, but rather a term in $D_{|\mu}\phi$ itself. In other words, we merely multiply by a new field,

$$D_k \phi \equiv e_k^\mu D_{|\mu}\phi. \tag{10.5.78}$$

Here, the e_k^μ are 16 new field variables with the transformation properties determined by Eq. (10.5.74) to be

$$\delta e_k^\mu = (\partial_\nu \xi^\mu) e_k^\nu - i\varepsilon_k^i e_i^\mu. \tag{10.5.79}$$

We note that the fields e_k^μ and A_μ^{ij} are independent and unrelated at this stage, although they will be related by the Euler–Lagrange equations of motion.

We find the invariant matter Lagrangian density $\mathcal{L}'_{\text{matter}}$ defined by

$$\mathcal{L}'_{\text{matter}} \equiv \mathcal{L}_{\text{matter}}(\phi, D_k\phi),$$

which is an invariant scalar. We can obtain the invariant matter Lagrangian density $\mathcal{L}''_{\text{matter}}$ which is an invariant scalar density by multiplying $\mathcal{L}'_{\text{matter}}$ by a suitable function of the new field variables,

$$\mathcal{L}''_{\text{matter}} \equiv \mathcal{E}\mathcal{L}'_{\text{matter}} = \mathcal{E}\mathcal{L}_{\text{matter}}(\phi, D_k\phi).$$

The invariance condition (10.5.73) for $\mathcal{L}''_{\text{matter}}$ is satisfied if a factor $\mathcal{E}$ itself is an invariant scalar density,

$$\delta\mathcal{E} + (\partial_\mu \xi^\mu)\mathcal{E} \equiv 0.$$

The only function of the new field variable e_k^μ which obeys this transformation law and does not involve the derivatives is

$$\mathcal{E} = [\det(e_k^\mu)]^{-1}, \tag{10.5.80}$$

where the arbitrary constant factor has been chosen such that $\mathcal{E} = 1$ when e_k^μ is set equal to δ_k^μ. The final form of the modified matter Lagrangian density $\mathcal{L}''_{\text{matter}}$ which is an invariant scalar density is given by

$$\mathcal{L}''_{\text{matter}}(\phi, \partial_\mu\phi, e_k^\mu, A_\mu^{ij}) \equiv \mathcal{E}\mathcal{L}_{\text{matter}}(\phi, D_k\phi). \tag{10.5.81}$$

As in the case of Weyl's gauge principle, we can define the modified current densities in terms of $\mathcal{L}_{\text{matter}}(\phi, D_k\phi)$ by

$$\mathcal{T}_\mu^k \equiv \frac{\partial \mathcal{L}''_{\text{matter}}}{\partial e_k^\mu} \equiv \mathcal{E}e_\mu^i \left[\frac{\partial \mathcal{L}_{\text{matter}}}{\partial(D_k\phi)} D_i\phi - \delta_i^k \mathcal{L}_{\text{matter}} \right], \tag{10.5.82}$$

$$\mathcal{S}_{ij}^\mu \equiv -2\frac{\partial \mathcal{L}''_{\text{matter}}}{\partial A_\mu^{ij}} \equiv i\mathcal{E}e_k^\mu \frac{\partial \mathcal{L}_{\text{matter}}}{\partial(D_k\phi)} S_{ij}\phi, \tag{10.5.83}$$

where e_μ^i is the inverse of e_i^μ, satisfying

$$e_\mu^i e_i^\nu = \delta_\mu^\nu, \quad e_\mu^i e_j^\mu = \delta_j^i. \tag{10.5.84}$$

In order to express the conservation laws of these currents in a simple form, we extend the definition of $D_{|\mu}\phi$. Originally, it was defined for $\phi(x)$, and is to be defined for any other quantity which is invariant under the ξ^μ transformations and transforms linearly under the ε^{ij} transformations. We extend $D_{|\mu}$ to any quantity which transforms linearly under the ε^{ij} transformations by ignoring the ξ^μ transformations altogether. Thus we have

$$D_{|\nu} e_i^\mu \equiv \partial_\nu e_i^\mu - A_{i\nu}^k e_k^\mu, \tag{10.5.85}$$

according to the ε^{ij} transformation law of e_i^μ. We call this the ε covariant derivative. We calculate the commutator of the ε covariant derivatives as,

$$[D_{|\mu}, D_{|\nu}]\phi = \frac{1}{2} i R_{\ \ \mu\nu}^{ij} S_{ij}\phi, \tag{10.5.86}$$

where $R_{\ j\mu\nu}^i$ is defined by the following equation,

$$R_{\ j\mu\nu}^i \equiv \partial_\nu A_{\ j\mu}^i - \partial_\mu A_{\ j\nu}^i - A_{\ k\mu}^i A_{\ j\nu}^k + A_{\ k\nu}^i A_{\ j\mu}^k. \tag{10.5.87}$$

This quantity is covariant under the ε^{ij} transformations. $R^{i}{}_{j\mu\nu}$ is closely analogous to the field strength tensor $F_{\alpha\mu\nu}$ of the non-Abelian gauge field. $R^{ij}_{\mu\nu}$ is antisymmetric in both pairs of indices.

In terms of the ε covariant derivative, the ten conservation laws of the currents, (10.5.82) and (10.5.83), are expressed as

$$D_{|\mu}(\mathcal{T}^{k}_{\nu}e^{\mu}_{k}) + \mathcal{T}^{k}_{\mu}(D_{|\nu}e^{\mu}_{k}) = \mathcal{S}^{\mu}_{ij}R^{ij}_{\mu\nu}, \tag{10.5.88}$$

$$D_{|\mu}\mathcal{S}^{\mu}_{ij} = \mathcal{T}_{i\mu}e^{\mu}_{j} - \mathcal{T}_{j\mu}e^{\mu}_{i}. \tag{10.5.89}$$

Now we examine our ultimate goal, the Lagrangian density $\mathcal{L}_{\mathrm{G}}$ of the "free" self-interacting gravitational field. We examine the commutator of D_k and D_l acting on $\phi(x)$. After some algebra, we obtain

$$[D_k, D_l]\phi = \frac{1}{2}iR^{ij}_{kl}S_{ij}\phi - C^{i}{}_{kl}D_i\phi, \tag{10.5.90}$$

where

$$R^{ij}_{kl} \equiv e^{\mu}_{k}e^{\nu}_{l}R^{ij}_{\mu\nu}, \quad C^{i}{}_{kl} \equiv (e^{\mu}_{k}e^{\nu}_{l} - e^{\mu}_{l}e^{\nu}_{k})D_{|\nu}e^{i}_{\mu}. \tag{10.5.91}$$

We note that the right-hand side of Eq. (10.5.90) is not simply proportional to ϕ but also involves $D_i\phi$.

The Lagrangian density $\mathcal{L}_{\mathrm{G}}$ for the "free" self-interacting gravitational field must be an invariant scalar density. If we set $\mathcal{L}_{\mathrm{G}} = \mathcal{E}\mathcal{L}_0$, then $\mathcal{L}_0$ must be an invariant scalar and a function only of the covariant quantities R^{ij}_{kl} and $C^{i}{}_{kl}$. All the indices of these expressions are of the same kind, unlike the case of the non-Abelian gauge field, so that we can take the contractions of the upper indices with the lower indices.

The requirement that $\mathcal{L}_0$ is an invariant scalar in two separate spaces is reduced to the requirement that it is an invariant scalar in one space. We have a linear invariant scalar which has no analogue in the case of the non-Abelian gauge field, namely, $R \equiv R^{ij}_{ij}$. There exist a few quadratic invariants, but we choose the lowest order invariant. Thus we are led to the Lagrangian density $\mathcal{L}_{\mathrm{G}}$ for the "free" self-interacting gravitational field,

$$\mathcal{L}_{\mathrm{G}} = \frac{1}{2\kappa^2}\mathcal{E}R, \tag{10.5.92}$$

which is linear in the derivatives. In Eq. (10.5.92), κ is Newton's gravitational constant.

So far, we have given neither any geometrical interpretation of the local ten-parameter Poincaré transformation, (10.5.69), nor any interpretation of the 40 new fields, $e^{\mu}_{k}(x)$ and $A^{ij}_{\mu}(x)$. We shall now establish the connection of the present theory with the standard metric theory of the gravitational field.

Under the ξ^{μ} transformation which is a general coordinate transformation, $e^{\mu}_{k}(x)$ transforms as a contravariant vector, while $e^{k}_{\mu}(x)$ and $A^{ij}_{\mu}(x)$ transform as covariant vectors. Then the quantity

$$g_{\mu\nu}(x) \equiv e^{k}_{\mu}(x)e_{k\nu}(x) \tag{10.5.93}$$

is a symmetric covariant tensor, and therefore may be interpreted as the metric tensor of a Riemannian space. It remains invariant under the ε^{ij} transformations. We shall abandon the convention that all the indices are to be lowered or raised by the flat-space metric $\eta_{\mu\nu}$, and we use $g_{\mu\nu}(x)$ instead as the metric tensor. We can easily show that

$$\mathcal{E} = \sqrt{-g(x)} \quad \text{with} \quad g(x) \equiv \det\bigl(g_{\mu\nu}(x)\bigr). \tag{10.5.94}$$

From Eq. (10.5.93), we realize that $e_k^{\mu}(x)$ and $e_{\mu}^{k}(x)$ are the contravariant and covariant components of a tetrad system in Riemannian space. The ε^{ij} transformations are the tetrad rotations. The Greek indices are the world tensor indices and the Latin indices are the local tensor indices of this system. The original generic matter field $\phi(x)$ may be decomposed into local tensors and local spinors. From the local tensors, we can form the corresponding world tensors by multiplying by $e_k^{\mu}(x)$ or $e_{\mu}^{k}(x)$.

For example, from a local vector $v^i(x)$, we can form the world vector as

$$v^{\mu}(x) = e_i^{\mu}(x)v^i(x), \quad \text{and} \quad v_{\mu}(x) = e_{\mu}^{i}(x)v_i(x). \tag{10.5.95}$$

We note that

$$v_{\mu}(x) = g_{\mu\nu}(x)v^{\nu}(x),$$

so that Eq. (10.5.95) is consistent with the definition of the metric $g_{\mu\nu}(x)$, Eq. (10.5.93).

The field $A^i{}_{j\mu}(x)$ is regarded as a local affine connection with respect to the tetrad system since it specifies the covariant derivatives of local tensors or local spinors. For a local vector, we have

$$\begin{cases} D_{|\nu} v^i &= \partial_{\nu} v^i + A^i{}_{j\nu} v^j, \\ D_{|\nu} v_j &= \partial_{\nu} v_j - A^i{}_{j\nu} v_i. \end{cases} \tag{10.5.96}$$

We notice that the relationship between $D_{|\mu}\phi$ and $D_k\phi$, (10.5.78), could be written simply as

$$D_{\mu}\phi = D_{|\mu}\phi, \tag{10.5.97}$$

according to the convention (10.5.95). We shall, however, make a distinction between D_{μ} and $D_{|\mu}$ for a later purpose. We define the covariant derivative of a world tensor in terms of the covariant derivative of the associated local tensor. Thus, we have

$$\begin{cases} D_{\nu} v^{\lambda} &\equiv e_i^{\lambda} D_{|\nu} v^i = \partial_{\nu} v^{\lambda} + \Gamma^{\lambda}{}_{\mu\nu} v^{\mu}, \\ D_{\nu} v_{\mu} &\equiv e_{\mu}^{i} D_{|\nu} v_i = \partial_{\nu} v_{\mu} - \Gamma^{\lambda}{}_{\mu\nu} v_{\lambda}, \end{cases} \tag{10.5.98}$$

$$\Gamma^{\lambda}{}_{\mu\nu} \equiv e_i^{\lambda} D_{|\nu} e_{\mu}^{i} \equiv -e_{\mu}^{i} D_{|\nu} e_i^{\lambda}. \tag{10.5.99}$$

We note that this definition of $\Gamma^{\lambda}_{\ \mu\nu}$ is equivalent to the requirement that the covariant derivative of the tetrad components vanish,

$$\begin{cases} D_{\nu}e^{\lambda}_{i} & \equiv \ 0, \\ D_{\nu}e^{i}_{\mu} & \equiv \ 0. \end{cases} \tag{10.5.100}$$

For a generic quantity α, transforming according to

$$\delta\alpha = \frac{1}{2}i\varepsilon^{ij}S_{ij}\alpha + (\partial_{\mu}\xi^{\lambda})\Sigma^{\mu}_{\lambda}\alpha, \tag{10.5.101}$$

the covariant derivative of α is defined by

$$D_{\nu}\alpha \equiv \partial_{\nu}\alpha + \frac{1}{2}iA^{ij}_{\nu}S_{ij}\alpha + \Gamma^{\lambda}_{\ \mu\nu}\Sigma^{\mu}_{\lambda}\alpha. \tag{10.5.102}$$

The ε covariant derivative of α, defined by Eq. (10.5.85), is obtained by simply dropping the last term in (10.5.102). We calculate the commutator of the covariant derivative of α with the result,

$$[D_{\mu}, D_{\nu}]\alpha = \frac{1}{2}iR^{ij}_{\ \mu\nu}S_{ij}\alpha + R^{\rho}_{\ \sigma\mu\nu}\Sigma^{\sigma}_{\rho}\alpha - C^{\lambda}_{\ \mu\nu}D_{\lambda}\alpha,$$

where $R^{\rho}_{\ \sigma\mu\nu}$ and $C^{\lambda}_{\ \mu\nu}$ are defined in terms of $R^{i}_{\ j\mu\nu}$ and $C^{i}_{\ kl}$ in the usual way. These quantities are the world tensors and can be expressed in terms of $\Gamma^{\lambda}_{\ \mu\nu}$ in the form,

$$R^{\rho}_{\ \sigma\mu\nu} = \partial_{\nu}\Gamma^{\rho}_{\ \sigma\mu} - \partial_{\mu}\Gamma^{\rho}_{\ \sigma\nu} - \Gamma^{\rho}_{\ \lambda\mu}\Gamma^{\lambda}_{\ \sigma\nu} + \Gamma^{\rho}_{\ \lambda\nu}\Gamma^{\lambda}_{\ \sigma\mu}, \quad C^{\lambda}_{\ \mu\nu} = \Gamma^{\lambda}_{\ \mu\nu} - \Gamma^{\lambda}_{\ \nu\mu}. \tag{10.5.103}$$

We see that $R^{\rho}_{\ \sigma\mu\nu}$ is the Riemann tensor formed from the affine connection $\Gamma^{\lambda}_{\ \mu\nu}$. From Eq. (10.5.100), we have

$$D_{\rho}g_{\mu\nu}(x) \equiv 0.$$

It is consistent to interpret $\Gamma^{\lambda}_{\ \mu\nu}$ as an affine connection in a Riemannian space. The definition of $\Gamma^{\lambda}_{\ \mu\nu}$, Eq. (10.5.99), does not guarantee that it is symmetric so that it is not the Christoffel symbol in general. In the absence of the matter field, however, $\Gamma^{\lambda}_{\ \mu\nu}$ is symmetric so that it is the Christoffel symbol. The curvature scalar has the usual form, $R \equiv R^{\mu}_{\ \mu}$, where $R_{\mu\nu} \equiv R^{\lambda}_{\ \mu\lambda\nu}$. The Lagrangian density $\mathcal{L}_{\mathrm{G}}$ for the "free" self-interacting gravitational field, Eq. (10.5.92), is the usual one,

$$\begin{aligned} &\mathcal{L}_{\mathrm{G}}\big(g^{\mu\nu}(x),\Gamma^{\lambda}_{\mu\nu}(x)\big) \\ &\qquad = \frac{1}{2\kappa^2}\sqrt{-g}g^{\mu\nu}(\partial_{\nu}\Gamma^{\lambda}_{\mu\lambda} - \partial_{\lambda}\Gamma^{\lambda}_{\mu\nu} + \Gamma^{\rho}_{\mu\lambda}\Gamma^{\lambda}_{\nu\rho} - \Gamma^{\lambda}_{\mu\nu}\Gamma^{\rho}_{\lambda\rho}). \end{aligned} \tag{10.5.104}$$

Unity of All Forces: Electro-weak unification of Glashow–Weinberg–Salam is based on the gauge group

$$SU(2)_{\text{weak isospin}} \times U(1)_{\text{weak hypercharge}}.$$

It suffers from the problem of the nonrenormalizability due to the triangular anomaly in the lepton sector. In the early 1970s, it was discovered that non-Abelian gauge field theory is asymptotically free at short distance, i.e., it behaves as a free field at short distances. Thus the relativistic quantum field theory of the strong interaction based on the gauge group $SU(3)_{\text{color}}$ is invented and is called quantum chromodynamics.

The standard model with the gauge group

$$SU(3)_{\text{color}} \times SU(2)_{\text{weak isospin}} \times U(1)_{\text{weak hypercharge}}$$

which describes the weak interaction, the electromagnetic interaction and the strong interaction, is free from the triangular anomaly. It suffers, however, from a serious defect; the existence of the classical instanton solution to the field equation in the Euclidean metric for the $SU(2)$ gauge field theory. In the $SU(2)$ gauge field theory, we have the Belavin–Polyakov–Schwartz–Tyupkin instanton solution which is a classical solution to the field equation in the Euclidean metric. A proper account for the instanton solution requires the addition of the strong CP-violating term to the *QCD* Lagrangian density in the path integral formalism. The Peccei–Quinn axion and the invisible axion scenario resolve this strong CP-violation problem. In the grand unified theories, we assume that the subgroup of the grand unifying gauge group is the gauge group $SU(3)_{\text{color}} \times SU(2)_{\text{weak isospin}} \times U(1)_{\text{weak hypercharge}}$. We now attempt to unify the weak interaction, the electromagnetic interaction and the strong interaction by starting from the much larger gauge group G which is reduced to $SU(3)_{\text{color}} \times SU(2)_{\text{weak isospin}} \times U(1)_{\text{weak hypercharge}}$ and further down to $SU(3)_{\text{color}} \times U(1)_{\text{E.M.}}$ as a result of the requisite sequences of the spontaneous symmetry breaking,

$$G \supset SU(3)_{\text{color}} \times SU(2)_{\text{weak isospin}} \times U(1)_{\text{weak hypercharge}} \supset SU(3)_{\text{color}} \times U(1)_{\text{E.M.}}.$$

By now, we are almost certain that the true underlining theory of particle interactions, including gravitational interaction, is superstring theory. Actually, phenomenological predictions have followed from superstring theory.

10.6 Problems for Chapter 10

10.1. (Due to H. C.) Find the solution or solutions $q(t)$ which extremize

$$I \equiv \int_0^T \left[\frac{m}{2}\dot{q}^2 - \frac{1}{6}q^6 \right] dt,$$

subject to

$$q(0) = q(T) = 0.$$

10.2. (Due to H. C.) Find the solution $q(t)$ which extremizes

$$I \equiv \int_0^T \left[\frac{m}{2}\dot{q}^2 - \frac{\lambda}{3}q^3 \right] dt,$$

subject to

$$q(0) = q(T) = 0.$$

10.3. The action for a particle in a gravitational field is given by

$$I \equiv -m \int \sqrt{g_{\mu\nu} \frac{dx^\mu}{dt} \frac{dx^\nu}{dt}}\, dt,$$

where $g_{\mu\nu}$ is the metric tensor. Show that the motion of this particle is governed by

$$\frac{d^2 x^\rho}{ds^2} = -\Gamma^\rho_{\mu\nu} \frac{dx^\mu}{ds} \frac{dx^\nu}{ds},$$

with

$$\Gamma^\rho_{\mu\nu} \equiv \frac{1}{2} g^{\rho\sigma} (\partial_\mu g_{\sigma\nu} + \partial_\nu g_{\sigma\mu} - \partial_\sigma g_{\mu\nu}),$$

which is called the **Christoffel symbol**.

10.4. (Due to H. C.) The space–time structure in the presence of a black hole is given by

$$(ds)^2 = \left(1 - \frac{1}{r}\right)(dt)^2 - \frac{(dr)^2}{\left(1 - \frac{1}{r}\right)} - r^2[\sin^2\theta (d\phi)^2 + (d\theta)^2],$$

and the motion of a particle is such that $\int ds$ is minimized. Let the particle move in the x-y plane and hence $\theta = \frac{\pi}{2}$. Then the equations of motion of this particle subject to the gravitational pull of the black hole are obtained by extremizing

$$s \equiv \int_{t_i}^{t_f} \sqrt{\left(1 - \frac{1}{r}\right) - \frac{\left(\frac{dr}{dt}\right)^2}{\left(1 - \frac{1}{r}\right)} - r^2 \left(\frac{d\phi}{dt}\right)^2}\, dt,$$

with initial and final coordinates fixed.

a) Derive the equation of motion obtained by varying ϕ. Integrate this equation once to obtain an equation with one integration constant.

b) Derive the equation of motion obtained by varying r. Find a way to obtain an equation involving $\frac{dr}{dt}$ and $\frac{d\phi}{dt}$ and a second integration constant.

c) Let the particle be at $r = 2$, $\phi = 0$, with $\frac{dr}{dt} = \frac{d\phi}{dt} = 0$ at the initial time $t = 0$. Determine the motion of this particle as best you can. How long does it take for this particle to reach $r = 1$?

10.5. (Due to H. C.) The invariant distance ds in the neigborhood of a black hole is given by

$$(ds)^2 = \left(1 - \frac{2M}{r}\right)(dt)^2 - \frac{(dr)^2}{\left(1 - \frac{2M}{r}\right)},$$

where r, θ, ϕ (we set $\theta = \phi =$ constant) are the spherical polar coordinates and M is the mass of the black hole. The motion of a particle extremizes the invariant distance.

a) Write down the integral which should be extremized. From the expression of this integral, find a first-order equation satisfied by $r(t)$.

b) If $(r - 2M)$ is small and positive, solve the first-order equation. How much time does it take for the particle to fall to the critical distance $r = 2M$?

10.6. (Due to H. C.) Find the kink solution by extremizing

$$I \equiv \int_{-\infty}^{+\infty} dt \left(\frac{1}{2}\phi_t^2 - \frac{m^2}{2}\phi^2 + \frac{\lambda}{4}\phi^4 + \frac{1}{4}\frac{m^4}{\lambda}\right).$$

10.7. Extremize

$$I \equiv \int_0^{x_0} dx \int d\Omega\, \tilde{f}(x,\theta)\left[\cos\theta \frac{\partial f(x,\theta)}{\partial x} + f(x,\theta) \right.$$
$$\left. - \frac{\kappa}{4\pi}\int w(\vec{n} - \vec{n}_0) f(x,\theta_0)\, d\Omega_0\right],$$

treating f and $\tilde{f}$ as independent. Here κ is a constant, the unit vectors, $\vec{n}$ and $\vec{n}_0$, are pointing in the direction specified by spherical angles, (θ, φ) and (θ_0, φ_0), and $d\Omega_0$ is the differential solid angle at $\vec{n}_0$. Obtain the steady-state transport equation for anisotropic scattering from the very heavy scatterers,

$$\cos\theta \frac{\partial f(x,\theta)}{\partial x} = -f(x,\theta) + \frac{\kappa}{4\pi}\int w(\vec{n} - \vec{n}_0) f(x,\theta_0)\, d\Omega_0,$$

$$-\cos\theta \frac{\partial \tilde{f}(x,\theta)}{\partial x} = -\tilde{f}(x,\theta) + \frac{\kappa}{4\pi}\int w(\vec{n}_0 - \vec{n}) \tilde{f}(x,\theta_0)\, d\Omega_0.$$

Interpret the result for $\tilde{f}(x,\theta)$.

10.8. Extremize

$$I \equiv \int dt\, d^3\vec{x}\mathcal{L}\left(\psi, \frac{\partial}{\partial t}\psi, \vec{\nabla}\psi, \varphi, \frac{\partial}{\partial t}\varphi, \vec{\nabla}\varphi\right),$$

where

$$\mathcal{L} = -\vec{\nabla}\varphi\vec{\nabla}\psi - \frac{a^2}{2}\left(\varphi\frac{\partial}{\partial t}\psi - \psi\frac{\partial}{\partial t}\varphi\right),$$

treating ψ and φ as independent. Obtain the diffusion equation,

$$\vec{\nabla}^2\psi(t,\vec{x}) = a^2\frac{\partial\psi(t,\vec{x})}{\partial t},$$

$$\vec{\nabla}^2\varphi(t,\vec{x}) = -a^2\frac{\partial\varphi(t,\vec{x})}{\partial t}.$$

Interpret the result for φ.

10.9. Extremize

$$I \equiv \int dt\, d^3\vec{x}\left\{-\frac{1}{2m}\left(\left(\frac{\hbar}{i}\vec{\nabla}-\frac{e}{c}\vec{A}\right)\psi\right)^*\left(\left(\frac{\hbar}{i}\vec{\nabla}-\frac{e}{c}\vec{A}\right)\psi\right)\right.$$
$$\left.+\frac{1}{2}\left[\psi^*\left(i\hbar\frac{\partial}{\partial t}-e\phi\right)\psi+\left(\left(i\hbar\frac{\partial}{\partial t}-e\phi\right)\psi\right)^*\psi\right]-\psi^* V\psi\right\},$$

$$\vec{\nabla}\vec{A}+\frac{1}{c}\frac{\partial\phi}{\partial t}=0,$$

treating ψ and ψ^* as independent. Obtain the Schrödinger equation,

$$\left(i\hbar\frac{\partial}{\partial t}-e\phi\right)\psi=\frac{1}{2m}\left(\frac{\hbar}{i}\vec{\nabla}-\frac{e}{c}\vec{A}\right)^2\psi+V\psi,$$

$$-\left(i\hbar\frac{\partial}{\partial t}+e\phi\right)\psi^*=\frac{1}{2m}\left(\frac{\hbar}{i}\vec{\nabla}+\frac{e}{c}\vec{A}\right)^2\psi^*+V\psi^*.$$

Demonstrate that the Schrödinger equation is invariant under the gauge transformation,

$$\begin{aligned}\vec{A} &\to \vec{A}' = \vec{A}+\vec{\nabla}\Lambda,\\ \phi &\to \phi' = \phi-\left(\frac{1}{c}\right)\left(\frac{\Lambda}{\partial t}\right), \quad \text{where} \quad \left(\vec{\nabla}^2-\frac{1}{c^2}\frac{\partial^2}{\partial t^2}\right)\Lambda=0.\\ \psi &\to \psi' = \exp\left[\left(\frac{ie}{\hbar c}\right)\Lambda\right]\psi,\end{aligned}$$

10.10. Extremize

$$I \equiv \int dt\, d^3\vec{x}\left\{-\left|\left(\frac{1}{i}\vec{\nabla}-e\vec{A}\right)\psi\right|^2+\left|\left(i\frac{\partial}{\partial t}-e\phi\right)\psi\right|^2-m^2|\psi|^2\right\},$$

$$\vec{\nabla}\vec{A}+\frac{\partial\phi}{\partial t}=0,$$

treating ψ and ψ^* as independent. Obtain the Klein–Gordon equation,

$$\left(i\frac{\partial}{\partial t}-e\phi\right)^2\psi-\left(\frac{1}{i}\vec{\nabla}-e\vec{A}\right)^2\psi=m^2\psi,$$

$$\left(i\frac{\partial}{\partial t}+e\phi\right)^2\psi^*-\left(\frac{1}{i}\vec{\nabla}+e\vec{A}\right)^2\psi^*=m^2\psi^*.$$

10.11. Extremize

$$I \equiv \int d^4x \mathcal{L}_{\text{tot}},$$

where $\mathcal{L}_{\text{tot}}$ is given by

$$\mathcal{L}_{\text{tot}} = \frac{1}{4}[\bar{\psi}_\alpha(x), D_{\alpha\beta}(x)\psi_\beta(x)] + \frac{1}{4}[D^{\text{T}}_{\beta\alpha}(-x)\bar{\psi}_\alpha(x), \psi_\beta(x)]$$
$$+ \frac{1}{2}\phi(x)K(x)\phi(x) + \mathcal{L}_{\text{int}}(\phi(x), \psi(x), \bar{\psi}(x)),$$

with $D_{\alpha\beta}(x)$, $D^{\text{T}}_{\beta\alpha}(-x)$ and $K(x)$ given by

$$D_{\alpha\beta}(x) = (i\gamma_\mu\partial^\mu - m + i\varepsilon)_{\alpha\beta},$$
$$D^{\text{T}}_{\beta\alpha}(-x) = (-i\gamma^{\text{T}}_\mu\partial^\mu - m + i\varepsilon)_{\beta\alpha},$$
$$K(x) = -\partial^2 - \kappa^2 + i\varepsilon,$$

and $\mathcal{L}_{\text{int}}$ is given by the Yukawa coupling specified by

$$\mathcal{L}_{\text{int}}(\phi(x), \psi(x), \bar{\psi}(x)) = -G_0\bar{\psi}_\alpha(x)\gamma_{\alpha\beta}(x)\psi_\beta(x)\phi(x).$$

The $\gamma^{\mu'}$s are the Dirac γ matrices with the property specified by

$$\{\gamma^\mu, \gamma^\nu\} = 2\eta^{\mu\nu},$$
$$(\gamma^\mu)^\dagger = \gamma^0\gamma^\mu\gamma^0.$$

The $\psi(x)$ is the four-component Dirac spinor and the $\bar{\psi}(x)$ is the Dirac adjoint of $\psi(x)$ defined by

$$\bar{\psi}(x) \equiv \psi^\dagger(x)\gamma^0.$$

Obtain the Euler–Lagrange equations of motion for the ψ field, the $\bar{\psi}$ field and the ϕ field.

10.12. Extremize the action functional for the electromagnetic field A_μ,

$$I = \int d^4x \left(-\frac{1}{4}F^{\mu\nu}F_{\mu\nu} + B\partial^\mu A_\mu + \frac{1}{2}\alpha B^2\right),$$
$$F_{\mu\nu} \equiv \partial_\mu A_\nu - \partial_\nu A_\mu.$$

Obtain the Euler–Lagrange equations of motion for the A_μ field and the B field. Can you perform the q-number gauge transformation after canonical quantization?

10.13. Extremize the action functional for the neutral massive vector field U_μ,

$$I = \int d^4x \left(-\frac{1}{4}F^{\mu\nu}F_{\mu\nu} + \frac{1}{2}m_0^2 U^\mu U_\mu\right),$$
$$F_{\mu\nu} \equiv \partial_\mu U_\nu - \partial_\nu U_\mu.$$

Obtain the Euler–Lagrange equation of motion for the U_μ field. Examine the massless limit $m_0 \to 0$ after canonical quantization.

10.14. Extremize the action functional for the neutral massive vector field A_μ,

$$I = \int d^4x \left(-\frac{1}{4} F^{\mu\nu} F_{\mu\nu} + \frac{1}{2} m_0^2 A^\mu A_\mu + B \partial^\mu A_\mu + \frac{1}{2} \alpha B^2 \right),$$

$$F_{\mu\nu} \equiv \partial_\mu A_\nu - \partial_\nu A_\mu .$$

Obtain the Euler–Lagrange equations of motion for the A_μ field and the B field. Examine the massless limit $m_0 \to 0$ after canonical quantization.

Hint for Problems 10.12, 10.13 and 10.14:

Lautrup, B.: Mat. Fys. Medd. Dan. Vid. Selsk. **35**. **No.11**. 29. (1967).

Nakanishi, N.: Prog. Theor. Phys. Suppl. **51**. 1. (1972).

Yokoyama, K.: Prog. Theor. Phys. **51**. 1956. (1974), **52**. 1669. (1974).

10.15. Derive the Schwinger–Dyson equation for the interacting scalar fields $\hat{\phi}_i(x)$ $(i = 1, 2)$ whose Lagrangian density is given by

$$\mathcal{L}(\hat{\phi}_1(x), \hat{\phi}_2(x), \partial_\mu \hat{\phi}_1(x), \partial_\mu \hat{\phi}_2(x))$$
$$= \sum_{i=1}^{2} \left\{ \frac{1}{2} \partial_\mu \hat{\phi}_i(x) \partial^\mu \hat{\phi}_i(x) - \frac{1}{2} m_i^2 \hat{\phi}_i^2(x) \right\} - g \hat{\phi}_1^2(x) \hat{\phi}_2(x).$$

Hint: Introduce the proper self-energy parts $\mathbf{\Pi}_i^*(x, y)$ $(i = 1, 2)$ and the vertex operator $\mathbf{\Lambda}(x, y, z)$, and follow the discussion in Section 10.3.

10.16. Derive the Schwinger–Dyson equation for the self-interacting scalar field $\hat{\phi}(x)$ whose Lagrangian density is given by

$$\mathcal{L}(\hat{\phi}(x), \partial_\mu \hat{\phi}(x)) = \frac{1}{2} \partial_\mu \hat{\phi}(x) \partial^\mu \hat{\phi}(x) - \frac{1}{2} m^2 \hat{\phi}^2(x) - \frac{\lambda_4}{4!} \hat{\phi}^4(x).$$

Hint: Introduce the proper self-energy part $\mathbf{\Pi}^*(x, y)$ and the vertex operator $\mathbf{\Lambda}_4(x, y, z, w)$, and follow the discussion in Section 10.3.

10.17. Derive the Schwinger–Dyson equation for the self-interacting scalar field $\hat{\phi}(x)$ whose Lagrangian density is given by

$$\mathcal{L}(\hat{\phi}(x), \partial_\mu \hat{\phi}(x)) = \frac{1}{2} \partial_\mu \hat{\phi}(x) \partial^\mu \hat{\phi}(x) - \frac{1}{2} m^2 \hat{\phi}^2(x) - \frac{\lambda_3}{3!} \hat{\phi}^3(x) - \frac{\lambda_4}{4!} \hat{\phi}^4(x).$$

Hint: Introduce the proper self-energy part $\mathbf{\Pi}^*(x, y)$ and the vertex operators $\mathbf{\Lambda}_3(x, y, z)$ and $\mathbf{\Lambda}_4(x, y, z, w)$, and follow the discussion in Section 10.3.

10.18. Consider the bound state problem for a system of two distinguishable spinless bosons of equal mass m, exchanging a spinless and massless boson whose Lagrangian density is given by

$$\mathcal{L} = \sum_{i=1}^{2} \left\{ \frac{1}{2}\partial_\mu \hat{\phi}_i(x)\partial^\mu \hat{\phi}_i(x) - \frac{1}{2}m^2 \hat{\phi}_i^2(x) \right\}$$
$$+ \frac{1}{2}\partial_\mu \hat{\phi}(x)\partial^\mu \hat{\phi}(x) - g\hat{\phi}_1^\dagger(x)\hat{\phi}_1(x)\hat{\phi}(x) - g\hat{\phi}_2^\dagger(x)\hat{\phi}_2(x)\hat{\phi}(x).$$

a) Show that the Bethe–Salpeter equation for the bound state of the two bosons $\hat{\phi}_1(x_1)$ and $\hat{\phi}_2(x_2)$ is given by

$$S'_{\mathrm{F}}(x_1, x_2; B) = \int d^4x_3 d^4x_4 \Delta_{\mathrm{F}}(x_1 - x_3)\Delta_{\mathrm{F}}(x_2 - x_4)$$
$$\times (-g^2) D_{\mathrm{F}}(x_3 - x_4) S'_{\mathrm{F}}(x_3, x_4; B),$$

where $\Delta_{\mathrm{F}}(x)$ and $D_{\mathrm{F}}(x)$ are given by

$$\Delta_{\mathrm{F}}(x) = \int \frac{d^4k}{(2\pi)^4} \frac{\exp[ikx]}{k^2 - m^2 + i\varepsilon}, \quad \text{and} \quad D_{\mathrm{F}}(x) = \int \frac{d^4k}{(2\pi)^4} \frac{\exp[ikx]}{k^2 + i\varepsilon}.$$

b) Transform the coordinates x_1 and x_2 to the center-of-mass coordinate X and the relative coordinate x by

$$X = \frac{1}{2}(x_1 + x_2), \quad \text{and} \quad x = x_1 - x_2,$$

and correspondingly to the center-of-mass momentum P and the relative momentum p,

$$P = p_1 + p_2, \quad \text{and} \quad p = \frac{1}{2}(p_1 - p_2).$$

Define the Fourier transform $\Psi(p)$ of $S'_{\mathrm{F}}(x_1, x_2; B)$ by

$$S'_{\mathrm{F}}(x_1, x_2; B) = \exp[-iPX] \int d^4p \exp[-ipx]\Psi(p).$$

Show that the above Bethe–Salpeter equation in momentum space assumes the following form,

$$\left[\left(\frac{P}{2} + p\right)^2 - m^2\right]\left[\left(\frac{P}{2} - p\right)^2 - m^2\right]\Psi(p) = ig^2 \int \frac{d^4q}{(2\pi)^4} \frac{\Psi(q)}{(p-q)^2 + i\varepsilon}.$$

c) Assuming that $\Psi(p)$ can be expressed as

$$\Psi(p) = -\int_{-1}^{1} \frac{g(z)\,dz}{[p^2 + zpP - m^2 + (P^2/4) + i\varepsilon]^3},$$

substitute this expression into the Bethe–Salpeter equation in momentum space. Carrying out the q integration using the formula,

$$\int d^4q \frac{1}{(p-q)^2+i\varepsilon} \cdot \frac{1}{[q^2+zqP-m^2+(P^2/4)+i\varepsilon]^3}$$
$$= \frac{i\pi^2}{2[-m^2+(P^2/4)-z^2(P^2/4)]} \cdot \frac{1}{[p^2+zpP-m^2+(P^2/4)+i\varepsilon]},$$

and comparing the result with the original expression for $\Psi(p)$, obtain the integral equation for $g(z)$ as

$$g(z) = \int_0^1 \varsigma\, d\varsigma \int_{-1}^1 dy \int_{-1}^1 dx \frac{\lambda g(x)}{2(1-\eta^2+\eta^2x^2)} \delta(z - \{\varsigma y + (1-\varsigma)x\}),$$

where the dimensionless coupling constant λ is given by

$$\lambda = \left(\frac{g}{4\pi m}\right)^2,$$

and the squared mass of the bound state is given by

$$M^2 = P^2 = 4m^2\eta^2, \qquad 0 < \eta < 1.$$

d) Carrying out the ς integration, obtain the integral equation for $g(z)$ as

$$g(z) = \lambda \int_z^1 dx \frac{1+z}{1+x} \frac{g(x)}{2(1-\eta^2+\eta^2x^2)} + \lambda \int_{-1}^z dx \frac{1-z}{1-x} \frac{g(x)}{2(1-\eta^2+\eta^2x^2)}.$$

e) Observe that $g(z)$ satisfies the boundary conditions,

$$g(\pm 1) = 0.$$

Differentiate the integral equation for $g(z)$ obtained in d) twice, and reduce it to a second-order ordinary differential equation for $g(z)$ of the form,

$$\frac{d^2}{dz^2} g(z) = -\frac{\lambda}{1-z^2} \frac{g(z)}{1-\eta^2+\eta^2z^2}.$$

This is the eigenvalue problem.

f) Solve the above eigenvalue problem for $g(z)$ in the limit, $1 \gg 1-\eta > 0$, and show that the lowest approximate eigenvalue is given by

$$\lambda \approx \frac{2}{\pi}\sqrt{1-\eta^2}.$$

Hint for Problem 10.18: The Wick–Cutkosky model is discussed in the following articles.

Wick, G.C.: Phys. Rev. **96**., 1124, (1954).

Cutkosky, R.E.: Phys. Rev. **96**., 1135, (1954).

10.19. Consider the bound state problem of zero total momentum $\vec{P} = 0$ for a system of identical two fermions of mass m, exchanging a spinless and massless boson whose Lagrangian density is given by

$$\mathcal{L} = \widehat{\bar{\psi}}(x)(i\gamma_\mu \partial^\mu - m + i\varepsilon)\hat{\psi}(x) + \frac{1}{2}\partial_\mu \hat{\phi}(x)\partial^\mu \hat{\phi}(x) - g\widehat{\bar{\psi}}(x)\hat{\psi}(x)\hat{\phi}(x).$$

Define the bound state wave function of the two fermions by

$$[U_{\vec{P}}(x)]_{\alpha\beta} = \ < 0|\,\mathrm{T}[\hat{\psi}_\alpha(\frac{x}{2})\hat{\psi}_\beta(-\frac{x}{2})]\,|B> ,$$

$$[u_{\vec{P}}(p)]_{\alpha\beta} = \int d^4x \exp\left[ipx\right] [U_{\vec{P}}(x)]_{\alpha\beta},$$

$$[u_{\vec{P}=0}(p)]_{\alpha\beta} = \frac{\delta_{\alpha\beta}\chi(p)}{p^2 - m^2 + i\varepsilon}.$$

Show that the Bethe–Salpeter equation for the bound state to the first-order approximation is given by

$$\chi(p) = ig^2 \int \frac{d^4q}{(2\pi)^4}\left[\frac{1}{(p-q)^2 + i\varepsilon} - \frac{1}{(p+q)^2 + i\varepsilon}\right]\frac{\chi(q)}{q^2 - m^2 + i\varepsilon}.$$

Solve this eigenvalue problem by dropping the antisymmetrizing term in the kernel of the above. The antisymmetrizing term originates from the spin-statistics relation for the fermions.

Hint for Problem 10.19: This problem is discussed in the following article.

Goldstein, J.: Phys. Rev. **91**., 1516, (1953).

Bibliography

Local Analysis and Global Analysis

We cite the following book for the local analysis and global analysis of ordinary differential equations.

[1] Bender, Carl M., and Orszag, Steven A.: "*Advanced Mathematical Methods For Scientists And Engineers: Asymptotic Methods and Perturbation Theory*", Springer-Verlag, New York, (1999).

Integral Equations

We cite the following book for the theory of Green's functions and boundary value problems.

[2] Stakgold, I.: "*Green's Functions and Boundary Value Problems*", John Wiley & Sons, New York, (1979).

We cite the following books for general discussions of the theory of integral equations.

[3] Tricomi, F.G.: "*Integral Equations*", Dover, New York, (1985).

[4] Pipkin, A.C.: "*A Course on Integral Equations*", Springer-Verlag, New York, (1991).

[5] Bach, M.: "*Analysis, Numerics and Applications of Differential and Integral Equations*", Addison Wesley, Reading, Massachusetts, (1996).

[6] Wazwaz, A.M.: "*A First Course in Integral Equations*", World Scientific, Singapore, (1997).

[7] Polianin, A.D.: "*Handbook of Integral Equations*", CRC Press, Florida, (1998).

[8] Jerri, A.J.: "*Introduction to Integral Equations with Applications*", 2nd edition, John Wiley & Sons, New York, (1999).

We cite the following books for applications of integral equations to the scattering problem in nonrelativistic quantum mechanics, namely, the Lippmann–Schwinger equation.

[9] Goldberger, M.L., and Watson, K.M.: "*Collision Theory*", John Wiley & Sons, New York, (1964). Chapter 5.

[10] Sakurai, J.J.; "*Modern Quantum Mechanics*", Addison-Wesley, 1994, Massachusetts. Chapter 7.

[11] Nishijima, K.: "*Relativistic Quantum Mechanics*", Baifuukan, 1973, Tokyo. Section 4–11 of Chapter 4. (In Japanese.)

Applied Mathematics in Theoretical Physics. Michio Masujima

ISBN: 3-527-40534-8

We cite the following book for applications of integral equations to the theory of elasticity.

[12] Mikhlin, S.G. et al.: "*The Integral Equations of the Theory of Elasticity*", Teubner, Stuttgart, (1995).

We cite the following book for the application of integral equations to microwave engineering.

[13] Collin, R.E.: "*Field Theory of Guided Waves*", Oxford Univ. Press, (1996).

We cite the following article for the application of integral equations to chemical engineering.

[14] Bazant, M.Z., and Trout, B.L.: Physica, **A300**, 139, (2001).

We cite the following book for physical details of the dispersion relations in classical electrodynamics.

[15] Jackson, J.D.: "*Classical Electrodynamics*", 3rd edition, John Wiley & Sons, New York, (1999). Section 7.10. p. 333.

We cite the following books for applications of Cauchy-type integral equations to dispersion relations in the potential scattering problem in nonrelativistic quantum mechanics.

[16] Goldberger, M.L., and Watson, K.M.: "*Collision Theory*", John Wiley & Sons, New York, (1964). Chapter 10 and Appendix G.2.

[17] De Alfaro, V., and Regge, T.: "*Potential Scattering*", North-Holland, Amsterdam, (1965).
We note that Appendix G.2 of the book cited above, [9], discusses Cauchy-type integral equations in the scattering problem in nonrelativistic quantum mechanics in terms of the inhomogeneous Hilbert problems with the complete solution.

We cite the following article for the integro-differential equation arising from Bose-Einstein condensation in an external potential at zero temperature.

[18] Wu, T.T.: Phys. Rev. **A58**., 1465, (1998).

We cite the following book for applications of the Wiener–Hopf method in partial differential equations.

[19] Noble, B.: "*Methods Based on the Wiener–Hopf Technique for the Solution of Partial Differential Equations*", Pergamon Press, New York, (1959).
We note that the Wiener–Hopf integral equations originated from research on the radiative equilibrium on the surface of the star.

We cite the following articles for the discussion of Wiener–Hopf integral equations and Wiener–Hopf sum equations.

[20] Wiener, N., and Hopf, E.: S. B. Preuss. Akad. Wiss. 696, (1931).

[21] Hopf, E.: "*Mathematical Problems of Radiative Equilibrium*", Cambridge, New York, (1934).

[22] Krein, M.G.: "*Integral equation on a half-line with the kernel depending upon the difference of the arguments*", Amer. Math. Soc. Transl. (2), **22**, 163, (1962).

[23] Gohberg, I.C., and Krein, M.G.: "*Systems of integral equations on the half-line with kernels depending on the difference of the arguments*", Amer. Math. Soc. Transl. (2), **14**, 217, (1960).

We cite the following article for discussion of the iterative solution for a single Wiener–Hopf integral equation and for a system of coupled Wiener–Hopf integral equations.

[24] Wu, T.T. and Wu, T.T.: Quarterly Journal of Applied Mathematics, **XX**, 341, (1963).

We cite the following book for the application of Wiener–Hopf methods to radiation from rectangular waveguides and circular waveguides.

[25] Weinstein, L.A.: "*The theory of diffraction and the factorization method*", Golem Press, (1969). pp. 66-88, and pp. 120-156.

We cite the following article and books for application of the Wiener–Hopf method to elastodynamics of the crack motion.

[26] Freund, L.B.: J. Mech. Phys. Solids, **20**, 129, 141, (1972).

[27] Freund, L.B.: "*Dynamic Fracture Mechanics*", Cambridge Univ. Press, New York, (1990).

[28] Broberg, K.B.: "*Cracks and Fracture*", Academic Press, New York, (1999).

We cite the following article and the following book for application of the Wiener–Hopf sum equation to the phase transition of the two-dimensional Ising model.

[29] Wu, T.T.: Phys. Rev. **149**., 380, (1966).

[30] McCoy, B., and Wu, T.T.: "*The Two-Dimensional Ising Model*", Harvard Univ. Press, Cambridge, Massachusetts, (1971). Chapter IX.

We note that Chapter IX of the book cited above describes practical methods to solve the Wiener–Hopf sum equation with full mathematical details, including discussions of Pollard's theorem which is the generalization of Cauchy's theorem, and the two special cases of the Wiener–Lévy theorem.

We cite the following articles for application of the Wiener–Hopf sum equation to Yagi–Uda semi-infinite arrays.

[31] Wasylkiwskyj, W.: IEEE Transactions Antennas Propagat., **AP-21**, 277, (1973).

[32] Wasylkiwskyj, W., and VanKoughnett, A.L.: IEEE Transactions Antennas Propagat., **AP-24**, 633, (1974).

[33] VanKoughnett, A.L.: Canadian Journal of Physics, **48**, 659, (1970).

We cite the following book for the historical development of the theory of integral equations, the formal theory of integral equations and a variety of applications of the theory of integral equations to scientific and engineering problems.

[34] Kondo, J.: "*Integral Equations*", Kodansha Ltd., Tokyo, (1991).

We cite the following book for the pure-mathematically oriented reader.

[35] Kress, R.: "*Linear Integral Equations*", 2nd edition, Springer-Verlag, Heidelberg, (1999).

Calculus of Variations.

We cite the following books for an introduction to the calculus of variations.

[36] Courant, R., and Hilbert, D.: "*Methods of Mathematical Physics*", (Vols. 1 and 2.), John Wiley & Sons, New York, (1966). Vol.1, Chapter 4. Reprinted in Wiley Classic Edition, (1989).

[37] Akhiezer, N.I.: "*The Calculus of Variations*", Blaisdell, Waltham, Massachusetts, (1962).

[38] Gelfand, I.M., and Fomin, S.V.: "*Calculus of Variations*", Prentice-Hall, Englewood Cliffs, New Jersey, (1963).

[39] Mathews, J., and Walker, R.L.: "*Mathematical Methods of Physics*", Benjamin, Reading, Massachusetts, (1970). Chapter 12.

We cite the following book for the variational principle in classical mechanics, the canonical transformation theory, the Hamilton–Jacobi equation and the semi-classical approximation to nonrelativistic quantum mechanics.

[40] Fetter, A.L., and Walecka, J.D.: "*Theoretical Mechanics of Particles and Continua*", McGraw-Hill, New York, (1980). Chapter 6, Sections 34 and 35.

We cite the following books for applications of the variational principle to nonrelativistic quantum mechanics and quantum statistical mechanics.

[41] Feynman, R.P., and Hibbs, A.R.: "*Quantum Mechanics and Path Integrals*", McGraw-Hill, New York, (1965). Chapters 10 and 11.

[42] Feynman, R.P.: "*Statistical Mechanics*", Benjamin, Reading, Massachusetts, (1972). Chapter 8.

[43] Landau, L.D., and Lifshitz, E.M.: "*Quantum Mechanics*", 3rd edition, Pergamon Press, New York, (1977). Chapter III, Section 20.

[44] Huang, K.: "*Statistical Mechanics*", 2nd edition, John Wiley & Sons, New York, (1983). Section 10.4.
The variational principle employed by R.P. Feynman and the variational principle employed by K. Huang are both based on Jensen's inequality for the convex function.

We cite the following book as a general reference for the theory of the gravitational field.

[45] Weinberg, S.: "*Gravitation and Cosmology. Principles and Applications of The General Theory of Relativity*", John Wiley & Sons, 1972, New York.

We cite the following book for the genesis of Weyl's gauge principle and the earlier attempt to unify the electromagnetic force and the gravitational force before the birth of quantum mechanics in the context of classical field theory.

[46] Weyl, H.: "*Space–Time–Matter*", Dover Publications, Inc., 1950, New York.
Although H. Weyl failed to accomplish his goal of the unification of the electromagnetic force and the gravitational force in the context of classical field theory, his enthusiasm for the unification of all forces in nature survived, even after the birth of quantum mechanics.

We cite the following articles and book for Weyl's gauge principle, after the birth of quantum mechanics, for the Abelian electromagnetic gauge group.

[47] Weyl, H.; Proc. Nat. Acad. Sci. **15**, (1929), 323.

[48] Weyl, H.; Z. Physik. **56**, (1929), 330.

[49] Weyl, H.; "*Theory of Groups and Quantum Mechanics*", Leipzig, 1928, Zurich; reprinted by Dover, 1950. Chapter 2, section 12, and Chapter 4, section 5.

We cite the following article as the first attempt to unify the strong and the weak forces in nuclear physics in the context of quantum field theory, without invoking Weyl's gauge principle.

[50] Yukawa, H.: Proc. Phys. Math. Soc. (Japan), **17**, 48, (1935).

We cite the following articles for the unification of the electromagnetic force and the weak force by invoking Weyl's gauge principle and the Higgs–Kibble mechanism and the proposal of the standard model which unifies the weak force, the electromagnetic force, and the strong force with quarks, leptons, the Higgs scalar field, the Abelian gauge field and the non-Abelian gauge field, in the context of quantum field theory.

[51] Weinberg, S.: Phys. Rev. Lett. **19**, 1264, (1967); Phys. Rev. Lett. **27**, 1688, (1971); Phys. Rev. **D5**, 1962, (1971); Phys. Rev. **D7**, 1068, (1973); Phys. Rev. **D7**, 2887, (1973); Phys. Rev. **D8**, 4482, (1973); Phys. Rev. Lett. **31**, 494, (1973); Phys. Rev. **D9**, 3357, (1974).

We cite the following book for discussion of the standard model which unifies the weak force, the electromagnetic force and the strong force with quarks, leptons, the Higgs scalar field, the Abelian gauge field and the non-Abelian gauge field by invoking Weyl's gauge principle and the Higgs–Kibble mechanism in the context of quantum field theory.

[52] Huang, K.: "*Quarks, Leptons, and Gauge Field*", 2nd edition, World Scientific, Singapore, (1992).

We cite the following article for the $O(3)$ model.

[53] Georgi, H. and Glashow, S.L.: Phys. Rev. Lett. **28**, 1494, (1972).

We cite the following articles for the instanton, the strong CP violation, the Peccei–Quinn axion hypothesis and the invisible axion scenario.

[54] Belavin, A.A., Polyakov, A.M., Schwartz, A.S., and Tyupkin, Y.S.; Phys. Letters. **59B**, (1975), 85.

[55] Peccei, R.D., and Quinn, H.R.; Phys. Rev. Letters. **38**, (1977), 1440; Phys. Rev. **D16**, (1977), 1791.

[56] Weinberg, S.; Phys. Rev. Letters. **40**, (1978), 223.

[57] Wilczek, F.; Phys. Rev. Letters. **40**, (1978), 279.

[58] Dine, M., Fishcler, W., and Srednicki, M.; Phys. Letters. **104B**, (1981), 199.

We cite the following article for the see-saw mechanism.

[59] Yanagida, T.; Prog. Theor. Phys. **64**, (1980), 1103.

We cite the following book for the *grand unification* of the electromagnetic, weak and strong interactions.

[60] Ross, G.G.; "*Grand Unified Theories*", Perseus Books Publishing, 1984, Massachusetts. Chapters 5 through 11.

We cite the following article for the $SU(5)$ grand unified model.

[61] Georgi, H. and Glashow, S.L.: Phys. Rev. Lett. **32**, 438, (1974).

We cite the following article for discussion of the gauge principle in the differential formalism originally due to H. Weyl and the integral formalism originally due to T.T. Wu and C.N. Yang.

[62] Yang, C.N.: Ann. N.Y. Acad. Sci. **294**, 86, (1977).

We cite the following book for the connection between Feynman's action principle in nonrelativistic quantum mechanics, and the calculus of variations; in particular, the second variation, the Legendre test and the Jacobi test.

[63] Schulman, L.S.; "*Techniques and Application of Path Integration*", John Wiley & Sons, New York, (1981).

We cite the following book for the use of the calculus of variations in the path integral quantization of classical mechanics and classical field theory, Weyl's gauge principle for the Abelian gauge group and the semi-simple non-Abelian gauge group, the Schwinger–Dyson equation in quantum field theory and quantum statistical mechanics, and stochastic quantization of classical mechanics and classical field theory.

[64] Masujima, M.: "*Path Integral Quantization and Stochastic Quantization*", Springer Tracts in Modern Physics, Vol.165, Springer-Verlag, Heidelberg, (2000). Chapter 1, Section 1.1 and 1.2; Chapter 2, Sections 2.3, 2.4 and 2.5; Chapter 3, Sections 3.1, 3.2 and 3.3; Chapter 4, Sections 4.1 and 4.3; Chapter 5, Section 5.2.

We cite the following book for discussion of the Schwinger–Dyson equation, and the Bethe–Salpeter equation from the viewpoint of the canonical formalism and the path integral formalism of quantum field theory.

[65] Huang, K.: "*Quantum Field Theory. From Operators to Path Integrals*", John Wiley & Sons, New York, (1998). Chapter 10, Sections 10.7 and 10.8; Chapters 13 and 16.

[66] Huang, K.: "*Quarks, Leptons, and Gauge Field*", 2nd edition, World Scientific, Singapore, (1992). Chapters IX and X.

[67] Huang, K.: "*Statistical Mechanics*", 2nd edition, John Wiley & Sons, New York, (1983). Chapter 18.

The Wick–Cutkosky model is the only exactly solvable model for the Bethe–Salpeter equation known to this day. We cite the following articles for this model.

[68] Wick, G.C.: Phys. Rev. **96**., 1124, (1954).

[69] Cutkosky, R.E.: Phys. Rev. **96**., 1135, (1954).

We cite the following books for canonical quantization, path integral quantization, the S matrix approach to the Feynman rule for any spin J, the proof of the non-Abelian gauge field theory based on BRST invariance and Zinn–Justin equation, the electroweak unification, the standard model, the grand unification of weak, electromagnetic and strong interactions, and the grand unification with the graded Lie gauge group.

[70] Weinberg, S.: "*Quantum Theory of Fields I*", Cambridge Univ. Press, New York, (1995).

[71] Weinberg, S.: "*Quantum Theory of Fields II*", Cambridge Univ. Press, New York, (1996).

[72] Weinberg, S.: "*Quantum Theory of Fields III*", Cambridge Univ. Press, New York, (2000).

Inclusion of the gravitational force in a unification scheme beside the weak force, the electromagnetic force and the strong force, requires the use of superstring theory.

Index